Biological Control Systems Analysis

Biological Control
Systems Analysis

John H. Milsum
Abitibi Professor of Control Engineering
McGill University

McGRAW-HILL BOOK COMPANY

New York   St. Louis   San Francisco
Toronto   London   Sydney

To M. L. M.

BIOLOGICAL CONTROL SYSTEMS ANALYSIS

Preface

The considerable challenge in writing this book about biological control systems has been mixed with some frustration. The frustration arises because although new knowledge and ideas are being generated very rapidly, yet the present overall state of quantitative knowledge is insufficient to permit development of a general control theory of biological systems. In consequence the proper title seems to be *Biological Control Systems Analysis* rather than *Theory*. Of course it can be rightly pointed out that the "balance of nature" and homeostasis have been basic concepts of biological thinking for a long time, but the relevant experimental data have often been concerned with the "steady-state" condition, and usually more qualitative than quantitative. Now an exact balance or steady-state condition is a very improbable state for a complex dynamic system, although, indeed, such balance may prevail on the average. Therefore the satisfactory functional analysis of control systems inevitably depends upon using dynamic testing procedures which, in turn, require the analyst to have a working mastery of the underlying mathematically based ideas. Thus, for example, we should not expect that establishing the qualitative information-flow pathways in patients having Parkinsonian tremors, or in an economy suffering boom-and-slump oscillations, would provide the knowledge necessary to bring these systems into satisfactorily stable operation, since it is the quantitative dynamic relations between elements that determine the system's overall behavior.

In consequence, there exists something of a "chicken-and-egg" type of initial-condition growth problem regarding biological control theory. The biological system's analyst needs a control theory to motivate and guide his designing of dynamic experiments, whereas the control theorist needs the resulting dynamic experimental data as a prerequisite to building up a general theory. Therefore my aim in this book is to stimulate the rate of growth by translating today's incomplete data and control theory into as much of a working textbook as possible, with full recognition that many of the models and even ideas discussed will shortly be outmoded. Parenthetically I would comment that the present need for good biological data is especially urgent since an increasing number of scientists from various outside disciplines [engineering, mathematics, (bio)physics, etc.] have become motivated to work in this area. If data are not provided in step with the growth of theory, there is always the danger that theory will be built up from an abstract philosophical viewpoint, based more on how biological control systems could (or should?) work than on how they do work.

Since my own point of approach has been from control engineering, I should emphasize that this discipline has a genuine practical reason for its concern with biological control systems. The engineer has always designed machines for man to operate, but in recent years performance on the machines' side has expanded so much that "man the operator" has tended to become a weak link in the dynamic chain, although he remains essential as an overall controller at the monitoring and purposeful level. Consequently, the control and system's engineer must integrate man's operating characteristics into his man-machine system for an optimum overall performance according to some human-oriented goal, but in order to do this he must have quantitative dynamic descriptions of man as the machine operator. Secondly, engineers are involved in producing many new artificial augmentations and replacements for man's subsystems, such as prostheses and heart-lung machines. These tasks are almost impossible unless a system's approach can be taken, which immediately requires quantitative dynamic descriptions of the natural systems themselves. Thirdly, scientists and engineers are significantly modifying the natural habitat and its populations through such effects as pollution, widespread use of chemicals to affect growth processes (both pro and con), and human death-rate control. Adequate knowledge of the ecosystems concerned is just as essential to rational progress in such matters as is the social motivation for them. Finally, mathematicians and engineers are developing a whole new class of machines based upon electronic computation. These machines exhibit rudimentary intelligence and, therefore, can relieve man of the many tedious mental jobs; examples are automatic language translation and automatic readers, "speakers," and "hearers" (especially for direct communication between man and his computers). Once again progress in such tasks is most rapid when there is fruitful interaction between the living and nonliving sciences.

In principle there is perhaps nothing new in such collaboration between disciplines since scientists have always depended heavily upon knowledge and techniques from outside their own fields. However the increasing physical and intellectual size of the systems which man creates or tolerates, together with the explosive splitting of science into so many specialties, have accelerated the need for groupings of functionally related disciplines. These groupings are not necessarily identical with the fewer disciplines of earlier days, and current examples are earth sciences, materials sciences, and—most relevant to this book—the variously described "cybernetics," "control and communications science," or "general system's theory."

When the biologist, social scientist, and indeed natural scientist collaborate with the engineer on these large new system's problems, their classical roles as analyzers of existing systems in contrast to the engineer's role as the synthesizer of previously nonexisting "hardware" systems needs reappraisal. The synthesis task in some practical senses is easier because hardware (such as control systems and computers) can be designed on the basis of abstract theories (such as linear system theory, counting, and logic), which man readily understands since he is their inventor. (In practice, the physical world forces some readjustment of such calculus and, for example, the control engineer recognizes that his amplifiers may saturate and the computer designer works with the easily implemented binary logic rather than the decimal.) On the other hand, the nature of the calculus by which living systems may be said to operate, and by which they should therefore be analyzed most easily, is not necessarily so simple. Herein therefore lies a challenge, since valid new theories will probably also provide new insight for designing better hardware systems; for example, one immediately thinks of the use of fallible, fragile components in redundant configurations and the liberal use of feedback.

The basic methodology advanced in this book stems from the relatively well-developed theory of control engineering which, in turn, is based on the modeling of physical systems by differential equations or derivatives thereof. Specifically it aims to show that mathematical modeling of biological systems on this basis is fruitful, being applicable especially to the analysis of homeostatic regulating systems, neuromuscular systems and multiorganism systems such as ecologies. One point which I have emphasized is that the need for dynamic analysis is fundamental, arising quite independently of whether the particular system or component being analyzed is of the closed-loop type. Therefore much of the book is devoted to developing the basic tools of linear dynamic analysis and control theory. The emphasis on linear theory arises because of the generalized, tractable, and easily comprehensible results which follow, but this is only meant to provide the introductory ideas. Models of the real world must usually be distributed and nonlinear, and their numerical solutions have to be obtained through the good works of electronic computers. On the other hand, I have not attempted to cover analysis of extremely complicated probabilistic systems such as the central nervous system or a national economy, which are usually included in the term *cybernetics*.

I have adopted a somewhat unconventional didactic method in view of the unsettled state of this subject, specifically by drawing freely upon a wide selection of models from the biological research literature, and incorporating these throughout the book in order to motivate each new necessary theoretical development. This entails the corresponding disadvantage, which I hope is outweighed, that some of the biological examples are treated at successively deeper levels in later chapters. Also in deference to those biologists who are supposedly not mathematically inclined, I have written the book so that certain sections of mathematical development (marked ▼) may be omitted at first reading without affecting comprehension of later material. Nevertheless the book will largely have failed in its purpose if such readers do not become motivated to return to these optional sections. I strongly recommend that these readers purchase at least one basic mathematical text at an appropriate level for themselves (see Bibliography). It is also presumed that some scientists and engineers may omit the optional sections because they already know the material involved. Clearly the presentation cannot be maintained at absolutely constant level because of these compromises, and indeed the level of presentation rises rapidly throughout the book on the assumption that the student will learn to consult the bibliography when necessary. I emphasize however that such techniques as that of Laplace transformation should be immediately accepted as useful working tools rather than as difficult mathematical concepts, the fine details of which must be conquered.

The book therefore is addressed to a mixed group of biologists, physicians, engineers, and indeed to all those interested in analyzing biological control systems. I have taught much of the material to mixed classes of such students, but this has required most of the emphasis to be placed on the earlier chapters, so that a prerequisite mathematics course has now been introduced which includes material from these early chapters. This latter course comprises about 50 hr and the remaining material is then taught in a 40-hr lecture course. I have made considerable effort to develop a set of worthwhile problems for the student to solve, and some of the biological ones are purposely left almost as open-ended as the subject itself in the hope that the student may pursue matters into the literature. My own students are asked in addition to undertake a project of analyzing more deeply some problem of their choice, and for this it is found valuable to pair off biologists and engineers.

In acknowledgments, the control theorist such as myself must first recognize his debt to the researchers of all disciplines who uncover the basic biophysical knowledge on which a general theory can be built. I can only hope that efforts such as this book may in

return be helpful to the researchers' efforts. Personally I acknowl-
edge the stimulation gained over a long period from many teachers
and colleagues, and also the constructive criticisms of the text or
its parts submitted ungrudgingly by a number of friends. Here I
can only name D. C. Baxter, B. D. Burns, M. Clynes, E. R. R.
Funke, G. Melvill Jones, R. W. Jones, G. L. d'Ombrain, H. M.
Paynter, F. A. Roberge, and N. Sugie. I also thank Mrs. T.
Hyland, Miss E. Gilday, and Mrs. A. Tempelman-Kluit, who typed
and retyped the manuscript without criticism, and finally my wife
Eileen, who edited, proofread, and otherwise cheerfully endured
this period.

John H. Milsum

Contents

4 *Energy Storage and Dissipation*   83

5 *The First-order Linear Lumped Model: The Operational Transfer Function*   99

6 *Transient Response Characteristics: Example of First-order System; Laplace-transform Techniques*   115

7 *Frequency Response Characteristics: Example of First-order System*   142

8   The Second-order Linear Lumped Model and Its Response Characteristics: The Pole-Zero Plot of Roots   179

9   High-order, Nonlinear, and Spatially Distributed Models   209

10   Computer Simulations and Numerical Solutions   240

11 Principles and Techniques of Feedback Control 274

12 The Special Challenge of Biological Control Systems 318

13 Biological Receptors 352

14 Statistical Signals 364

15 Biological Performance Criteria and Adaptive Control 401

1 *Introduction*

All living systems, from the simplest cell to the most complex ecology, maintain their viability in continual challenge to the basic law of the nonliving world, by which regimes of high order tend to be degraded into regimes of uniformly low order. Living systems can only achieve such high order by their ability to obtain energy externally and to control its expenditure internally. The basic energy source is the sun, of course, and photosynthesis is the technique whereby its energy is trapped. Since many living species have evolved without photosynthetic ability they must consume those species that retain it, in order to survive.

Naturally, therefore, it is to be expected that all living systems should efficiently control their internal environment. Even the unicellular creature is a complex biochemical "factory" transforming under careful control the various input foods for its internal economy. Then, because nature has evidently found specialization of cells beneficial organizationally, living systems such as man have evolved with a whole hierarchy of regulatory control systems, for example, blood-pressure control, temperature control, postural control. In addition, man, out of all his biological relatives, has developed the complex organ called the *cerebral cortex*, with its thinking, learning, and emotive powers. These powers introduce new control abilities whose mechanisms and structure we find very hard to comprehend, although by using them mankind has developed great control over his physical universe. Unfortunately, the complex sociological, economic, and industrial systems which have resulted do not yet exhibit the same desirable performance and stability as our own autonomous internal regulating systems.

These comments are intended to emphasize that all living systems comprise a complex of interacting processes and components. When they exhibit satisfactory performance, it is as a result of internal control mechanisms, which in turn depend upon adequate internal communication channels. Stability is one particular feature which is universally considered necessary for satisfactory performance, but the dynamic nature of the system's components often makes it difficult to achieve. Negative feedback is the necessary technique by which long-term stability in complex systems can be obtained, but its presence is not a sufficient condition for stability.

This description is recognizable by the control engineer as defining the servomechanisms and regulating control systems which he designs and builds. It was this structure similarity which Wiener pointed out in 1948, when he defined *cybernetics* as "the science of control and communication in man and machine" [1.1].*
As such, cybernetics is clearly an all-embracing term and therefore includes everything that is said within this book. However, cybernetics is really developing most fruitfully as the unifying theory behind very complex systems, for example, the central nervous system (CNS), a national economy, an industrial firm [1.2]. Since such complexities are not tackled in this book, the word *cybernetics* has not been used in its title. Many readable introductions to cybernetics have recently been written and are listed in the Bibliography.

Biophysics is also an apparently all-embracing title; thus, for example, Casey [1.3] remarks in introducing his book: ". . . since the name, 'Biophysics' means so many different things to so many different people, the big difficulty has been to decide what not to write." Etymologically, biophysics must mean the study of physical processes incorporated in living systems. Thus, biophysics should completely overlap the half of cybernetics concerned with living systems. In practice, however, biophysics has tended to specialize in the various physical phenomena of forces, radiation, energetics, and structure at the molecular and cellular levels [1.3–1.7].

This book concentrates upon the area left relatively untouched by contemporary biophysics and cybernetics—that of the dynamics and control of biological systems of intermediate complexity, such as the range of homeostatic or regulatory control systems. The results of biophysics are the basic laws which govern the functioning of the system's components, but the main emphasis is placed on the ways in which these components act in concert as a system.

* Numbers in brackets refer to the references at the end of the book.

At the same time, the cybernetic problem of the cerebral cortex is largely ignored since we deal with reflex and autonomous regulatory systems which are not primarily mediated from the cortex, at least not continually in normal operation. Autonomous and neuro-muscular control systems within man and other animals usually comprise several organs, have neural or hormonal effector and feed-back channels, and have as their function the regulation of physical variables of the internal body or its limbs. Some examples are:

Body temperature
Cardiac output
Water and electrolyte balance
Respiration
Blood chemistry; sugar level, red and white cells, etc.
Audition
Control of posture and locomotion
Oculomotor system—tracking and stabilization
Pupillary reflex system

In addition, the population dynamics of ecological systems have essentially the same features except that there is no active con-trolling mechanism; instead, control is achieved passively by mutual interaction of populations with each other and their plant food sources.

This book attempts to apply in biology the general theory for the dynamic analysis of control systems which the control engineer has developed and applied so successfully within the last 20 years to his technological problems. Of course, Watt built his famous fly ball governor for control of steam engine speed in the late eighteenth century, but little significant theory was developed (with the notable exception of Maxwell's work [1.8]) until just before World War II, when feedback amplifiers in electronic equip-ment represented the state of the art. During and since the war, a great expansion has occurred in both control theory and its appli-cations. The automatic control of gun aiming by radar was the major wartime achievement, and today there are such diverse applications as computer control of oil refineries and control of space probes during interplanetary travel. Furthermore, present research work in machine learning and artificial intelligence promises much relief for man from tedious, repetitive mental tasks. At the same time, however, such applications are producing corresponding social and economic problems which will require much new under-standing of their dynamics before they can be solved satisfactorily.

The developments of control engineering have been facilitated because the dynamics of the systems can be described by differential equations. Upon this mathematical modeling, or representation of systems, has developed an impressive and highly useful body of theory and techniques, especially for linear systems. However, this fast development would have been impracticable without electronic analog and digital computers, because these machines enable the engineer to obtain satisfactory numerical answers to his design problems whenever the complexity of the system model makes the search for analytical answers unhelpful. The analog computer especially has been an important conceptual tool in simulating systems by an easily manipulatable electronic analog.

In autonomous biological systems many of the physiological processes to be controlled can also be satisfactorily modeled by differential equations. Consequently, biologists and engineers have found collaboration valuable in utilizing control theory and computers as conceptual tools for building up mathematical models which will duplicate, in some significant respects, the biological system under study. Although model building cannot by itself uncover new fundamental knowledge, it may provide new insight for the investigator who can see for the first time how the various system components interact dynamically in producing the whole response. Also, it is often a cheap and painless way of predicting the outcome of inconvenient biological experiments, and it is consequently useful to suggest crucial new biological experiments which should be undertaken. As a result, the model itself is modified, and the iterative process is repeated. Therefore, the mathematical model programmed onto a computer can be viewed as a fast but approximate exploratory tool which quickens the biological discovery process. It is worth repeating that for any realistically complex model it will almost always be impossible to predict system responses without the aid of a computer. A cheap computer of low accuracy which the biologist can readily master can often suffice, however.

The biologist may reasonably question the value of linear differential equation models for his inevitably complex biological systems, especially since nature is so frequently nonlinear. As a partial answer, we must first understand the elegance and generalized advantages which are obtainable from a linear modeling whenever it seems even faintly reasonable. Then, when the discrepancies appear, one can start to postulate nonlinearity, nonstationarity, and other modifications. The fact that essential nonlinearities are actually so abundant in living systems constitutes much of their challenge to one interested in control theory. For the control engineer there is also the probability that any worthwhile advances of theory made in the biological area will be valuable in engineering.

A major feature of autonomous control systems is that although they seem to have rather simple overall performance requirements (constant body-core temperature, etc.), the control loop may involve neural paths with a large number of parallel neurons. This multipath and statistical method of communication clearly defies detailed analysis, but since the output requirement of the channel is simple and probably analog in quantity (heat production in muscle, etc.), one can reasonably expect that a simple modeling may suffice for overall systems analysis. In this respect, the comparison with molecular versus continuum gas behavior is superficially very relevant. Thus, as is well known, the large number of gas molecules makes it impracticable to predict the force exerted upon any finite-sized surface as a result of individual molecular collisions; instead, however, the gas is treated as a continuum and the universal gas laws then render the calculation trivial.

This book is written both for the biologically trained scientist who wishes to apply the principles of dynamic analysis and control to his systems and for the engineer who is interested in the area often called biomedical engineering. The theory is developed around and applied to biological systems almost in entirety, except where it seems useful to point out engineering-biological analogies. Work in this area is in its infancy and therefore no claim is made that particular models and techniques used in the text will not shortly be superseded. However, it is hoped that the approach and underlying principles will continue to be valuable. In the text are modeled as many different biological systems as possible consistent with clarity, and these are taken from the pioneering work of a widely scattered group of biologists, clinicians, engineers, physicists, and mathematicians. This fact alone attests the interdisciplinary stimulation which is arising from the study of biological control systems.

2 Dynamic Systems and Their Control

A Qualitative Introduction

2.1 Introduction

This chapter introduces some of the basic principles and phenomena met in dynamic systems, and more particularly in dynamic control systems. The account is largely qualitative in order to emphasize ideas without involving undue mathematical manipulation. After some systems definitions in Sec. 2.2, the information-flow diagram for the man-machine system of driving a car is prepared in Sec. 2.3. Section 2.4 develops some necessary general ideas for the system's modeling which are illustrated quantitatively for a very simple population-growth model in Sec. 2.5. This is essentially an uncontrolled system, and Sec. 2.6 considers the modeling of a feedback-control system, the reflex system for the regulation of light impingement upon the retina. The basic structure of active feedback-control systems, as implied from this example, is then developed more generally in Sec. 2.7. The nature of the steady-state and transient responses of control systems is explored in Sec. 2.8 with the use of the pupil example, and some differences between engineering and biological systems are pointed out in Sec. 2.10 by using the analog system of automatic aperture control in cameras. Section 2.9 discusses some of the experimental difficulties and possibilities in analyzing intact biological systems,

in particular that of opening biological loops; the nature of instability is also introduced.

2.2   *Some Systems Definitions*

(A) SYSTEM

A *system* in this context is defined as any collection of "communicating" materials and processes which together perform some function in which the investigator is interested.   The system's behavior
is determined (Fig. 2.1a) by:

1. The characteristics of the components or subsystems.
2. The structure of communication between components, which usu-
    ally involves feedback paths.
3. The *input* signals or variables to the system.   These are initially
    assumed to be independent variables under the investigator's con-
    trol but some may in fact be controlled by the outputs of other
    systems.

The results of the system's operation upon the inputs are the
outputs.   Any variable of the system may be considered an output,
dependent only upon the particular interest of the system's investigator.   Note that neither input nor output variables need particularly be material flows; variables such as voltage, temperature, and
pressure qualify equally well as the liters per second of blood flow.

In consequence, a system may be represented symbolically as
an input-output device; one conventional symbol is a block with

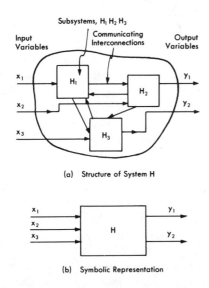

(a)   Structure of System H

(b)   Symbolic Representation

FIG. 2.1   System.

incoming and outgoing arrows to represent the relevant variables (Fig. 2.1*b*). Any system may always be broken down into smaller, connected subsystems if one so wishes, as in Fig. 2.1*a*.

The living cell is a convenient example of a system, its boundary being formed by the membrane. Internally the cell has many subsystems, such as nucleus and mitochondria, and communication occurs between them through the actions of the various biochemical materials. The input variables relevant to a particular study are selected from the large number of possible ones which the investigator could manipulate, including environmental concentrations of the various inflowing chemicals, environmental temperature, and pressure. The output variables relevant to the study could include internal concentrations or output flow rates of internally synthesized chemicals, metabolic rate, and transmembrane potential.

In its turn the living cell is a subsystem and a basic building block for a hierarchical set of larger systems, of which a representative few members are given in order of increasing physical size and complexity:

Cell
Organ (e.g., heart, liver, lung)
Organ system (e.g., cardiac output-control system)
Organism (e.g., man)
"Tribe"
Community (e.g., sociological, economic, ecological)

(B) DYNAMIC SYSTEM

The adjective *dynamic* appears ubiquitously in this book. Etymologically it is derived from the Greek word *dunamis*, meaning "power," but it is used here in the accepted current sense of *time varying;* this change in definition is only slight since normally a system only varies in time as a result of forces acting upon it (where force is generalized to include mechanical, chemical, electromagnetic, etc.). In general all nontrivial physical systems must be dynamic, owing to their internal energy-storing or energy-dissipating nature. Thus, for example, no known biological transducer will output a new constant level of action potential (spike) frequency immediately after a sudden "step" change of input condition (pressure, length, temperature, etc.). Instead, there is a perceptible delay followed by a transient of some sort, such as the spike frequency sketched in Fig. 2.2 for a stylized biological receptor.

Mathematically, the result of this time variability is that the systems are described by differential equations, which constitute a methodical way of writing down dynamic interactions. If the variables are functions of space as well as of time then ordinary differential equations no longer suffice but the more complex partial differential equations must be used. Differential equations are

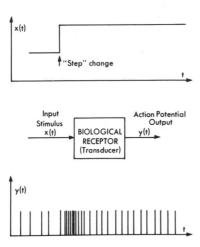

FIG. 2.2 *Stylized step response of transducer.*

particularly suitable for systems with smoothly changing variables, such as the temperature-control system, but they cannot directly be applied to systems with essentially discontinuous states, such as neurons and in particular the neuronal components of the CNS. However, it seems that the neurons in such a system as temperature control are essentially transmitting analog information (that is, they are continuously variable) so that the essential functioning of the whole system can be represented by a model using continuous variables. The action potentials represent a type of pulse frequency modulation, which is pursued in Sec. 2.6.

(C) CONTROL SYSTEM

The term *control system* is used in this book synonymously with *negative-feedback control system*. The simplest possible representation of such an *active* control system is given in Fig. 2.3a. The *reference input* is the basic input to the system causing it to behave in a desired manner, and therefore it is synonymously called the *desired output* of the system. The *actual output* of the dynamic system being controlled (the controlled process) is transmitted in a *feedback* path for comparison with the desired output in an active control element. If there is *error*, or mismatch, between actual and desired outputs, the active control element produces an *actuating* or *control effort* which acts upon the controlled process to force the actual output toward the value of desired output, which thus in turn reduces the error. Negative feedback exists here because the output is subtracted from the input, and consideration shows that potential controllability is always greater if one can compare current performance achieved with that desired. In particular, this *closed-loop* action permits the system to compensate actively for the

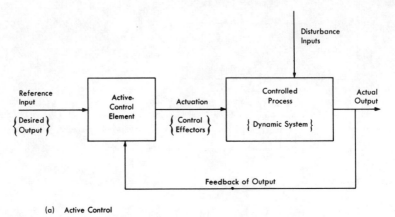

(a)   Active Control

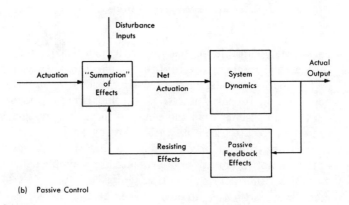

(b)   Passive Control

FIG. 2.3   Feedback-control-system configuration.

inevitable *disturbance inputs* of environmental variables.   The latter
compensation effect provides one of the main reasons for closed-loop
control of the many regulatory systems in the body, for example, in
temperature control to compensate for varying rates of environmental heat loss.   Another basic reason is that with closed-loop
control, a desirable dynamic response can be maintained even when
the parameters of the controlled process change.   In engineering
systems the comparison of input with output usually implies direct
differencing of the quantities, in which the direct result is the system error; however, in biological control systems this direct sub-
traction may often be replaced by some form of parameter modification (Sec. 2.7).

In general both blocks of Fig. 2.3a are themselves comprised of
further subsystems, but the configuration shown is the simplest
which can explain the essential features of active feedback control.
This configuration can be termed *active control* because a superficially similar phenomenon exists which can be termed *passive*

control [2.1]. Thus, in most physical and physiological systems any change of input inherently produces effects which tend to resist such changes, and thereby to reestablish an equilibrium. This is illustrated schematically in Fig. 2.3*b*, which in its entirety represents merely an expansion of the box marked "Controlled Process" in Fig. 2.3*a*. Although the configurations of Fig. 2.3*a* and *b* are apparently almost identical, there are in fact important differences in design and functioning, which will become more obvious as further examples are developed. One working distinction is to term a feedback control system *active* when the feedback pathway has been incorporated with a control "purpose," but otherwise *passive*. Thermoregulation in the mammal is then active, whereas the control in a cold-blooded animal is passive since it is entirely dependent upon the environmental conditions.

2.3 Man-Machine Example

A simple example from the class of man-machine systems, namely, driving a car, illustrates some of these phenomena and distinctions. Consider a car being driven in equilibrium (at constant speed) on a level road. For some accelerator pedal deflection x_1 an equilibrium speed (velocity) V_1 exists when the engine's propulsive road force F_e equals the car's resistance force F_r due to road and air friction (Fig. 2.4*a*). F_r is some increasing function of the speed V and is therefore labeled $F_r(V)$. F_e is assumed to depend on accelerator deflection x_1 alone and is therefore labeled $F_e(x_1)$. Actually, F_e typically decreases with increasing V, as shown dashed, but this does not affect the principles of this presentation. If the operator wishes to increase car speed and suddenly depresses the accelerator pedal position at time t_1 to x_2, then acceleration ensues until a new equilibrium speed V_2 is reached at which engine and resistive forces again balance (Fig. 2.4*a*). The following causal chain of events exists:

1. On depression of the accelerator, the propulsive force immediately increases by F.
2. F is initially a net excess of driving force over resistive force and by Newton's second law of motion produces an initial acceleration a, satisfying the relation

$$F = ma \qquad (2.1)$$

where m is the mass of the car.

3. The effect of acceleration is to increase the velocity $V(t)$ as some function of time (Fig. 2.4*b*), whose mathematical nature is not pursued at present. Note that by definition of the system under study, the velocity is the natural variable to select as output in this system.
4. The effect of increasing velocity is to increase the resistive force,

$F_r(V)$, which feeds back "passively" to reduce the net accelerating force F (Fig. 2.4c). Thus, the acceleration decays and the new equilibrium velocity becomes established (Fig. 2.4b).

Unless the driver is very expert it is unlikely that he could predict at all accurately his accelerator position x to produce a speed V_2. In fact he is only subconsciously aware of accelerator position at all since it is a subsidiary variable in his closed-loop chain of driving control, as discussed later.   Figure 2.4c shows the specific version of the passive control configuration of Fig. 2.3b that is relevant to this example, where a new symbol, a circle with the letter sigma (Σ), has been introduced to show algebraic summation of input variables, each with the sign indicated beside its arrival at the circle.   This type of figure indicates the flow of information among the various components of the system, and is called an *information-flow diagram*. It is often a preliminary to the *block diagram* which incorporates the mathematical models of the various components, as discussed later.

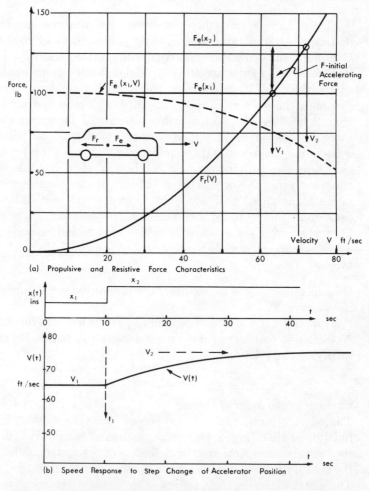

FIG. 2.4   Speed control of car.

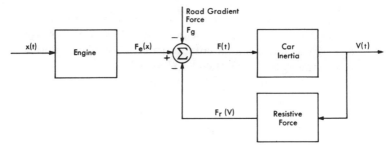

(c) Passive Control Configuration-Information Flow

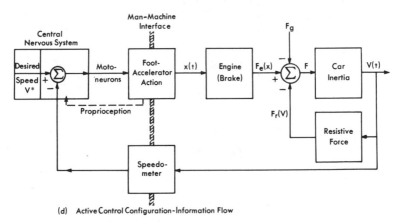

(d) Active Control Configuration-Information Flow

FIG. 2.4 *For descriptive legend see opposite page.*

Active closed-loop control of car speed arises in the general case where the driver decides that he will maintain a certain desired speed V^*; this becomes the reference input to the system and therefore is equally the desired output of the system, as shown in the information-flow diagram of Fig. 2.4d. This man-machine system has several interfaces where the two interact. The desired speed V^* exists in the driver's CNS, and the actual speed must be fed back to him by some such transducer as the speedometer (although in practice the driver obtains other visual clues to his speed from the environment). As a result of comparing actual with desired speed the CNS decides on some actuation command, which is then transmitted via motoneurons to the appropriate leg muscles. In turn the leg muscles depress or relax the accelerator position $x(t)$, or apply the brake for large negative force. The remainder of the information-flow diagram is identical with that given in Fig. 2.4c for passive control. There is some local feedback between accelerator position and CNS via leg-muscle proprioception, since skeletal-muscle control is of the closed-loop type.

Figure 2.4d shows that the disturbance effects of changing road gradient F_g can be compensated for by the feedback of speed information. Again, however, as so often happens in biological control systems, a sensible shortcut is made in this regard by all good drivers; that is, the driver can achieve better control by observing any forthcoming changes of gradient and using predictive actuation to change F_e to match the new F_g before any change in speed occurs. This is an important feature to incorporate in any control system whenever the disturbance effects can be measured; however, the inherent advantage of active closed-loop negative-feedback control is that compensation against disturbances is available even if these latter are not measurable.

Having now explored qualitatively the distinction between active and passive control, we will in future reserve the term *control system* (or *control*) for the active type alone, namely, incorporating a purposeful closed-loop action with negative feedback. It is also worth emphasizing that although the information flow in hybrid systems such as man-machine can easily be presented qualitatively on these diagrams, there may well be more difficulty when one wishes to write mathematical models.

2.4   Modeling and Block Diagrams

In the car example a qualitative verbal modeling has been performed which has been represented schematically by the information-flow diagrams. The next step in performing a quantitative analysis is to make *mathematical models* for the blocks constituting this information-flow diagram; when this has been done the diagram contains only mathematical symbols and is generally then called a *block diagram*. Since the use of blocks and interconnecting lines to represent physical systems and their intercommunications may well represent an unfamiliar concept to some readers, we will now present the relevant principles in some detail.

(A) INPUT–OUTPUT CONCEPT

Every block represents some conveniently bounded subsystem or process, for which one or more input and one or more output signals can specify the functional connections with the other subsystems. Such a model is made as simple as possible consistent with reality, and in particular only the information-flow paths are emphasized, rather than the detailed energetic phenomena. Thus, in the car engine representation, the single input of accelerator position is shown as determining uniquely the output of propulsive road force. However, there are many other variables which might be considered as inputs since they do affect in some measure the engine output, for example, ambient temperature and altitude, gasoline quality, valve leakage, and the important variable already mentioned, car speed. The point is that in "nonpathological" cases none of these variables *significantly* affects the output (except for speed, which would normally be included in the model). However, if one of these variables is changing very greatly in a particular study, it should be included.

These points are equally relevant for an analogous biological force-producing system, the muscle. Thus, the tension produced depends upon the number of motoneurons firing and their spike frequency, as well as upon contraction velocity (comparable to car speed), but one could usually ignore external temperature and pressure, food which has been eaten recently, and breathing rate, to name only a few other possible variables.

(B) INFORMATION VERSUS ENERGY AND MATERIAL FLOWS

It would be natural to include as inputs the material flows of gasoline (and oxygen) and of biochemical fuel (and oxygen) to the two force-producing transducers discussed, and as outputs their respective products of combustion. It must be emphasized that such a representation would usually be confusing and inaccurate. To repeat a major point, the input-output (and black-box) concept is concerned with mapping the causal flow of information within the system. Under normal conditions the flows of gasoline and biochemical fuel are not independent variables, since they are either "oversupplied" or manipulated as dependent variables to provide the power supply demanded by the accelerator position and motoneuron excitation, respectively. Consequently they are of no further concern in understanding and predicting the dynamic behavior of their systems. However, for clarification, the inflow of necessary but dependent energy (or power) and materials may be shown on the block by some such symbolism as the broad arrow shown in Fig. 2.5. The criterion for omitting an input variable is therefore whether it is dependent upon other input variables. In some cases energy flow would be a prime independent input, for example, when gasoline, biochemical fuel, or oxygen could not be supplied fast enough under some condition of concern to the analyst. It should also be noted that material flows of fuel become energy flows within

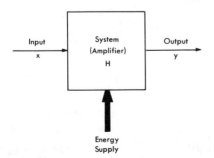

FIG. 2.5 The amplifier component.

their appropriate engine (transducer); on the other hand, some transducers use nonmaterial energy flows, for example, electric motors. Hence the broad arrow of Fig. 2.5 may represent either material or energy flow.

(C) AMPLIFIER

An amplifier is a device which increases (amplifies) the level of a signal, in either amplitude or power. Thus, amplifiers for electrophysiological use accept signals at very low voltage and power level (say 10^{-4} volt, 10^{-10} watt) and output at convenient recording levels (10^1 volts, 10^{-1} watt). The amplification is achieved because the input is able to *modulate* or *vary under control* a considerable electric energy supply represented by the broad arrow. This electrophysiological amplifier uses the same form of energy for all three variables, but such is not generally necessary. Thus, both the car engine and the muscle are amplifiers; the first has a low-power-level displacement-type input and a high-power-level mechanical-force output, derived by converting and modulating a chemical energy supply; the muscle, on the other hand, has a low-power-level electric-type (or biochemical-type, dependent upon one's viewpoint) input and a high-power-level mechanical-force output, derived by utilizing a biochemical energy supply.

(D) MODELING—TRANSFER FUNCTIONS

So far, all discussion has been qualitative in that no differential or other equations have been written to describe the phenomena occurring within the systems, the subsystems, or their symbolic blocks. The basic method of modeling is to write equations, usually of the differential type, to describe these input-output phenomena, and the art of modeling consists in reducing these equations to the least complexity which still accounts for the significant phenomena with which the analyst believes himself concerned. The differential equations must be derived from the various laws of physics, chemistry, etc., and from experimentally observed data.

Thus, Newton's second law of motion states that with constant mass the net accelerating force equals the product of mass and the acceleration in the direction of the force. Equation (2.1) is here repeated:

$$F(t) = ma(t) \equiv m\frac{dV}{dt} \qquad (2.1)$$

The alternative versions of Eq. (2.1) arise because acceleration physically equals the rate of change of velocity with respect to time. Both these variables and also the accelerating force vary in time, and are therefore termed *functions of time*. The causality

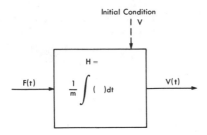

FIG. 2.6 *"Transfer function" for car inertia.*

indicated on the figure is that the force $F(t)$ is transformed by the car inertia system into a velocity $V(t)$ as output. To be able to write the required form

$$V(t) = f[F(t),m]$$

where $f[\quad]$ means "function of . . . ," it is sufficient in this case to integrate both sides of Eq. (2.1) and rearrange:

$$\int a(t)\, dt = \frac{1}{m} \int F(t)\, dt$$

that is,

$$V(t) - V_1 = \frac{1}{m} \int_{t_1}^{t} F(t)\, dt \qquad (2.2)$$

The term V_1 represents the initial value of $V(t)$ when the change occurred at t_1. Hence, the fundamental nature of the dynamic process described as *car inertia* is that the integral of accelerating force over an interval of time, divided by the mass, equals the change of velocity in that interval. Initial conditions affect only the actual magnitude, not the nature of the dynamic response. If, therefore, the analyst remembers as a general rule that all integrating actions are subject to an initial condition, he can omit explicit reference to it and simplify Eq. (2.2) as

$$V(t) = \frac{1}{m} \int F(t)\, dt \qquad (2.3)$$

The words *car inertia* may now be replaced by the symbolic form

$$\frac{1}{m} \int (\quad)\, dt \qquad (2.4)$$

where it is understood that the integration is to be performed upon the input variable of force to produce the output variable of velocity (Fig. 2.6). In addition, the initial condition may be explicitly shown, if desired, as a secondary input quantity.

Expression (2.4) is the precursor of what is termed the *transfer function*, that is, the ratio of output to input in "operational" form, for which the generalized symbol H will be used, namely,

$$\frac{V}{F} = H \qquad (2.5)$$

This technique is justified in Chap. 3, and at present should be considered only as a convenient shorthand. Thus, for car inertia we have

$$H = \frac{1}{m} \int (\quad) \, dt \tag{2.6}$$

(E) BLOCK DIAGRAM

The block diagram thus consists of a set of blocks interconnected by directed lines to indicate unidirectional cause-effect relationships.

The *lines* represent variables which are relevant to the desired analysis and which are physically measurable in the real system, at least in principle. When two blocks reciprocally interact, two interconnecting lines are necessary.

The *summing points* arise whenever the variables must be added or subtracted, for example, in the expression for net accelerative force (Fig. 2.4c)

$$F = F_e - F_r - F_g$$

The *blocks* represent those transfer functions of the various subsystems which can conveniently be separated from each other. In general, multiple inputs and outputs can exist for any block, although one of each often suffices in practice. An output variable may act as input to several other blocks, for example, $V(t)$ in Fig. 2.4d.

The *configuration* of block diagram for a given system is not usually unique, since there are multiple ways in which the differential equations may be represented in a block diagram. However, the cause-effect directions of some blocks will probably require inversion, and intuitively the analyst often feels that a particular configuration presents the causal relations of his physical system most usefully. For example, it seems physically meaningful that a change in velocity results from a force being applied over some period of time, thus justifying the cause-effect direction already used.

In principle the block diagram adds no information to that available in the describing differential equations. In practice, however, its graphical display of information-flow patterns can clarify the analyst's thinking about the system; furthermore, various simple techniques are available to "collapse" block diagrams into simpler patterns. Although this latter process is formally identical to the successive elimination of variables in a set of equations, the graphical one is less tedious and provides more physical "feel" for the process.

(F) COMPUTER SIMULATIONS

Mathematical models of physiological systems usually need to be studied on computers because they are inevitably too complicated for convenient analytical solution. Furthermore, even when the

analytical solution is known, it is usually sufficiently complicated that numerical solutions must be obtained on a computer,
and in this case it is often more economical to program the original
describing equations rather than the analytical solution.   On
an analog computer the electric circuit is specified directly by
the block diagram, and therefore corresponds to the real system
being modeled since both models satisfy the same differential equations.   Because of this physical correspondence, and because the
analyst can manipulate various parameters directly, the analog
computer is said to provide a *simulation* of the real system.   Since
analog computers are relatively cheap and moderately accurate,
they are ideal for studying such dynamic systems as the many
autonomous regulating systems, but for complex systems involving
logic, algebra, or high accuracy the digital computer often becomes
essential.

2.5   Population-growth Model

An extremely simplified model of population growth is now presented, giving rise to the simple first-order differential equation
whose solution is the classic exponential growth or decay curve.
Consider a population whose statistics are measured at suitable
unit intervals of time Δt; at time $(n\,\Delta t)$ the population is conveniently written as $P(n\,\Delta t)$, or more compactly P_n (Fig. 2.7).
The rate of growth of population dP/dt at time $(n\,\Delta t)$ is then
approximately equal to the recent increment of population ΔP,
divided by the time interval Δt, viz.,

$$\frac{dP}{dt} \approx \frac{P_n - P_{n-1}}{\Delta t} \tag{2.7}$$

Figure 2.7 shows that this formula can only be exact as Δt becomes
very small, and indeed it then constitutes a simple definition for the
differentiation of a continuous function.   In principle, populations

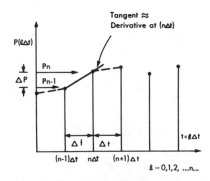

FIG. 2.7   Approximation to derivative.

are not continuous functions, and furthermore a time increment which is small relative to gestation period, etc., would not normally be very useful in studying population growth. Provided, however, that large populations are being studied, Eq. (2.7) allows the rate of growth to be evaluated from experimental data. It is the most elementary of various approximation formulas possible.

The simplest growth model is obtained by noting that, to a reasonable accuracy, the rate of growth is proportional to a constant multiplied by the population:

$$\frac{dP(t)}{dt} = kP(t) \qquad t > 0 \tag{2.8}$$

where k is the net birth rate, which is the birth rate minus the death rate. In fact the net birth rate is a function of the distribution of population ages, availability of food supplies, etc.

Equation (2.8) is mathematically a *first-order, linear, ordinary differential equation* with *constant* coefficients (k and the unity constant for dP/dt). It is, in fact, the simplest dynamic model which can be written. Very few differential equations are analytically solved in this book, but it is done for this simple case. First Eq. (2.8) is rewritten as

$$\frac{dP(t)}{P(t)} = k \, dt$$

and then both sides are integrated over the time interval 0 to t:

$$\log_e P(t) - \log_e P_0 = kt \qquad t > 0 \tag{2.9}$$

where $\log_e$ ( ) equals logarithm of ( ) to base e, or natural logarithm, that is,

$$P(t) = P_0 e^{kt} \equiv P_0 \exp{(kt)} \qquad t > 0 \tag{2.10}$$

Equation (2.10) gives the classic exponential curve of the population explosion (Fig. 2.8), so direly predicted by Malthus [2.2] in the eighteenth century and now so urgent a problem in the twentieth. The positive value of coefficient k implies a net excess of births over deaths. If the reverse is the case, k is negative and a decaying exponential curve results (Fig. 2.8). This decay curve is a fundamental component of the response of stable dynamic systems and frequently recurs in this book. Note that the curves for k negative and positive are mirrored continuations of each other.

The time derivative of an exponential curve is of some interest; Eq. (2.10) yields, after differentiating,

$$\frac{dP}{dt} = kP_0 e^{kt} \qquad t > 0$$

and using Eq. (2.10) again

$$\frac{dP}{dt} = kP \qquad t > 0 \tag{2.11}$$

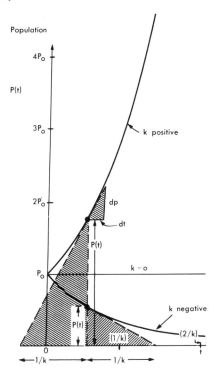

FIG. 2.8 *Exponential population curves.*

In fact, this is merely the differential equation with which we started, Eq. (2.8), but the procedure provides a useful check that the solution of Eq. (2.10) is correct. Because the derivative is graphically the tangent to the curve, the similar triangles drawn on Fig. 2.8 may be used to show that the time required for the tangent of the exponential curve at any point $P(t)$ to pass through zero is $1/k$. This quantity $1/k$ is an important parameter of the system's dynamics, having the dimensions of time (actually k has the following dimensions: number of people per unit time divided by number of people). The quantity kt is therefore dimensionless and provides a convenient abscissa scale in many cases; since its magnitude is proportional to time, it is usually termed *dimensionless time*. Therefore, if we define

Dimensionless time $\qquad\qquad \tau = kt$

Dimensionless population $\qquad \rho = \dfrac{P}{P_0}$

Eq. (2.10) becomes

$$\rho(\tau) = e^{\tau} \qquad \tau > 0 \tag{2.12}$$

There are now no free parameters (coefficients) to this expression and hence a unique growing exponential curve is obtained for all positive k, and a unique decaying exponential for all negative k.

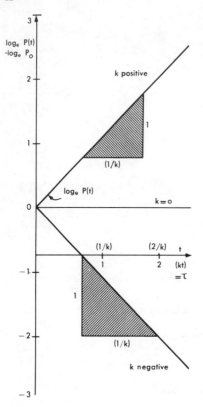

FIG. 2.9   *Semilogarithmic plotting of exponential curves.*

Hence Eq. (2.12) is called the standard form of the first-order system's response.

The analyst must often search for a law to fit experimental data, and a straight line is obviously easier to distinguish by eye than any other curve (by lining it up so that the eye looks down it). The well-known technique for producing this when the suspected law is exponential is to use a linear abscissa and a logarithmic ordinate. The result is a straight line, as shown in Fig. 2.9, for positive, negative, or zero k values, which Eq. (2.9) shows to be expected. Note that the initial population P_0 merely provides an offset to the ordinate scale which is otherwise independent of P_0. When dimensionless time τ is used, the slope becomes ± 1 with natural-base logarithms; if base-10 logarithms are used, then the slopes are smaller, obeying the relation

$$\log_{10} x \equiv (\log_{10} e)(\log_e x) = 0.434 \log_e x \qquad (2.13)$$

Block Diagrams.   Figure 2.10a shows the information-flow diagram for this example and essentially repeats the relations derived in the text. Figure 2.10b presents the corresponding symbolic block diagram. There is an immediate difficulty regarding input-output, however, since this system has no independent input variable, but only the initial condition P_0. On the other hand, the simple positive-feedback closed loop which exists when k is positive makes clear that the regenerative circulating effect produces an exponential rise in population. Of course, the population cannot increase indefinitely at an increasing rate; instead, k typically depends upon some function of the actual population P and available food supplies, so that an external input would be involved in a more realistic mathematical modeling. This system is not actively controlled, for when k is negative it is passively controlled, and when k is positive its control is unstable in any meaningful sense. As a general rule, positive feedback tends to produce unstable systems, whereas negative feedback tends to stabilize systems, at least within certain dynamic limits.

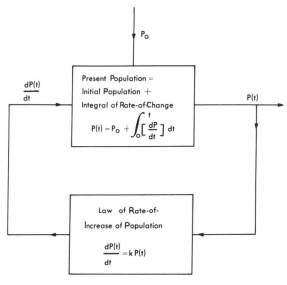

(a) Information Flow Diagram

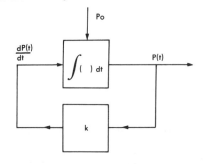

(b) Symbolic

FIG. 2.10 Block diagrams for population model.

This elementary example has been developed in some detail, because it represents one of the basic dynamic phenomena on which understanding of more complex systems can often be built.

2.6   The Pupil-control System

A typical biological control system, the reflex regulation of pupil area by the iris, is examined qualitatively in this section.   The eye represents the largest sensory, or data-gathering, instrument for many classes of animals.   It must have excellent dynamic response in order to transmit, and perhaps resolve, the rapid changes which occur both spatially and temporally.   The perception and processing of visual information partially occur in the retinal cells and their associated second- and third-order neurons.   The third-order neurons particularly are considered part of the brain, and their "computing system" is difficult to understand, although great progress is currently being made.   This aspect of the system certainly fits into the "complex" category defined in Chap. 1 as cybernetics, and therefore will not be dealt with in this book. However, in order to gather its relevant data the eye has several control systems which do permit useful modeling by the techniques of this book.

Eye Positioning.   Each eyeball must be positioned in its orbit for any given target position, and three antagonistic pairs of muscles work in coordinated fashion to achieve this (Sec. 12.2).   When a target is moving, the eye must track, which is the classic function of a servomechanism such as an automatic gun-aiming device.   The analogy applies further in that just as we have rejected consideration of the complex data-processing system by which the retinal image provides the necessary tracking commands, so also do we not need to consider the details of the radar system by which the gun-aiming tracking commands are obtained in order to understand how the gun-aiming system works.

In binocular vision, target tracking at varying distance requires also that the eye movements be slightly different in order that the visual axis of each eye may be on target.   This feature is called *vergence* (convergence and divergence).

Focusing.   The corneal lens complex (Fig. 2.11) constitutes an effective, but optically complicated, convex lens to focus the image of a perceived object upon the retina.   *Accommodation* is the process of changing the lens power until the image is in focus, when the object is at varying distances, and in mammals is achieved by action of the ciliary muscle which modifies the convexity of the lens.   In most bony fishes, accommodation is achieved by moving the lens with respect to the retina [2.3], which of course is the method engi-

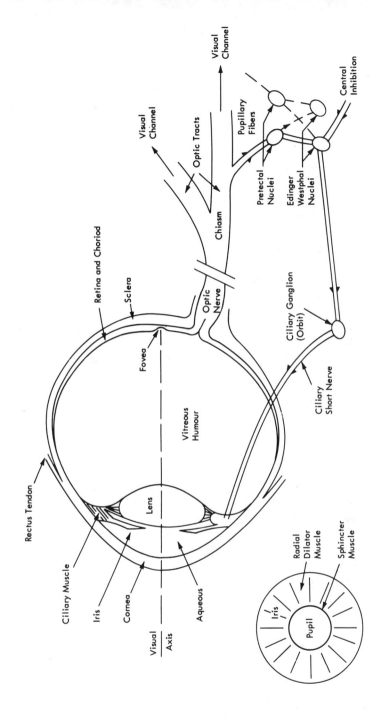

FIG. 2.11 *Pupil-control pathways.*

neers have found easiest to provide in their equivalent focusing system for the camera.

Pupil-control System. The iris (Fig. 2.11) provides an opening for the passage of light, called the *pupil*, while the iris itself contains black pigment which effectively occludes all passage of light elsewhere. There are two antagonistic muscles in the iris for variation of pupil area, the radial dilator muscle and the sphincter muscle. The sphincter muscle is more developed and is apparently the most significant effector in providing the working basis of an autonomous, reflex, closed-loop system for control of pupil diameter.

One major function of the pupil-control system is to help produce clear retinal images for close work by excluding the lens periphery where optical aberrations are greatest, and by increasing the depth of focus [2.3, p. 421]. It is interesting to note that the control systems for convergence, accommodation, and pupil constriction must all work together to produce clear images for near vision. This raises questions concerning interactions and performance criteria for complex systems, which are discussed briefly in Sec. 15.2. Another major function of the pupil-control system seems to be that of regulating the illumination reaching the retina, and this aspect is now explored in more detail.

(A) INFORMATION–FLOW DIAGRAM FOR PUPIL [2.4–2.7]

In considering the pupil's reflex dynamic response to changing light intensities, the sphincter muscle seems to be the main effector as already mentioned. The physiological information-flow diagram is simplified if the sympathetic innervation of the dilator muscle is ignored, but the "push-pull" type of operation involving both muscles can certainly be included if such seems necessary. One further restriction is made—ignoring the cross coupling of neural pathways between the two eyes. This coupling results, for example, in the consensual reaction of both pupils being contracted together when extra illumination is applied to one eye only. The closed-loop chain to be studied is not, however, affected by this omission. Figure 2.11 shows the essential physiological pathways by which the retinal illumination is fed back to control the pupil diameter. Information from all over the retina is used in this afferent or feedback path, although the necessary data processing is not understood, and the resulting information travels in pupillary fibers which are conveyed within the optic nerve to the chiasm. They leave the optic tract and synapse in the pretectal nucleus. Fibers then pass to both the Edinger-Westphal nuclei, enter the ciliary ganglion, and finally reach the sphincter muscle via the short ciliary nerve. This is the efferent or outflowing pathway from the CNS.

The physiological reflex arc is now closed because retinal illumination may pass through a reflex arc to modify the pupil area and

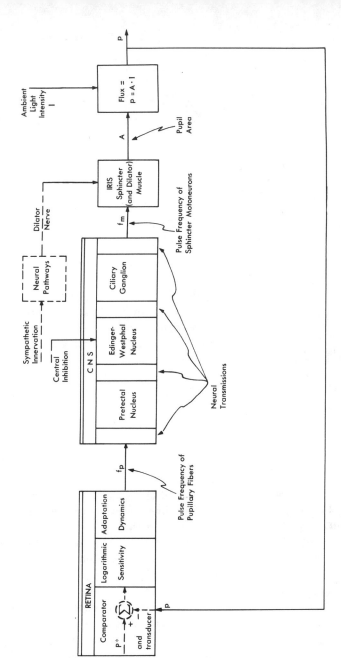

FIG. 2.12 Information-flow diagram for pupil-control system.

27

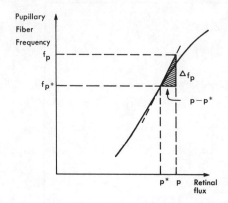

FIG. 2.13   *Equivalent comparator action of nonlinearity.*

hence the retinal illumination itself. This physiological diagram can be converted into an information-flow diagram rather easily (Fig. 2.12), except that the variable being controlled by this system must be defined, and any known neurophysiological phenomena should be included in the blocks.

Retinal Blocks—Comparator. It is postulated that the pupil reflex exists to maintain the total value of retinal light flux p near to, or constant at, some desired value p^*,

where $p = AI$ *(2.14)*
   A = pupil area
   I = intensity of illumination

(This formula assumes uniform illumination and undistorted transmission within the eye.) The basic requirement for a negative-feedback stabilizing action is merely that when retinal flux exceeds p^*, increased stimulation of the sphincter muscle should ensue so as to reduce the pupil area correspondingly, and vice versa. This function does not explicitly require that the desired flux p^* appear in an arithmetic-differencing comparator to produce an error signal, although this is often a conceptually useful device. Since the device may appear contrived to the physiologist, however, a more natural graphic alternative follows. Let us presume that a possible steady-state curve relating retinal flux p as input to pupillary fiber frequency f_p as output is as shown in Fig. 2.13. If the regulatory system is stable, then at some given conditions of illumination, etc., the pupil area is A^*, the flux is p^*, and the neural frequency is f_p^*. If one approximates this arbitrary curve as a straight line about the operating point,

$$\Delta f_p = k(p - p^*) = -k(p^* - p)$$ *(2.15)*

where Δf_p is conveniently defined as the "error signal" or difference in neural frequency from the equilibrium condition, namely,

$$\Delta f_p \triangleq f_p - f_p^*$$ *(2.16)*

Equation (2.15) does not directly implement comparator action because the sign inversion of negative feedback is not yet fulfilled. In the pupil system, sign inversion occurs in the actuator block, however, because increasing muscle motoneuronal pulse frequency f_m *decreases* pupil area A. Together, therefore, the retina and the sphincter muscle produce comparator action equivalent to that of Fig. 2.12. The problem of linearizing such nonlinearities as shown in Fig. 2.13 is considered in Sec. 3.5, and for the present we need only state that linearization is often permissible for small signal variations provided that the curve does not change slope abruptly.

Retinal Blocks—Logarithmic Sensitivity. The retina is able to accept light intensity variations over a range of 10^8 at least, with the corresponding pupil area changing by a factor of only about 5 after suitable time has elapsed for retinal adaptation. Over a central range of intensity of about 10^6 there is a linear law relating A and $\log I$:

$$A = A_0 - k_1 \log I \qquad (2.17)$$

which is the well-known Weber-Fechner law. This is illustrated in Fig. 2.14, which is adapted from [2.8] [see also 2.3, p. 421].

Retinal Blocks—Adaptation Dynamics. The logarithmic sensitivity of the retina is the steady-state result of a neurochemical process whose details are not discussed here. However, when the light intensity changes there is a slow transference from the old

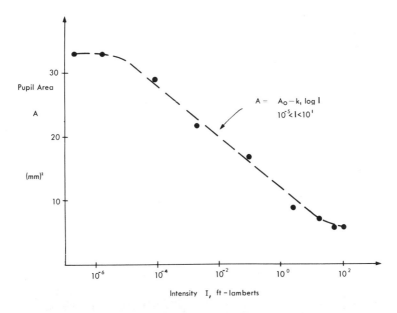

FIG. 2.14 *Logarithmic sensitivity of pupil area to light intensity.* *(Adapted from Wagman and Nathanson.)*

adapted pupil area to the new one. The time required for adaptation varies with the direction of light change; thus *light* adaptation is relatively fast and is completed in about 1 min, whereas dark adaptation occupies up to 30 min [see 2.3, p. 430, for experimental data].

Complete Retinal Block. The complete retina has been shown as three directly joined blocks, comprising the functions of comparator, logarithmic sensitivity, and adaptation dynamics. Physiologically it is very difficult to separate them into different processes, because they are inextricably mingled in the same neurochemical operation. A simple criterion for separation is the ability to show causal input-output variables between successive blocks, and this certainly cannot be done here. Thus the ordering of the three blocks is to some extent arbitrary and only the overall retinal block can satisfactorily show input-output variables; these latter are the total retinal flux p and the pupillary fiber pulse frequency f_p, respectively, with the assumed reference input flux p^* being an internally generated quantity.

Transmission through the CNS. The information flow through the CNS en route to the sphincter muscle can superficially be modeled rather simply. Neurons can be considered as simple transmission lines characterized by a time delay due to the finite speed of transmission. This speed is 70 to 120 m/sec in the larger myelinated fibers common in skeletal-muscle control, so that the "cable" transmission delay in the pupil system is typically less than 1 msec, which is negligible with respect to others in the system. (To some extent smaller and slower fibers are also involved, as shown in Sec. 12.4.)

The synaptic junctions in the pretectal nucleus, the Edinger-Westphal nucleus, and the ciliary ganglion give rise to delays of the order of 1 msec, due to the synaptic release and diffusion of neurochemicals. In general, the extremely large number of synaptic connections for each neuron in nuclei, ganglia, etc., give rise to poorly understood transfer characteristics. However, the pupil system seems to be relatively directly controlled so that perhaps not very much modification of signals occurs at the named junctions. In passing we should note, however, that the Edinger-Westphal nucleus is the junction for crossover innervation from the other pupil and inhibition from central areas such as the cortex and diencephalon.

Information Content of Neural Signals. In almost all the systems considered in this book, continuous analog-type variables are being controlled (pressure, temperature, muscle tension, etc.). From this and other reasoning it seems probable that the phenomenon of discrete neural pulses (spikes, action potentials), which char-

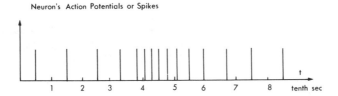

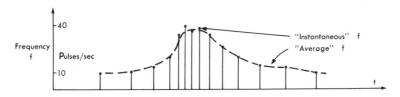

FIG. 2.15 Simple pulse frequency coding in neuron.

acterize neural transmission, does not imply a "digital-computer" type of information processing by nerves, but rather that the reciprocal of the average time between pulses is the significant variable. This reciprocal is the average pulse frequency and is essentially continuous (Fig. 2.15). It is therefore ideal for representing continuous physical variables, although of course the functional relationship may be nonlinear, for example, the logarithmic sensitivity already discussed. If this model were sufficient, neural transmission would be equivalent to continuous analog-type transmission such as the representation of physical quantities in the analog computer by varying voltages. In fact, nature has added a second technique for increasing total signal strength, that of the recruitment of more neurons. Thus, the term *pulse frequency* is henceforth intended to encompass the total signal strength of both frequency and recruitment.

Pupil Effector—The Sphincter Muscle. The output of the CNS block is the sphincter motoneuron pulse frequency f_m, and this is the direct input to the sphincter muscle, the primary effector mechanism that produces pupil-area changes. The dilator muscle condition is obviously also relevant and is therefore included as an auxiliary input. Finally, the output of the iris is pupil area. The operation of this block involves force and mass, and is therefore inevitably dynamic.

Controlled Process—Retinal Light Flux. The retina is subject to an uncontrollable and unpredictable ambient light intensity. We have presumed that the retina attempts to regulate, or to maintain essentially constant, the total light flux, at least until the slower adaptation process has occurred. The total flux is the product of this intensity and pupil area [Eq. (2.14)], so that regulation is achieved if A decreases as a result of I increasing. Thus the controlled process is the algebraic relation among A, I, and p, and is a little unusual in that there are no dynamics involved; that is, p is propagated from pupil to retina with the speed of light, resulting in a time delay which is negligible with respect to the physiological lags.

The information-flow diagram for this closed-loop system is now complete, and its dynamic behavior will shortly be investigated. At this point it will prove valuable to generalize the structure or "functional anatomy" of closed-loop control systems as much as possible, so that an ordered approach to the analysis of many analogous systems may be possible. For such a generalization, we will turn to the well-developed methodology of control engineering.

2.7   The Generic Structure of Control Systems

(A) GENERAL

Control-systems methodology has developed around the generic structure shown in Fig. 2.16. Even in engineering this is only a stylized model, and in practice there may be many differences of detail, especially the inclusion of extra internal loops and multiple inputs. Nevertheless, the model does reveal most essential information-flow features, and this is probably also true for physiological systems. This structure is consistent with the very simple form given in Fig. 2.3a, but in addition shows the features necessary in the active control element.

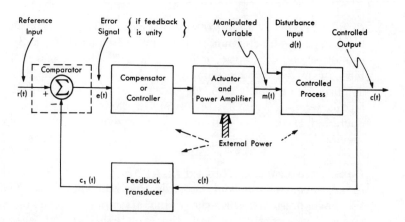

FIG. 2.16   Generic structure of control systems.

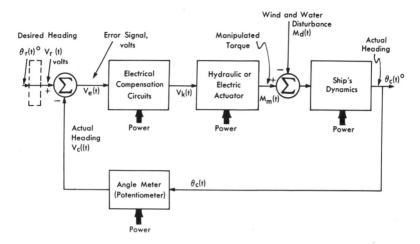

FIG. 2.17 *An engineering servomechanism—power steering of ship.*

(B) SERVOMECHANISM AND REGULATOR

Control systems may usefully be divided into two classes—servomechanisms and regulators. The *regulator* (or *homeostat*) has the less ambitious criterion of maintaining the controlled output $c(t)$ at some constant desired reference level $r(t)$, and the system error $e(t)$ is therefore the difference between the desired and achieved levels. The regulator's dynamic problem arises in correctly programming "manipulated variable" action $m(t)$ to compensate for the effect of the disturbance input $d(t)$ and for changes of parameters in the loop's subsystems. At intervals the reference input (set point) may change to some new level, but usually it is more important that the transient response to this change be stable than that it be as fast as possible.

The *servomechanism* has the extra and primary objective of "tracking," or following slavishly, a variable reference input $r(t)$ as well as the regulating function just described. Thus, as an example, the power steering of a ship (Fig. 2.17) operates mostly as a regulator in transoceanic passage, but then as a servomechanism during rapid in-port maneuvering. Clearly, the difference between servomechanism and regulator is only one of performance emphasis, but this may well lead to important changes in design philosophy.

Typically the designer's task is to assemble components to achieve the desired overall function under satisfactory control and economy. The adjective *satisfactory* implies a performance criterion by which the goodness of performance can be judged, and these are discussed in Chaps. 11 and 15. The dynamic problems involved in control arise most typically because of "inertia" or "sluggishness" in the controlled process, although other components of the control loop often produce dynamic lags also. Dynamic components usually involve energy storage, which is a subject discussed in Chap. 4.

(C) STRUCTURAL COMPONENTS OF CONTROL SYSTEM

Controlled Process. The controlled process is the logical starting point in considering control systems. It is the basic process (pupil, heating of body, ship dynamics, catalytic cracking tower) which must transform some material or information inputs into desired output states under good control despite the presence of disturbances and parameter changes. The process may often involve the controlled utilization of much power. The sluggish nature of the process usually requires description by differential equations, and often the system designer must provide or evolve phase-leading elements elsewhere which will provide "antisluggishness." Even in the unusual pupil case where the process has no sluggishness, it does inevitably appear elsewhere in the loop, particularly in the sphincter muscle.

Comparator. The comparator is directly instrumented in engineering systems as a voltage difference, pressure difference, etc. Its basic input-output characteristics are shown graphically in Fig. 2.18, where the negative slope of the output-versus-feedback lines is the essential characteristic of negative feedback. The comparator action is directly instrumented in some biological systems also, such as by the eye and CNS in the servomechanisms for visual target tracking and manual manipulation.

However, the comparator probably does not exist as a separate functional component in most biological homeostats. Thus, some loops maintain stable operating levels because an equivalent comparator and negative-feedback action exists in some loop compo-

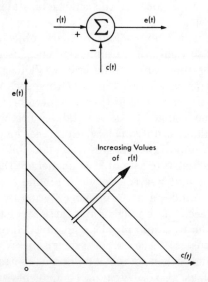

FIG. 2.18 Basic comparator characteristics.

nent, as illustrated in Fig. 2.13; that is, the correcting signal change is generated as a function of the difference between the stable operating level of the system and the present value.

Another biological alternative is a nonlinear function of two variables, or a parameter modifier, of which one example would be the output pulse frequency f_c of a neuron subject to one excitatory input f_a and one inhibitory input f_b. If we postulate the extremely simplified division relation [2.9]

$$f_c = \frac{f_a}{f_b} + f_0 \qquad (2.18)$$

where f_0 is the quiescent pulse rate, then the two plots for f_0 versus f_a and f_b, respectively, are as given in Fig. 2.19. In Sec. 3.5 it is shown that a linearized relation for small changes about an operating point is

$$\Delta f_c = k_1 \Delta f_a - k_2 \Delta f_b \qquad (2.19)$$

That is, for small changes about some operating conditions f_{a0}, f_{b0} the change Δf_c is as given in Eq. (2.19). In this equation the constants k_1 and k_2 equal the partial slopes shown in Fig. 2.19, and for empirical curves these constants may be evaluated numerically by graphical means. This linearized relation shows that differential comparator action also exists for such a case, that is, when f_a is the reference input and f_b the feedback of controlled output; compare Fig. 2.19*b* with Fig. 2.18.

As we have shown that the linear error-activated comparator action can be abstracted from nonlinear and arbitrary control functions, it should now be emphasized that this simplification is not necessary in practical modeling; on the contrary, in computer modeling of biological systems one should incorporate such relations as Eq. (2.18) or Fig. 2.19*b* whenever these seem important. However, the conceptual ability to abstract such linearized error relations should justify the use of the latter technique throughout the book; or stated with changed emphasis, although the mathematical forms and the phenomena may look different, their function may still be expressible by the comparator concept.

Compensator or Controller. The stylized control system uses the system error to correct its performance. In practice the feedback circuits may include signals from other points, such as the ship's turning rate and acceleration in Fig. 2.17, but such refinements do not significantly affect the principle of error actuation.

The system error is fed into a compensation or controller block, which the engineer usually designs to "speed up" the signal and hence reduce system lags, and also to cancel out long-term errors. This improvement is usually brought about at a low power level,

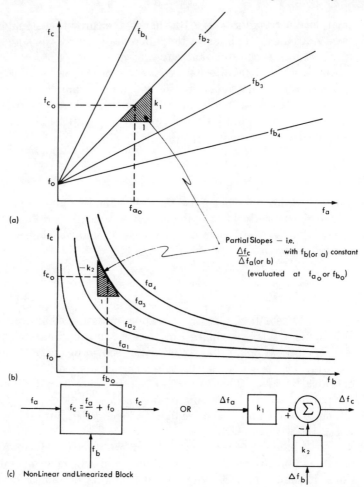

FIG. 2.19 *Equivalent comparator action of nonlinear dual-input process.*

by circuits which may be electrical, mechanical, hydraulic, etc. On the other hand, biological systems have not often developed with such liberality of components, so that this block may often be omitted. The improvement of the biological system's dynamic response more typically seems to be implemented by the feedback transducer, as indeed the comparator action is also. The retina is a case in point and other examples will be given.

Actuator. The actuator is necessary to provide the manipulated variable input to the controlled process. Almost invariably, amplification of power level is involved, and often also transduction from one energy domain to another. There is not usually any need to provide dynamic improvement at this stage, and unfortunately some deterioration of dynamic performance may often result in practice because of the high power level of this final control element.

In engineering systems the actuator usually can provide both positive and negative values of its manipulated variable output, for example, voltage from amplifier, thrust from aircraft engines. However, there are exceptions, such as the car engine for propulsive force and the brake for retarding force. In biological control, actuators are typically paired for bidirectional control; antagonistic muscle pairs provide an obvious example; shivering and sweating provide another pair (Sec. 3.3). In the pupil-control system only one actuator seems needed by virtue of normal operation around a nonzero value of dilator muscle tension (Fig. 2.13), but it is significant that the antagonist pair exists.

Feedback-measuring Device. A feedback-measuring device is obviously a very important link in the closed loop. It is nearly always a separate functional unit in engineering systems, for example, the angular potentiometer of Fig. 2.17. In biological systems it is also often a separate receptor organ, for example, pain and pressure receptors and muscle proprioceptors. In other cases, however, the "measured" variable may bring about a parameter modification in already existing organs, for example, the reception of temperature in the hypothalamus. In other cases again, reception may be an auxiliary function of an organ which mainly exists for another purpose; for example, in the pupil-control system the retina measures the total flux as a secondary function.

In poorly designed systems the dynamic response of the transducer may be very slow; a classical example of this is perhaps the mercury-in-glass thermometer for monitoring of patients, whereas thermocouples and other currently available devices are essentially instantaneous. Nature, on the other hand, seems to have developed feedback receptors which "speed up" the response information by including "rate sensitivity" or derivative action; in large measure this component then also doubles for the engineering function of the controller or compensation circuit, as briefly discussed already. This subject will recur in further biological examples.

Feedback receptors are often transducers in that they convert energy from one form to another, and their power level is usually low. There is some conceptual advantage in defining the energetic nature of different types of elements, and this is done below.

(D) ENERGETIC DEFINITION OF SYSTEM COMPONENTS

Information flows have been established as the important input-output measures of a system, with the necessary but dependent energy flows not normally being important in determining system dynamics. However, all system components involve energy flow,

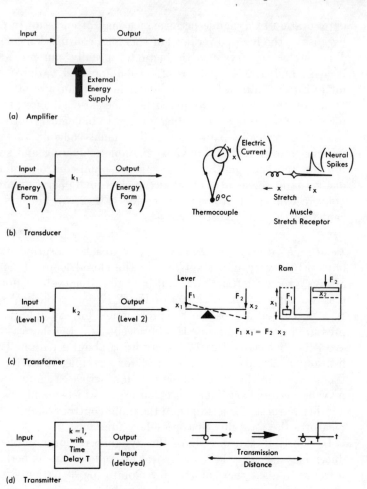

FIG. 2.20   *Energetic definition of system components.*

and some useful nonexclusive operational definitions can be made on this basis.

Amplifier (Fig. 2.20*a*).   An amplifier increases either the effort, the flow, or the power level of the information-flow process under study.   Muscles are examples of biological amplifiers, and in general most actuating components of control systems are amplifiers. *Amplifying transducers* also satisfy the following definition of a transducer.

Transducer (Fig. 2.20*b*).   A transducer converts energy from one form to another; for example, a thermocouple converts thermal energy into electrical energy.   An amplifying transducer converts energy but in addition raises it to a higher level by utilizing another

controlled-energy source.   Thus, the insect's olfactory organ detects
a few molecules of given biochemical by using a controlled internal
energy source to increase the energy level of its afferent neural signal.

Transformer (Fig. 2.20c).   A transformer changes the effort level,
without changing the power level or the form of energy; examples
are levers, electric transformers, hydraulic rams, and gears.   In biological systems the lever is widely used to apply muscular tensions
to the skeleton.   Sometimes a mechanical advantage or amplification of force level is obtained, but more frequently the reverse is
true and the muscle must exert more force than is achieved at the
point of useful action, for example, the jaw muscles [1.3].

Transformers do not amplify the power level, since they are
passive systems; in fact, there is always a small energy loss in
friction.   Apart from this the conservation-of-energy principle is
obeyed, so that, in the lever example,

$$F_1 x_1 = \text{work done} = F_2 x_2 \qquad (2.20)$$

(see Fig. 2.20c).   This equation shows that if the force level is
amplified, a smaller displacement is achieved by the output.

Transmitter (Fig. 2.20d).   The transmitter transmits signals over
a distance, ideally without loss of fidelity or change of energy level.
The neuron is an element which closely approaches this definition
except that in detail it utilizes controlled energy sources along the
way and is therefore an amplifying transmitter.   Hormonal transmission in the blood stream does not amplify, and is even subject
to some dispersion.   The particular process of radio transmission
of instrumental information is often called *telemetering*.

2.8   The Dynamic Response Characteristics
of the Pupil-control System

(A) PURPOSE

An information-flow diagram has been prepared in some detail for
the pupil-control system (Sec. 2.6) and compared with the generalized structure of control systems (Sec. 2.7).   Such information-flow
diagrams have value in themselves, since the causal flows of information can be readily appreciated.   However, the information-flow
diagram and more particularly the block diagram are also useful
for the mathematical manipulations and computer simulations
which are generally necessary in order to obtain useful information
about the system's dynamic response characteristics.   Dynamic
characteristics are very important because control systems are normally not at rest.   Thus, it is unusual for steady-state conditions
to exist, as exemplified by the pupil system since its area normally

fluctuates randomly and continuously over a small range, quite apart from major variations due to lightin-tensity changes.

In the methods of analysis which are developed in succeeding chapters the output of a system is calculated from a knowledge of both the input and the system characteristic. The biologist, however, is usually faced with an existing but unknown system, into which there is an input and from which an output results. His problem is then to infer the system's characteristics, that is, to make a model. The control engineer also faces this problem with systems such as oil refineries, where the complexity defies simple a priori model making. The problem becomes less difficult when the input may be manipulated by the experimenter than when it must be left to vary according to the system's normal operation, but in either case a study of the input-output behavior may enable one to argue back to some postulation of the system's dynamics. In the present section the qualitative natures of both steady-state and transient pupil responses are presented, in order to illustrate several control-system phenomena without mathematical manipulation.

(B) STEADY–STATE RESPONSE

A steady-state response clearly exists when inputs and system variables are constant. In the general case this response can be calculated by solving a set of algebraic equations, because in the steady state there are by definition no rates of change; hence, all derivatives of the differential equations are zero, for example, the acceleration $dV/dt = 0$ in the steady-state speed condition of the car. If the steady-state relations are nonlinear, a graphical solution technique may offer advantages, as illustrated in Fig. 2.21 for the pupil [2.7]. For exemplification, some rather arbitrary but plausible relations have been drawn between the four system variables, pupil area A, retinal flux p, afferent neural frequency f_p, and efferent neural frequency f_m (Fig. 2.21a). In addition, the $p:A$ plot is shown for several arbitrary values of ambient light intensity I. Note that the assumption of steady-state conditions allows the retinal adaptation and other component dynamics to be ignored. If the curves are conveniently redrawn in the back-to-back fashion of Fig. 2.21b, the steady-state solution constitutes those values of the four variables that allow a traversal around the diagram to close back to the starting value. In practice one guesses a value of pupil area, say A_1, and traverses the "loop"; if the result does not close on to the starting value A_1 then one "iterates" successively until a solution is reached. It should be noted that it is the closed-loop nature of the system that prevents one from immediately reading off the steady-state values, and also that it is the negatively sloped characteristic of f_m on A that provides a negative-feedback closed loop and hence ensures that the iteration will converge.

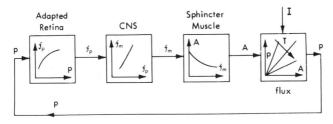

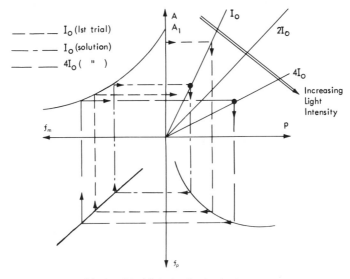

(a) Steady–State Information Flow Diagram

(b) Graphical Solution for A, p etc.

FIG. 2.21   Steady-state solution of pupil system.   (*After Jones.*)

(C) DYNAMIC RESPONSE

A dynamic system can be characterized usefully by its response to certain stylized inputs, of which the most commonly used are:

1. The transient inputs of impulse and step (Fig. 2.22), for which the outputs are the impulse response and step response, respectively

2. The sinusoidal input which produces the steady-state frequency response

Random inputs are also used, and all three types reveal the same mathematical information about linear systems, at least in principle. Frequency response and random signal analyses are not considered until later chapters, however, since some mathematical background is needed.

Figure 2.22 shows the four different pupil responses typically obtained [2.6] when *pulse* and *step* inputs of light intensity are used, of both "light" and "dark," that is, the eye is subjected to either

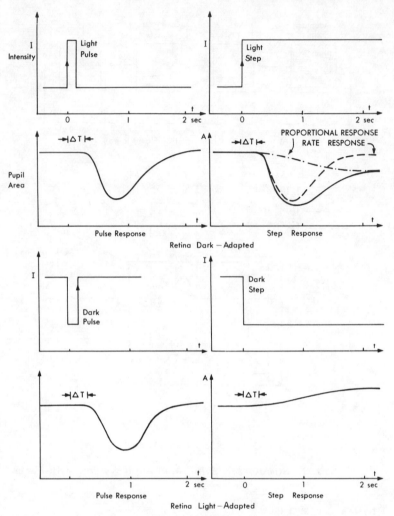

FIG. 2.22   *Transient responses of pupil system.*   (*From Clynes.*)

brief or sustained increases and decreases in light intensity from some retinally adapted medium-intensity level. An *impulse* is defined to be infinitely short, and therefore is only a mathematical ideal, but input *pulses* can usually be made sufficiently short so that no response is exhibited by the system until after the pulse is completed. In this case the realizable pulse is indistinguishable from the unrealizable impulse for all practical purposes. This practical equivalence is important because for mathematical manipulations the impulse is much more tractable than the pulse.

These responses are obtained with the eye bathed by the light-intensity source, and since the response is completed within several seconds the retinal-adaptation characteristic plays no significant part. A pupillometer is used to monitor the pupil diameter, or area, continuously. This instrument [2.4] irradiates the eye with unperceived infrared light, utilizing the fact that the pupil and iris reflectances are different, to obtain an electrical voltage measure of the pupil size.

A somewhat variable time delay of approximately 250 msec occurs before any responses are noticeable to these input stimuli. It is generally thought to be due mostly to the retinal transduction process from light energy to nerve pulses [2.5, 2.7], since neural transmission delays are two orders of magnitude smaller. The four illustrated responses present a challenging proposition to the model-making analyst, for they clearly cannot be produced by a linear system, in which reversal of input polarity merely reverses the polarity of the response without affecting its pattern. However, the pupil responds to both light and dark pulses with a rapid contraction followed by a slower dilation back to the original area. This response is complete within about 2 sec, during which the adaptation of retinal sensitivity can be ignored. The apparent paradox can be resolved by postulating that the pupil's neural signal from the retina has some component dependent upon the rate of change of light intensity, and in particular dependent only upon positive rates of change of intensity. The total dark pulse response is then constituted essentially of zero response to the dark "step component" followed by a standard contraction-dilation response to the closely following light "step component" (see arrows on the pulses). The difference between the dark and light step responses is also resolved by this model. Thus, the dark step response is essentially a relatively slow compensating dilation without overshoot, and this is often defined as a proportional response. The light step, however, excites both the proportional response and the rate-sensitive response, so that the total response overcontracts rapidly and then dilates back to a new steady state, until adaptation subsequently occurs.

It should be emphasized that even though these curves do not present the full story, they do already reveal unexpected and fundamental phenomena which the model maker must explain. Actually these generic forms of response are subject to quantitative differences due to a number of other variables, including input intensity level, amplitude of change, and different sequential patterns [2.10, 2.11]. The experimental data are well documented in [2.6], [2.10], and [2.11] together with appropriate mathematical models which we ignore in this qualitative chapter.

The emerging picture is that this control system provides rapid compensatory pupil-area changes against changes of light intensity; in this action a faster (and overshooting) response occurs to increasing intensity changes than to decreasing ones, which may perhaps be viewed teleologically as a desired protective device. In addition, retinal adaptation, again with more rapid change to light increase than decrease, adjusts the retinal sensitivity to achieve regulation over a longer subsequent interval of the order of minutes [2.3, 2.12].

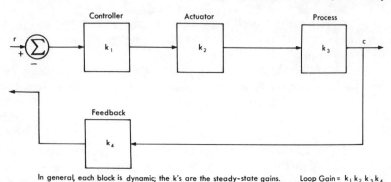

In general, each block is dynamic; the k's are the steady-state gains.   Loop Gain = $k_1 k_2 k_3 k_4$

FIG. 2.23   *Opening the closed loop—loop gain.*

2.9   *Open- and Closed-loop Systems—Closed-loop Instability*

(A) THE OPENING OF CLOSED–LOOP SYSTEMS

A closed-loop system is said to be opened, or open-loop, if a break is made in information flow anywhere within the loop.   In engineering systems this is normally simple to do, and the feedback path is the one most frequently broken by disconnecting it from the summation point (Fig. 2.23).   This procedure allows the engineer to study a simpler system and in particular to test different blocks in the system separately.   Furthermore, certain powerful control theorems allow the closed-loop dynamic behavior to be predicted from knowledge of open-loop behavior (Chap. 11).   It is the biologist's misfortune that he cannot normally open the loop in his experimental living systems, at least not reversibly and without harm.

Loop Gain.   The reader may have wondered whether the responses of Fig. 2.22 were for the open or the closed loop.   In fact, the two sets are not greatly different because of the low static loop gain in this closed loop (Fig. 2.23).   The static loop gain of a loop is the product of all component "steady-state" gains within the loop and is the amount by which a signal is multiplied on returning to itself; it is therefore dimensionless.   The pupil's loop gain is about 0.2, which does in fact produce a very stable system, as is shown in Sec. 2.9*B*.

The pupil system permits a very neat trick which opens the loop without physiological damage, because there is a residual pupil area even when the sphincter muscle is maximally excited.   If a smaller beam of light than this residual area is focused through the pupil [2.5], then when its light intensity is increased the flux increases and a compensating action is propagated around the loop, as already discussed, except that the resultant constriction of the pupil can no longer occlude any light and therefore reduce the flux (Fig. 2.24). The loop gain can be calculated from measurements of these changes, but not completely directly in this case because the perturbation is

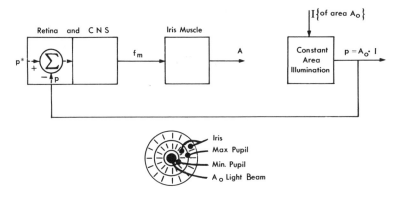

FIG. 2.24 Opening the pupil's closed loop.

made upon a disturbance variable of the loop I rather than on the internal loop variable of pupil area A which actually breaks the loop [2.5]. Thus the change of illumination intensity ΔI produces a perceived flux change

$$\Delta p_1 = A_0(\Delta I) \tag{2.21a}$$

where A_0 is the beam area. Although the pupillary reflex system is actually prevented from nullifying this change because of the beam's narrowness, the flux change that would normally have resulted can be calculated as

$$\Delta p_2 = (\Delta A)I_0 \tag{2.21b}$$

where I_0 is the average intensity of pupil illumination and ΔA is the change of pupil area which results from ΔI. The loop gain K is then the dimensionless ratio of these two changes:

$$K = \frac{\Delta p_2}{\Delta p_1} = \frac{\Delta A}{A_0} \frac{I_0}{\Delta I} \tag{2.22}$$

It is this ratio which Stark reports as approximating to 0.2 for the pupil [2.5].

(B) MANIPULATING THE CLOSED LOOP—INSTABILITY

We have seen that closed-loop control can provide compensation against system disturbances; in addition, as shown in Chap. 11, dynamic performance can be improved if the controller is correctly designed, and this usually means increasing the loop gain, among other effects. However, a basic upper limit of gain always exists under any particular circumstance, in that the closed loop eventually becomes unstable, that is, breaks into sustained or even destructive oscillations. Thus, in his control system designs the engineer typically makes some sort of trade-off between stability and speed of response.

Therefore, when one is faced with an unknown system to analyze, it is valuable to be able to manipulate loop gain. Although this is often not possible in studying biological systems, it can be per-

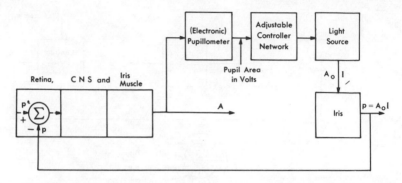

FIG. 2.25   *Electronic reclosing of pupil's opened loop.*   (*After Stark.*)

formed with the pupil system. The pupillometer continuously measures the pupil area, providing as output a convenient voltage signal. If the thin-light-beam technique is retained in order to open the physiological loop, the latter may be reclosed electronically by arranging for the pupillometer signal to drive the light intensity of the thin beam (Fig. 2.25) [2.5]. The advantage for the analyst of this rearrangement is that the electronic part of the loop may now be experimentally manipulated by the insertion of an appropriate electric controller network with adjustable parameters. As one example, a continuous oscillation of the pupil similar to that known clinically as *hippus* can be artificially induced. However, the necessary dynamic characteristics of the electric network can now be measured so that one can infer something of the pathological conditions for hippus from this experimental knowledge.

At this point it is necessary to describe some of the sufficient conditions for production of sustained oscillations in closed-loop systems. Consider first the simple negative-feedback system (Fig. 2.26) in which the only dynamic effect is a pure time delay, that is, an element which outputs the input unchanged in shape except that it is uniformly delayed in time. If this system is subjected to a step reference input, then one can trace the signals around the loop and appreciate that the output must be a square wave of half-period equal to the time delay. This idealized example is perhaps only useful in pointing out that a steady input level can become an oscillating output. Next, then, consider inputting either a rectangular pulse or half-period sine-wave pulse of the same duration as the time delay. This will generate a continuing square wave or sine wave of zero average level.

A feature of this last example is that at a specific frequency the time delay has the same effect as another device which simply inverts the input signal, namely, has a transfer function of

$$H = -1 \tag{2.23}$$

In fact this equation specifies the basic condition for instability in a linear negative-feedback closed loop (Chap. 11). Consider the

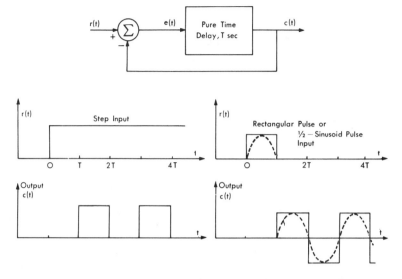

FIG. 2.26   Oscillations in pure-time-delay circuit.

inputting of a sine wave (Fig. 2.27) into such a closed loop and again follow signals around the loop. Initially the error equals the input, but it is then inverted across H and reinverted at the error comparator so as to add to the effect of the reference input. Actually this inversion does involve a half-cycle delay, so that the regenerative effect circulates exactly as for the time delay, but because the input is also oscillating there is a continuous buildup in the output oscillation. A quantitative form of this result may also be derived nonrigorously from Fig. 2.27, where the magnitude of H is allowed to differ from unity by an arbitrarily small amount δ at the particular sinusoidal frequency. Thus

$$e = r - c$$

and

$$c = -(1 - \delta)e$$

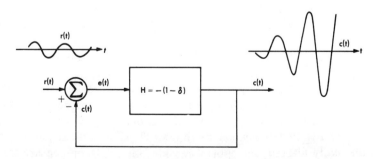

FIG. 2.27   Closed loop when $H \to -1$.

Solving these equations yields

$$c = -\frac{1-\delta}{\delta} r \approx -\frac{r}{\delta} \quad \text{if} \quad \delta \ll 1 \tag{2.24}$$

Thus as δ approaches zero the output oscillation builds up toward infinite amplitude at the same frequency as the input, although in practice saturation or other nonlinearities prevent infinite oscillations being attained.

Returning briefly to the induced hippus of the pupil, the experimental technique requires the controller network to be adjusted until the whole system satisfies Eq. (2.23). The oscillation is induced by some arbitrary or accidental pulse and thereafter maintains itself; if the system were driven with a sine wave at this frequency the oscillation would quickly grow until saturation occurred. One interesting subjective feature of this forced pupil oscillation is that the subject perceives only an oscillating light intensity without appreciating that he is engaged in a mutually excitatory interaction with the light source. Of course, this ignorance is due to the autonomous nature of the pupil system.

2.10 Automatic Aperture Control in Cameras— An Engineering Analog of the Pupil System

In recent years cameras have become available with automatic control of aperture area. The film is the photosensitive equivalent of the retina, for which the system regulates the incident light (or the time-integrated flux in this case). The camera system differs in detailed function because taking a picture is a one-shot affair, and it is controlled most easily if the correct exposure time can be calculated in advance. In addition, there are further degrees of freedom in the system since both film speed and exposure time are arbitrarily under the control of the camera operator. The resulting information-flow diagram for some practical systems (Fig. 2.28) differs from that for the pupil system in that there is no negative feedback to close the loop; that is, the system is open loop. The light intensity is conveniently measured by a photosensitive light meter, but if it is behind the aperture which is normally closed, then no useful measurement is available until the split second of exposure. Instead, therefore, the light meter is positioned so as to measure intensity continuously and to adjust the actuator and linkage for correct aperture area, subject to the reference input settings of film speed and exposure time; this configuration is often designated *feed-forward*. Note that for this system the desired performance measure is not total flux, but rather integrated flux—namely, flux multiplied by exposure time. Thus, if the exposure time is manually altered, the control mechanism must compensate for this with altered aperture. Closed-loop control has also been imple-

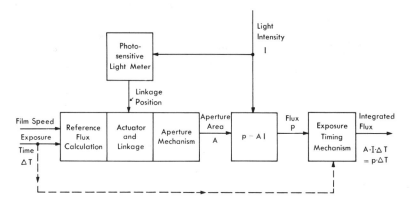

FIG. 2.28 Automatic control of camera aperture.

mented by placing the light meter behind the aperture mechanism so that aperture flux is actually measured, but this obviously places a much higher demand on the system's dynamic performance. Exposure time has also been brought under automatic control in some designs by using an electronic circuit to integrate the light flux, but only in the open-loop configuration already discussed.

In view of these technical considerations, what are the penalties of open-loop control? The open loop is not self-checking, and if such malfunctions as partially operative photocells or increased linkage friction occur, there is no feedback of actual flux to indicate this and take corrective action. However, the designer is faced with a fairly specific performance objective and only limited environmental changes. Furthermore, precision components can be designed to give almost infallible action over long periods of time, and finally when failure does occur the part can be taken out and repaired or replaced. In these circumstances open-loop control is more economical. On the other hand, nature's components must combat a large variability of environmental disturbances, and at the same time the evolutionary "design technique" is slow, as well as limited with regard to materials. Therefore the best system-design strategy has evidently been the redundant (parallel) use of fallible components and liberal utilization of feedback. However, it is also worth noting that the same necessary compromise between speed of response and stability is found in biological systems as in engineering systems. This is evidenced by the fact that in both disease and evolutionary processes a number of oscillatory malfunctions occur, such as hippus, epilepsy, Parkinson's disease, clonus, and predator-prey population cycles.

3 *Mathematical Modeling of the System*

Thermoregulation

3.1 Introduction

In this chapter, emphasis is laid upon the techniques used in obtaining a mathematical model for a dynamic system, the art of modeling. A mathematical model is an abstract representation of physical phenomena, and constitutes a basic creative step of the scientific method. For a physiological system the model is a concise way of quantifying the interacting behavior of those physiological mechanisms which have a more or less clearly defined functional goal. There are many practical and conceptual uses to which this model may be put, but perhaps the overriding motivation arises because the functioning of a dynamic system cannot be understood by merely establishing causal information pathways. Thus, knowledge of the relative phasing of such information is critically important, which requires in turn knowing both the pathways and the numerical values of the parameters. Consequently, mathematical model building involves the usual combination of abstract hypothesizing and experimentation. However, the model's interactions are usually too complex to work out in one's head, so that electronic computers become essential additional tools. The analog computer is particularly valuable because it sets up an electronic analog of the system via the equations of the mathematical model, and because this analog provides a flexible tool for parameter manipulation, etc. By this ability to "play with" the system's model, the possible presence and values of different defining parameters may be elicited which would otherwise remain undetected. If the model successfully defines the physiological system's dynamic behavior with such parameters, presumably there should be physiological counterparts whose presence and values can be sought experimentally. In this special sense, mathematical model making does increase physiological knowledge rather directly, and is not the sterile theory-oriented tool which the biologist might initially suspect. In summary, the overall method involves an iterative process between physical experimentation and model simulation. Philosophically it is exactly the classic method of science, except that the model-making (hypothesizing) stage now involves computer simulations because of the system's complexity. Perhaps it should be added that mathematical modeling does not imply that relations are necessarily deterministic, for in a real sense statistical models can also be subjected to rigorous mathematical treatment.

The development of a model is illustrated here by consideration of a particular system, the autonomic thermoregulatory system of mammals. It is hoped that this technique will produce an increasingly intuitive feel for the nature of biological control systems, concurrently with the development of formal control-theory tools.

Modeling comprises at least two major approaches. The first consists in writing equations for well-established physical processes from a priori knowledge; examples of this are the light-flux equation for the retina and the laws of heat transfer for the thermoregulatory system. Although the physical laws may be well known, this phase is not usually trivial because considerable art is needed to simplify the geometric system which actually exists into one which yields simple relations. One such simplification in the retinal model has been to assume constant spatial efficiency in trapping photons; many and drastic simplifications are necessary in the thermoregulatory system.

How simple the model may be depends upon whether the expected resulting errors exceed some reasonable tolerances, which the analyst must predetermine from knowing his purpose in the modeling. In general, these tolerances should not be stricter than the expected errors of real measurements upon the system.

The second approach is the *empirical* one of deriving cause-effect relations from experimental data, because the actual phenomena are not understood. In such cases the experiments must be carefully designed so that closed-loop chains of cause-effect relations and noncausal correlations do not invalidate the results.

With our present knowledge of biological control systems, neural and hormonal phenomena must generally be modeled by the empirical approach, and often there is not satisfactory experimental data even for this. For example, some empirical data are becoming known about the transfer characteristics of sensory neural receptors, but a priori models cannot yet be made with confidence. When these afferent pathways are followed to CNS nuclei, the empirical data and a priori knowledge become even scarcer, and in general the CNS integrative sites represent the realm of cybernetic systems which we ignore. In some particular cases, however, experimental evidence may suggest very simple input-output relations such as was postulated for the pupil system in the pretectal nucleus and the Edinger-Westphal nucleus (Fig. 2.21). Finally, when the efferent signals are followed from CNS sites to effectors such as muscles and glands, experimental data become more satisfactory again, although there is still little biophysical knowledge allowing a priori modeling.

Thus a priori, or biophysical, modeling will generally be restricted to the sluggish *controlled processes* (for example, the vascular system) and possibly their *effectors* (for example, the heart as a pump). Empirical modeling with scanty data must usually suffice for the rest of the closed-loop chain—namely, feedback transducers, comparators, controllers, and often effectors.

3.2   On Cold-bloodedness and Warm-bloodedness

Every living creature metabolizes, and in the process produces heat which must be dissipated to the environment at the same average rate if temperature is to remain essentially constant. *Homothermous*, or "warm-blooded," animals (birds and mammals) have evolved an *active* control system to achieve this, whereas the remaining "cold-blooded," or *poikilothermous*, creatures have only *passive* control. The basis for evolving thermoregulation is presumably the survival advantage gained when the animal's metabolic activity is essentially independent of ambient temperature conditions, since otherwise the underlying nature of cellular metabolism produces a strong dependence upon temperature.

Consider, then, the cold-blooded animal's response when it moves into a warmer environment after previously reaching equilibrium of internal temperatures. The heat loss to the environment decreases, and hence some of the internally generated metabolism must be stored internally, which in turn raises the internal temperature; as a result the metabolic rate increases and we have the exponential growth situation again. However, heat production and dissipation are nonlinear functions of temperature so that a new equilibrium condition does eventually result. As a rough rule the metabolic rate approximately doubles for each increase of 10°C, at least over a limited range [3.1, chap. 49]. In general, therefore, the internal temperature in cold-blooded animals tends to follow the ambient temperature, and the animal cannot achieve desired activity levels at arbitrary times and places.

The warm-blooded animal actively works toward a constant-temperature internal environment, despite varying ambient conditions and varying desired metabolic rates. It achieves this regulation by actively varying both the system's heat-loss and heat-generation characteristics. As one direct result, the overall metabolism increases with decreasing ambient temperature, rather than the reverse as in poikilothermy. If the core of the body, with its many organs, is viewed as a biochemical processing plant, regulation of temperature ensures its constant efficiency. The regulation required is very strict by biological standards, since enzyme activity is greatly reduced if the temperature falls a few degrees below the normal 37°C, and since CNS cells are irreversibly damaged above about 41°C; that is, a maximum tolerance of about ±4°C only is permitted (and usually bettered) over a possible ambient temperature range of about ±50°C, and also over considerable variations in wind speed.

The general nature of this system's requirements should convince one that a feedback-control system is necessary. Specifically, the system seems to be a regulator, as defined in Chap. 2, having as its desired performance a constant temperature. However, the need for constant temperature seems largely limited to the deep-core region of the body, which provides an essentially analogous problem to that of maintaining constant temperature in buildings.

Certain interesting questions arise about this system during such a widespread phenomenon as fever. At the onset of fever the usual reactions to a cold environment are exhibited, although the environment is not cold, as a result of which the heat-flow balance is disturbed and the patient becomes hotter; the reverse occurs when the fever breaks. The obvious question which a control engineer asks is: Has the reference temperature (thermostat) been set to a higher level in order to aid the organism in its fight against some disease, or is the change tolerable by the system and produced merely as a side reaction, or finally does the change represent a breakdown of regulation which is essentially harmful to the system? or more shortly: Is the temperature rise beneficial, neutral, or harmful? Unfortunately, there seems to be insufficient knowledge to answer the question satisfactorily at present, although the clinical treatment desirable presumably depends on this answer. It is pertinent to note that pyrogens, which are released in the body by organisms such as bacteria, probably do raise the effective set point.

Hibernation is another interesting thermoregulatory phenomenon, because mammals which enter this condition apparently reset their temperature-regulation set point (to around 6°C). However, temperature control is maintained and in particular the animal increases heat production if necessary to avoid having the body temperature drop to 0°C [3.2]. The condition is presumably entered to reduce the high thermoregulatory heat production necessary at low ambient temperatures, which in turn raises the question of why the normal set point is as high as 37°C. Probably this is in order that the

animal's limited heat-dissipating capacity in hot ambients may be
sufficient, for clearly the required heat dissipation would increase if
the normal set point were reduced.   The usual problem of trading
off various requirements is illustrated by these points, as well as by
nature's flexible solutions.

3.3   The "Anatomy" of Thermoregulation

The body is a material mass with a corresponding heat capacity—
that is, at any time an amount of thermal energy is stored which is
approximately proportional to temperature.   When there is temperature equilibrium the body must steadily dissipate heat at a rate
equal to the internal generation rate.   If these rates become mis-
matched the internal temperature starts a corresponding rate of
change.   The gross information-flow diagram for the complete control system is shown in Fig. 3.1, and the incomplete modeling which
follows largely incorporates the recent results of several workers in
this area [3.3–3.11].

(A) CONTROLLED PROCESS AND ACTUATOR—THE BODY

For thermoregulatory modeling purposes, there seem to be at least
three different regions of the controlled process, and these necessarily incorporate the following actuating mechanisms:

1.   The *deep body*, or *core*, comprises all the body except the skeletal
     muscles and skin, but in particular includes the viscera and the CNS.
     The fact that the material is definitely not homogeneous adds to
     the difficulty of mathematical modeling.   Much of the basal meta-
     bolic rate (BMR) is generated here and must be dissipated to the

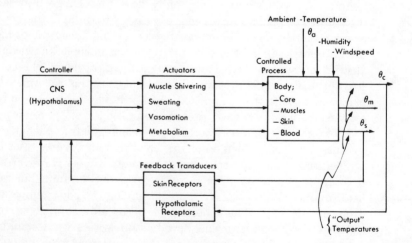

FIG. 3.1   Human thermoregulation—crude information-flow diagram.

environment. The BMR is controlled by the endocrine system and thus to some extent can provide an actuator for thermoregulation.

2. The *skeletal muscles* generally surround the core and comprise rather more than one-third of man's body weight. They are of particular interest in temperature regulation because they shiver when the system cools; this shivering is uncoordinated (normal tension-producing operation of an antagonistic pair is coordinated) so that all the energy input appears as a heat flow instead of some as mechanical work. The economy of component design is worth noting here; that is, the muscle serves a second useful purpose as actuator in the temperature-control system in addition to its presumed primary but different one of actuator in skeletal posture and movement. The tonic contraction of skeletal muscles, combined with their inherent inefficiency in converting chemical energy to mechanical energy, also makes the muscle region one of continuous BMR.

3. The *skin* provides external cover to the muscles and core. Disregarding other functions, it is therefore a thermal insulator for the temperature-control system, although a very actively variable one. The insulation effect is directly varied by the *vasomotor* effect, which decreases blood flow to the surface (vasoconstriction) when heat loss is to be reduced, and vice versa (vasodilation). In addition, there is the mechanism of *sweating* through the skin pores to produce water evaporation and therefore increased heat loss. Apart from evaporation the skin also loses heat by convection and radiation, when the skin is hotter than the environment. Evaporation allows the organism to survive against an adverse temperature gradient (ambient higher than skin surface), when the other modes produce a net gain of heat to the body.

For furry animals, hair erection is another surface function which reduces heat loss against cold weather. However, such animals cannot sweat profusely in hot weather (for example, the dog has no sweat glands except on the paw pads), and hence the panting phenomenon has been evolved to increase evaporation from the respiratory region, especially the tongue. However, since the present mathematical modeling is performed for a nude man, the last two effects are henceforth excluded.

The circulating blood plays a large part in achieving heat transfer among these three regions, and therefore should ideally be considered a fourth. However, this would complicate the model more than presently necessary.

(B) FEEDBACK TRANSDUCERS—TEMPERATURE MEASUREMENT

House-heating control (Fig. 3.2) utilizes an analogous configuration, with the thermostat constituting the reference input, feedback transducer, and controller combined (the transducer is typically a "bimetal strip" which bends as a function of temperature, thus making or breaking an electrical contact). It is located deep in the temperature-controlled part and achieves regulation within about $\pm 1°F$, which is normally considered satisfactory. Finer control may easily be achieved by placing a temperature transducer in the ambient air, which involves the same feed-forward of disturbance information that was discussed with regard to the car driver and aperture control in cameras. If the outside temperature falls, for example, heat production can immediately be increased to match the increased heat-loss rate *before* the internal temperature has had time to fall and so signal its need more slowly through the feedback loop. The overall effect is to quicken the response and reduce oscillation in the system.

It has been satisfactorily determined [3.8] that the body contains both deep central and skin thermoreceptors, as well as some others, for example, in the respiratory tract [3.11]. The central receptors are obviously analogous to the thermostat, and the skin receptors provide as close information to ambient air conditions as can conveniently be obtained physiologically.

Cutaneous Thermoreceptors. Two types of thermoreceptors have been identified in the skin [3.8]. The *cold receptor* responds particularly to decreasing skin temperatures, but there is a steady-state tonic discharge which varies with temperature (in a rather curious way, discussed later). The skin is more profusely supplied with these cold receptors than with warmth receptors. The *warmth receptor* responds particularly to increasing skin temperatures, but also exhibits steady-state discharge.

Hypothalamic Thermoreceptors. These receptors are near the site of the controller for the temperature-regulation system. Their

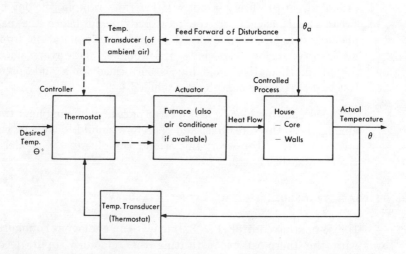

FIG. 3.2 House thermoregulation—crude information-flow diagram.

mode of action is not well understood but a good correlation has been demonstrated between anterior hypothalamus temperature and the level of a local "slow potential" in this region [3.8].

In Fig. 3.3 the gross information-flow diagram of Fig. 3.1 has been expanded to include these individual feedback paths.

(C) CONTROLLER AND COMPARATOR—THE HYPOTHALAMUS

The hypothalamus is thought to be the "controller" for thermoregulation, having two complementary sites: the *heat-maintenance center* situated in the posterior hypothalamus, and the *heat-loss center* situated in the anterior hypothalamus. By experimental calorimetric work, Benzinger et al. [3.5] have deduced that the heat-loss center uses core-temperature information (hypothalamic thermoreceptors) to drive the actuating variable of heat loss. Sweating is the primary mechanism for augmenting heat loss, but this is aided by vasodilation, which increases skin temperature and also provides the necessary fluid for evaporation. The heat-maintenance center, however, controls muscle metabolism on the basis of both skin- and core-temperature information. Figure 3.4 shows the corresponding neurological configuration suggested in [3.5], and Fig. 3.3 incorporates the pathways in an information-flow diagram (the place of termination of the warmth receptors seems in some doubt). It should be commented that there may be some anticipatory control pathways also; for example, increased sweating occurs very rapidly after the initiation of work in a warm room, apparently independently of thermoreceptor information [3.12].

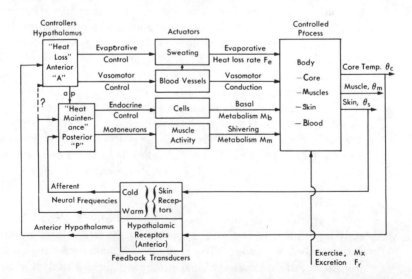

FIG. 3.3 *Human thermoregulation—detailed information-flow diagram.*

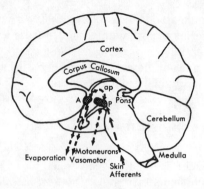

FIG. 3.4 Thermoregulatory pathways of hypothalamus. (*From Benzinger et al.*)

Although Fig. 3.3 provides a fairly detailed diagram of information flows at the qualitative level, the controlled process itself has still not been broken down into its functional relations. This breakdown, of course, is essential to a quantitative model, and furthermore is the major topic of this chapter. The particular functional aspect of present concern is heat transfer in its various forms of conduction, convection, and radiation, with evaporation as a special form also involving mass transfer. These matters are considered in the next two sections.

3.4 Lumping and Partial Differential Equations

Thus far, all emphasis has been upon the dynamic behavior of systems, that is, upon the time-domain response to various inputs or disturbances such as initial conditions and steps. Implicitly we have assumed that each system variable being considered is constant all over the system at any given instant of time—that is, it is constant in *space* but not in *time*. In practice, however, all matter and energy are distributed in both space and time. As a simple example, the driver and his car do not always have the same instantaneous acceleration; furthermore, the driver's tissues and organs in general have different accelerations under such conditions. Thus the only *exact* method of modeling is to write each variable θ_j of any system as a function of both time and all relevant space variables, $\theta_j(x,y,z,t)$. In practice, however, one recognizes that for safe driving conditions the driver's organs and the car essentially move as one rigid point mass. Thus Newton's second law need only be applied to the total mass rather than to a large number of interconnected masses. The result is a tremendous modeling simplification, at no expense of inaccuracy within the analyst's requirements. On the other hand, if one is examining the effects on the body of safety belts during accidents, then car and body masses must be separately considered. Furthermore, if one considers, for example, the vertical movement of a modern aircraft's flexible wing in gusty air, the wing's mass must be considered as smoothly distributed along the wing. The mathematical result of considering such spatial distribution is that partial differential equations apply instead of ordinary ones, because spatial derivatives are involved as well as those of time. Partial-differential-equation models are not treated in detail in this book because they do involve significantly greater difficulty, and because many biological servomechanisms and regulators can fortunately be treated by modeling with ordinary differential equations alone. This is also true of many engineering systems, from which field the descriptive term *lumping* is taken; a lumped system may comprise several, or many, different discrete masses, condensers, etc., but each discrete unit is a "lump" having a uniform value of its describing variable at any one time. As a particular example of this lumping concept, we shall represent the core temperature in thermoregulation by one homogeneous lump; that is, the lungs, liver, intestines, etc., are all assumed to be at the same temperature at any given time, and furthermore their physical properties of density, specific heat, and thermal conductivity are assumed to be uniform in space. Different lumped temperature variables are specified for each of two other lumps: the skeletal muscles and the skin. Although these may seem wildly inaccurate assumptions to the uninitiated reader, contemporary modeling based on them has produced reasonable agreement with physiological measurement.

Since we are vitally concerned with heat conduction in the modeling of thermoregulation, the process of lumping is now examined for this case. In the house-heating example let us postulate that a given wall is homogeneous in its thermal properties of conduction and energy storage. If the internal and external wall temperature are uniform across their respective surfaces, and if a steady-state temperature difference $\theta_1 - \theta_2$ exists across the wall, there is a steady flow of heat, q cal/sec (Fig. 3.5a), defined by

$$q = \frac{kA}{L} (\theta_1 - \theta_2) \qquad \text{cal/sec} \qquad (3.1)$$

where k = thermal conductivity, cal/(sec)(°C)(cm)
   A = surface area, cm²
   L = heat-flow path length, cm

This equation is really the solution of a physically derived differential equation with distance x as variable, often called the Fourier law of heat conduction.

$$q = -kA \frac{d\theta}{dx} \qquad \text{or} \qquad d\theta = -\frac{q}{kA} dx \qquad (3.2)$$

and after integration from $x = 0$ and $\theta = \theta_1$ to any x, θ condition there results

$$\theta(x) = \theta_1 - \frac{q}{kA}\, x \tag{3.3}$$

as shown in Fig. 3.5a.

The above equations are for steady-state conditions in time, but in practice neither the house wall nor the animal's skin is likely to remain for long in the steady state. Consider then, for example, the response when the temperature of the inner face θ_1 is suddenly raised, while θ_2 of the outer face remains unchanged. Intuitively one knows that the inner surface of the wall starts to warm up first, but that ultimately a new and larger steady-state heat flow is established, with a correspondingly larger temperature gradient along the x axis. Some representative distribution curves at various increasing times are sketched in Fig. 3.5b, showing that the temperature θ is a function of both x and t, and is therefore written $\theta(x,t)$.

After the sudden change in input temperature, the new steady-state temperature distribution cannot be immediately established because extra energy must first be stored in the wall in local amounts corresponding to the local temperature. This process can be analyzed easily by lumping the wall into N homogeneous thin slabs of thickness Δx (Fig. 3.6a), where

$$\Delta x = \frac{L}{N} \tag{3.4}$$

Each slab is then conceptually described by a single temperature variable $\theta_n(t)$, which is a function of time only. Although this is a

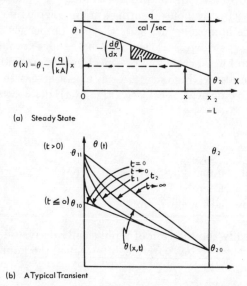

(a) Steady State

(b) A Typical Transient

FIG. 3.5   Heat conduction; temperature distribution in homogeneous wall.

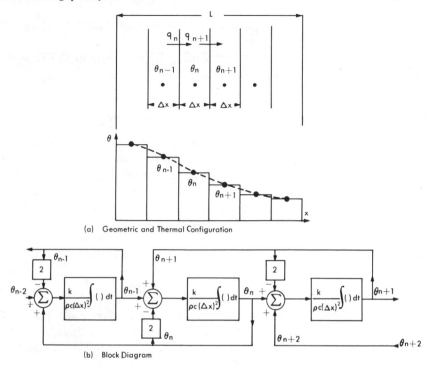

(a)   Geometric and Thermal Configuration

(b)   Block Diagram

FIG. 3.6   *The lumping procedure; applied to one-dimensional homogeneous heat conduction.*

convenient concept from the energy-storage viewpoint, one can still use the local-temperature-gradient concept to define the local heat-flow rates.   Thus the heat inflow and outflow rates of the nth slab are obtained from Eq. (3.1) as

$$\textit{Influx} \qquad q_n = \frac{kA}{\Delta x}(\theta_{n-1} - \theta_n)$$

$$\textit{Outflux} \qquad q_{n+1} = \frac{kA}{\Delta x}(\theta_n - \theta_{n+1}) \tag{3.5}$$

The formal statement of energy conservation in this case comes from the first law of thermodynamics:

$$Q = E + W \tag{3.6}$$

That is, the net quantity of heat entering the system Q equals the energy stored internally E plus external work done W, all in heat units.   In our case $W = 0$ and furthermore we work in heat-flow rates rather than in absolute quantities.   The rate of storage of heat q_{En} in the nth slab is proportional to the rate of temperature change:

$$q_{En} = \rho c A \, \Delta x \, \frac{d\theta_n}{dt} \qquad \text{cal/sec} \tag{3.7}$$

where c equals the specific heat of slab material, cal/(g)(°C), and ρ is the density, g/cm³.

Hence, by combining Eqs. (3.7) and (3.5) in the rate-of-change version of Eq. (3.6),

$$\theta_{n-1} - 2\theta_n + \theta_{n+1} = \frac{\rho c}{k}(\Delta x)^2 \frac{d\theta_n}{dt} \qquad (3.8)$$

This represents a recurrence set of difference-differential equations, whose block-diagram form is shown in Fig. 3.6b. The form is directly applicable for analog-computer simulation, but for digital-computer solution the $d\theta_n/dt$ term must also be converted into difference form. In general, errors due to lumping may be made arbitrarily small as the number of lumps N is increased so that the slab width Δx is decreased. It is not necessary for the lumps to be the same size, and in practice a small number of lumps suffice for most modeling purposes.

▼

For the interested reader the partial differential equation of heat conduction is now derived by passing to the limit as Δx becomes increasingly smaller; others may now jump to the paragraph following Eq. (3.14) without sacrificing their understanding of later sections.* We consider heat flow across an infinitesimally thick slab defined by walls at x and $x + dx$ (Fig. 3.6). The relations corresponding to Eqs. (3.5) are now derived directly from either Eq. (3.2) or Eqs. (3.5) by noting that as $\Delta x \to 0$,

$$\begin{aligned} q_x &= -kA\frac{d}{dx}\theta_x \triangleq -kA\frac{d\theta}{dx} \\ q_{x+dx} &= -kA\frac{d}{dx}\theta_{x+dx} \end{aligned} \qquad (3.9)$$

The problem of finding an analytical form for θ_{x+dx} causes us to introduce conventional partial derivative notation because θ is a function of the independent variables x and t, and derivatives are needed with respect to each; these are called *partial derivatives* and written $\partial\theta/\partial x$ and $\partial\theta/\partial t$. With regard to the value of θ at $x + dx$, it is plausible that this value is approximated by

$$\theta_{x+dx} = \theta_x + \frac{\partial\theta}{\partial x}dx \qquad (3.10)$$

That is, since $\partial\theta/\partial x$ is the spatial slope of θ at x, multiplying it by dx provides the incremental value for θ. Hence, from Eq. (3.10), Eq.

---

* Sections marked with a triangle may be omitted during a rapid first reading without excessive loss of comprehension. However, they generally provide the mathematical basis for various theoretical aspects and will need to be studied in order to achieve a thorough understanding of control theory.

(3.9) becomes

$$q_x = -kA \frac{\partial \theta}{\partial x}$$

$$q_{x+dx} = -kA \frac{\partial}{\partial x}\left(\theta + \frac{\partial \theta}{\partial x} dx\right) = -kA\left(\frac{\partial \theta}{\partial x} + \frac{\partial^2 \theta}{\partial x^2} dx\right)$$

(3.11)

and in particular the net incremental heat flux into the element is given by their difference

$$dq = q_x - q_{x+dx} = kA \, dx \frac{\partial^2 \theta}{\partial x^2}$$

(3.12)

The net incremental heat storage in the element is rewritten from Eq. (3.7) as

$$dq_E = c\rho A \, dx \frac{\partial \theta}{\partial t}$$

(3.13)

and hence by combining Eqs. (3.12) and (3.13) there results

$$\frac{\partial^2 \theta}{\partial x^2} = \frac{\rho c}{k} \frac{\partial \theta}{\partial t}$$

(3.14)

This is the partial differential equation for one-dimensional transient heat conduction in a homogeneous medium (ρ, c, and k all constants). It can be directly generalized for three spatial variables if required.

In the thermoregulatory example we use a lumped model because of the great difficulty in working with partial differential equations, especially when the system is so patently nonhomogeneous; in particular, we represent each of the three zones (core, muscle, and skin) by a single lump each defined by a single temperature variable: $\theta_c(t)$, $\theta_m(t)$, $\theta_s(t)$. This raises the question of the actual geometric lumping, because considerable simplification is also necessary in this regard. The simplest model is that of one-dimensional conduction through three parallel homogeneous slabs (Fig. 3.7a), and this is the model used by Crosbie et al. [3.4] and earlier workers. Note that muscle and skin layers need not be shown on each side of the core because lateral symmetry can be presumed. The thicknesses of the slabs do not necessarily bear a direct relation to the corresponding body regions, but instead are determined from steady-state physiological data so that the actual and modeled core-to-skin temperature drops are equal. In making this calculation the measured thermal conductivity of wet tissue is used, namely,

$$k = 0.4 \text{ kcal}/(\text{m})(\degree\text{C})(\text{hr})$$

In using this model under non-steady-state conditions, the effect of vasomotor action upon thermal conductivity must be considered, especially in the peripheral region. Crosbie et al. therefore made their k value dependent also upon body temperature and derivatives, a relation which is considered more relevantly later. Stolwicjk

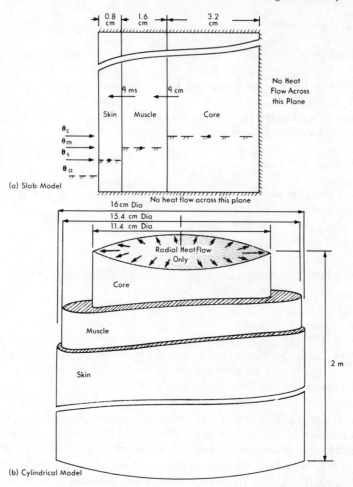

FIG. 3.7   Lumped models of body for temperature regulation. [(a) *From Crosbie et al.; (b) from Stolwicjk.*]

[3.3] and others have proposed a concentric-cylinder model (Fig. 3.7b) which is more satisfying intuitively. In this case the criterion has been to maintain the same surface-volume ratio in the model as in man himself. The heat flow is assumed to be independent of sector location but dependent upon radial distance, and it is therefore still one-dimensional, although the heat-conduction equation must be rewritten for a radial variable.

It was noted earlier that thermal conductivity must be varied in these models in order to account for vasomotor activity. Actually, vasomotor activity affects blood-flow patterns rather than tissue conductivity directly. What should we say then of the blood-flow phenomenon? Clearly the blood has heat capacity and therefore should in principle be separately modeled. Its method of heat transfer is usually defined as *convection;* that is, the heat-transfer process occurs simultaneously with a mass transfer of fluid spatially. The rate of convective heat transfer depends upon local temperature differences which typically vary in both space and time within the body, so that inclusion of the blood separately in the model involves more complexity. Fortunately, the effect can be included in the heat-transfer equation if a simple model is desired because the rapid circulation time of the blood within the body (of the order of 1 min) is short compared with the relatively sluggish behavior of the controlled system. In consequence, the blood can always be considered to be transferring heat at a rate proportional to zonal temperature difference, which therefore allows this effect to be considered as conduction. However, the blood can transfer heat essentially directly from core to skin without passing through muscles, so that a new equivalent conduction pathway must be incorporated in the model. This pathway also is the one most directly affected by vasomotor action, and hence vasomotor activity is shown in the model as controlling the core-skin conductivity.

The original aim of this section was to show how the controlled process, the body, might be modeled. Certain lumping approximations have now been demonstrated and as a result an information-flow diagram for the body can be drawn (see Fig. 3.11). The heat transfer at the skin is due to other effects than conduction; since these effects also involve the modeling of nonlinear functions, it is convenient to consider them separately.

3.5 Nonlinearity and Linearization; Heat-transfer Examples

(A) MORE MODES OF HEAT TRANSFER

At the skin surface, heat is lost by convection, evaporation, and radiation.

Convection. Convection is defined as the transfer of heat within a fluid, gas, or liquid, and is usually caused by a heat flow at some boundary, such as from a solid surface, for example, the skin. The conventional relation is Newton's law of cooling, which defines the convective heat-flow rate F_c from the surface as, for example,

$$F_c \triangleq q = hA(\theta_s - \theta_a) \qquad \text{kcal/hr} \qquad (3.15)$$

where h = heat-transfer coefficient, $\text{kcal/(hr)(°C)(m}^2)$
  A = surface area, m^2
  θ_a = ambient temperature, °C
  θ_s = skin temperature, °C

Thus heat-flow rate depends linearly upon the forcing temperature difference as in heat conduction. However, the heat-transfer coefficient h is in practice a complicated function of the temperatures and other variables, although it can often be considered constant over limited ranges.

Evaporation. Evaporation involves mass transfer of fluid from skin to air, and also a change of state from liquid to vapor. In general, evaporation is a complex function of such variables as relative air velocity, wetted skin area, and relative humidity, and it is certainly nonlinear. Crosbie et al. [3.4] defined vaporization rate as the relevant variable in thermoregulation, rather than sweating rate, since the aim of the heat-loss center is only to achieve a given heat-loss rate. They were able to express this vaporization rate as a nonlinear function of essentially one variable only, the body temperature. Since this empirical relation is somewhat similar to the physically based law of radiation, detailed attention is deferred until the next heading.

Radiation. In general a given body surface engages in a heat-flux interchange with the surrounding enclosure for which the net effect is expressed by the Stefan-Boltzmann law

$$q = F_{\rm rad} = \sigma A'(\theta_s{}^4 - \theta_e{}^4) \qquad {\rm kcal/hr} \qquad\qquad (3.16)$$

where θ_s = temperature of skin, °K

θ_e = temperature of radiant enclosure, °K

A' = effective radiating area, which must be less than actual skin area, due to body topology and posture

≈ 2 m² in man

σ = Stefan-Boltzmann constant for black-body radiation (approximated by human body)

= 0.49×10^{-7} kcal/(m²)(hr)(°K⁴)

°K = degrees Kelvin $\triangleq$ °C + 273

Radiation heat loss can be considerable from a nude body with a cooler enclosure; thus, for example, in a room with 70°F (21°C, 294°K) walls the heat-loss rate would be about 100 kcal/hr for a

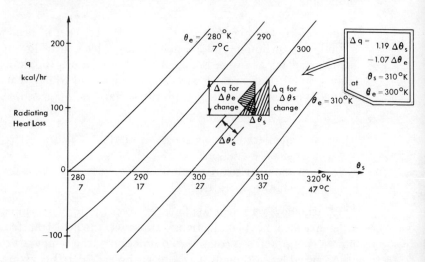

FIG. 3.8   Radiating heat-transfer rates.

skin temperature 10°C higher (Fig. 3.8), which is larger than the BMR ($\approx$ 70 kcal/hr). When the body is clothed, the skin temperature can remain high even though its effective emitting temperature, that of the clothes, is much less. The feeling of discomfort, familiar to everyone, in a room whose air temperature is comfortable (70 to 75°F) but whose "enclosure" temperature is low, is thus explained on the basis of large radiant heat loss from exposed skin areas. Casey [1.3] presents interesting data concerning the proportional heat-loss rates from conduction and convection, radiation, skin evaporation, and respiration under various bodily exercise and clothing conditions.

The nonlinear nature of Eq. (3.16) is evident. Some simplifications, including linearization, are examined for this case in the next two sections, but we first establish a basic mathematical result.

▼

(B) TAYLOR'S SERIES EXPANSION

Evaluation of Eq. (3.16) requires the differencing of two temperature terms after each has been raised to the fourth power. In biological situations the temperature difference is normally small, although the absolute temperatures are themselves large. A computational simplification is made possible in this case by Taylor's theorem [for example, (3.13)], which allows any continuous, differentiable function $f(x)$ to be expressed as a relatively simple polynomial function in a small region. It has many applications apart from that of present interest, in particular in obtaining series expansions for sine, exponential, log, and other functions (see Prob. 3.3). The theorem states that

$$f(x + \Delta x) = f(x) + f'(x) \frac{\Delta x}{1!} + \cdots + f^{(n)}(x) \frac{(\Delta x)^n}{n!} + \cdots \qquad (3.17)$$

where $f^{(n)}(\dot{x}) \triangleq \dfrac{d^n}{dx^n} f(x)$

That is, the value of $f(x + \Delta x)$ is approximated by a series of terms comprising $f(x)$ and the sum of n derivatives each weighted individually by the term $(\Delta x)^n/n!$. Convergence is assured provided Δx is "small" and in many practical cases the first few terms only are needed. In particular as $\Delta x \to 0$ one obtains the classical definition of a derivative by ignoring quadratic and higher terms in Δx:

$$f'(x) = \frac{d}{dx} f(x) = \lim_{\Delta x \to 0} \frac{f(x + \Delta x) - f(x)}{\Delta x} \qquad (3.18)$$

One particular result is the binomial series, here expressed in unitary form as

$$(1 + \delta)^n = 1 + n\delta + \frac{n(n - 1)}{2} \delta^2 + \cdots \qquad (3.19)$$

and it is this expression which is now utilized in the radiation example. Note that if n is a positive integer the number of terms is $n + 1$, but otherwise it is infinite. In Eq. (3.16) define $\theta_s = \theta_e + \Delta\theta$ and therefore write

$$\theta_s{}^4 - \theta_e{}^4 = \theta_e{}^4[(1 + \delta)^4 - 1] \tag{3.20}$$

where $\delta = \dfrac{\Delta\theta}{\theta_e}$

For typical biological situations $\Delta\theta$ is of order 10°C (or less) so that $\delta \leq \frac{1}{30}$. Therefore δ^2 becomes negligible and Eq. (3.20) can be rewritten

$$\theta_s{}^4 - \theta_e{}^4 = 4\delta\theta_e{}^4 = 4(\theta_s - \theta_e)\theta_e{}^3 \tag{3.21}$$

This result may not seem particularly valuable, although some useful simplification has been achieved; however, the potential advantage of expansion techniques in numerical evaluations and computer simulations should be stressed. In this case one has in fact performed a linearization for θ_s in a small region about an operating point $\theta_s = \theta_e$; that is, the function is here linear in θ_s, as shown in Fig. 3.8. Linearization is examined below in more detail.

▼

(C) NONLINEARITY AND LINEARIZATION

The process of linearizing an arbitrary continuous function by drawing a tangent to the curve at some desired operating point was illustrated graphically in Fig. 2.13 for a presumed pupillary fiber frequency relation. The corresponding analytical procedure has just been discussed for the particular case of radiation. In general, the linearization formula for a function of one variable is obtained by truncating the relevant Taylor's series after the first derivative, namely

$$f(x + \Delta x) = f(x) + k \Delta x \tag{3.22}$$

where k is the value of df/dx at the chosen x value. Often it is convenient to work only with delta variables, that is, variables representing small changes from a fixed operating point. Thus in this case we define

$$\Delta f(x) \triangleq f(x + \Delta x) - f(x) \tag{3.23}$$

whence the linearized input-output relation of Eq. (3.22) becomes

$$\Delta f = k \Delta x \tag{3.24}$$

Let us now return to the radiation expression (3.16). If we first consider q as a function of the variable $\Delta\theta_s$ only, Eq. (3.24) provides the linearized relation as

$$\Delta q = \frac{dq}{d\theta_s} \Delta\theta_s = \sigma A' 4\theta_s{}^3 \Delta\theta_s \tag{3.25}$$

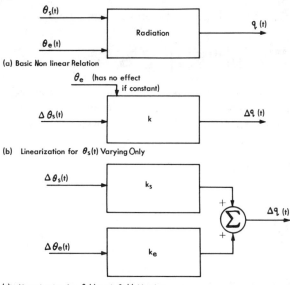

(a) Basic Non linear Relation

(b) Linearization for $\theta_s(t)$ Varying Only

(c) Linearization for $\theta_s(t)$ and $\theta_e(t)$ Varying

FIG. 3.9   Nonlinear functions and linearization.

This result has a different purpose than that of Eq. (3.21), although each has been obtained through Taylor's theorem.   Equation (3.21) merely provides an easy way of calculating the net heat flux when the temperature difference is small, whereas Eq. (3.25) represents a linearized input-output relation, or transfer function, between θ_s as input and heat-flow rate as output.   This linear relation is only true for small changes from any given operating point θ_s (Fig. 3.8), and the *gain* k in fact depends upon the cubed value of θ_s about which the small change is considered, but is independent of θ_e, although this last fact is not obvious graphically.   Numerically the value of k at a skin temperature of 37°C is about 1.18 kcal/ (hr)(°C) for a nude man, using the A' and σ values quoted.

The block-diagram representation of the results so far are summarized in Fig. 3.9.   Figure 3.9a shows the basic nonlinear relation of Eq. (3.16) for $\theta_s(t)$ and $\theta_e(t)$ as the time-varying inputs.   Figure 3.9b shows the linearized relation of Eq. (3.25) when θ_e is assumed not to vary in time.   Actually the latter's absolute value is immaterial in this case, and θ_e need not be shown as an input except for conceptual clarity.

In practice θ_e is just as likely to vary in time as is θ_s and therefore we now consider the general problem of linearizing a function with more than one independent input variable.   We first define an output variable $z(t)$ as a general function of two input variables $x(t)$ and $y(t)$, viz.

$$z = f(x,y) \tag{3.26}$$

A generalization of Taylor's theorem allows one to modify Eq. (3.24):

$$dz = \left[\frac{\partial f}{\partial x} (x,y) \right]_{y=\text{const.}} dx + \left[\frac{\partial f}{\partial x} (x,y) \right]_{x=\text{const.}} dy \qquad (3.27)$$

In evaluating each partial derivative, note that the independent variable not involved must be held constant. As in Eq. (3.24), it may be assumed that Eq. (3.27) is valid for small finite changes, so that the linearized relation becomes

$$\Delta z = k_x \, \Delta x + k_y \, \Delta y \qquad (3.28)$$

where

$$k_x \equiv \left[\frac{\partial f(x,y)}{\partial x} \right]_{y=\text{const.}}$$
$$k_y \equiv \left[\frac{\partial f(x,y)}{\partial y} \right]_{x=\text{const.}}$$

As pointed out in the neuron example of Eq. (2.18), these partial derivatives correspond to tangents drawn to the appropriate curves, at appropriate values of the input variables (see Fig. 2.19). However, analytical expressions can also be obtained directly if the describing relation is known. In the radiant heat-transfer case, the partial derivatives are essentially of the same form as given in Eq. (3.25):

$$\begin{aligned} k_s &= \sigma A' 4\theta_s{}^3 \\ -k_e &= \sigma A' 4\theta_e{}^3 \end{aligned} \qquad (3.29)$$

(the minus sign on k_e is merely for conceptual convenience). Figure 3.9c shows the result of this linearization, namely, that a nonlinear relation has been replaced by two linear "multiplications by a constant" plus one summation; of course the variables of the block diagram can now only refer to small changes about some selected operating point. In Fig. 3.8 the "geometric" construction is shown; the evaluation of $[\partial q/\partial \theta_s]_{\theta_e=\text{const.}}$ is direct, but that of $[\partial q/\partial \theta_e]_{\theta_s=\text{const.}}$ is perhaps not so obvious unless the curves are replotted on θ_e as abscissa and θ_s as parameter (compare with Fig. 2.19). However, this is not actually needed, and instead one can interpolate a small θ_e change with θ_s constant (a vertical line); the partial slope is then found by dividing the resulting Δq change by the $\Delta \theta_e$ perturbation.

When the function has more than two input variables Eq. (3.27) can be appropriately generalized. Although linearization has classically been an important tool of the control specialist because of the body of linear theory to which it gives access, such a simplification is not vital to progress, especially since electronic computers have become available. Certainly one should not linearize in situations where some fundamental dynamic phenomena would thereby be eliminated. As a tentative general rule, "soft" nonlinearities can

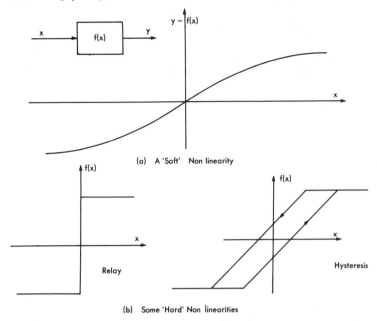

(a) A 'Soft' Non linearity

Relay

Hysteresis

(b)   Some 'Hard' Non linearities

FIG. 3.10   *"Soft" and "hard" nonlinearities.*

usually be linearized without affecting qualitative behavior, whereas this may not be true for "hard" nonlinearities.

A soft nonlinearity is continuous and differentiable, such as the *saturation* effect (Fig. 3.10a) which is inevitable in all transducers, amplifiers, etc., when the input magnitude increases sufficiently. A hard nonlinearity, on the other hand, is often discontinuous and even multivalued; the ideal relay and hysteresis are examples (Fig. 3.10b), and various dynamic phenomena such as subharmonics may result from feedback systems which incorporate such nonlinearities. In biological systems, hard nonlinearities are perhaps not too abundant, although the threshold characteristic of neurons is one. Hysteresis is an energy-dissipating phenomenon since the area within the loop on the sketch represents the energy which is dissipated as heat during each traversal of the loop. Animal joints exhibit hysteresis, but only to a small degree, presumably because evolution has found low-friction joints advantageous.

3.6   *Mathematical Model of the Controlled Process, the Body*

The last two sections have diverged from the thermoregulatory system in order to explore two important problems of modeling. However, we are now in a position to write the appropriate heat-balance equations for each of the three lumps (core, muscles, and skin) comprising the controlled process of the body. From the first law of thermodynamics [Eq. (3.6)], with no useful work output, we can write

Core $\qquad m_c c_c \dfrac{d\theta_c}{dt} = M_b - F_r - q_{cm} - q_{cs}$ $\qquad\qquad$ *(3.30)*

Muscles $\qquad m_m c_m \dfrac{d\theta_m}{dt} = M_m + M_x + q_{cm} - q_{ms}$ $\qquad$ *(3.31)*

Skin $\qquad m_s c_s \dfrac{d\theta_s}{dt} = -(F_c + F_e + F_{\text{rad}}) + q_{cs} + q_{ms}$ $\qquad$ *(3.32)*

and $\qquad q_{cm} = \dfrac{k A_{cm}}{L_{cm}} (\theta_c - \theta_m)$ $\qquad\qquad\qquad$ *(3.33)*

$$q_{ms} = \frac{k A_{ms}}{L_{ms}} (\theta_m - \theta_s)$$ *(3.34)*

$$q_{cs} = \frac{k_v A_{cs}}{L_{cs}} (\theta_c - \theta_s)$$ *(3.35)*

where m = mass, kg
$\qquad c$ = specific heat, kcal/kg
$\qquad k$ = thermal conductivity, kcal/(cm)(hr)(°C)
$\qquad A$ = effective area for heat conduction between lumps; also convective area, cm^2
$\qquad L$ = effective path length between lumps, cm
$\qquad M_m$ = muscle shivering metabolism, kcal/hr
$\qquad M_b$ = BMR, kcal/hr
$\qquad M_x$ = muscle exercise metabolism, kcal/hr
$\qquad F_c$ = convective heat-transfer rate from skin, kcal/hr
$\qquad h_s$ = convective heat-transfer coefficient—skin to air, kcal/(m^2)(hr)(°C)
$\qquad F_e$ = evaporative heat-transfer rate from skin, kcal/hr
$\qquad F_r$ = respiratory-excretory heat-loss rate, kcal/hr
$\qquad F_{\text{rad}}$ = radiant heat-transfer rate from skin, kcal = kcal/hr

and where subscripts c, m, and s refer to core, muscle, and skin conditions, respectively, and v refers to vasomotor action. These equations are valid for either the slab or the concentric model, since in either case the A and L constants must be idealized to fit available physiological data. A further simplification is directly available in the slab model, for which

$m = \rho L A$ $\qquad\qquad\qquad\qquad\qquad\qquad\qquad\qquad\qquad$ *(3.36)*

where ρ is the density of the tissue, g/cm^3, $\rho c \approx 1$ cal/(cm^3) (°C), and A is constant.

The reader interested in further numerical details of this example is referred to [3.4]. For the present purpose we proceed with the block-diagram representation of Eqs. (3.30) to (3.35), (3.15), and (3.16), which is shown in Fig. 3.11. This figure is certainly more

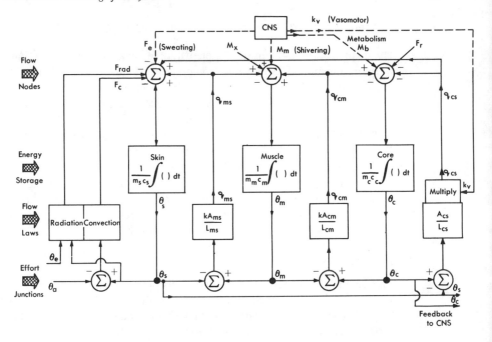

FIG. 3.11 *Block diagram for controlled process, the body.*

complex than previous examples, but it only involves a logical connecting of the several relations. Thus it should be noted immediately that the three first-order differential equations give rise to three energy-storage elements; on the other hand, the several modes of heat transfer give rise to more than three heat-flux paths, some of which are nonlinear. No attempt has been made at this stage to show the details of completing the loops back through the CNS; rather, the emphasis is on showing the dynamics of the controlled process. It should by now also be clear to the reader that obtaining the numerical values for constants in a model is not a trivial task.

The configuration of this block diagram has been carefully chosen so as to illustrate common or symmetric features of the model. Thus, *heat storage* is shown in descending paths, and *flow laws* in ascending paths; in addition, the features of *flow nodes* and *effort junctions* which are common to many circuits (Sec. 4.3) are shown.

In the next section some recent results from studies of the thermoregulatory system are presented. All of such quantitative efforts to simulate this system, using experimental physiological data, have been performed on analog computers, and although it is not convenient to introduce analog-computer programming at this stage, it is worth pointing out that the block diagram of Fig. 3.11 represents quite closely the manner in which the various elements of an analog computer are joined together to simulate a mathematical model.

3.7   Performance Characteristics of the Thermoregulatory System

Experimentally determined performance characteristics of the
human thermoregulatory system are now reported.   One purpose
is to complete the case study a little more satisfactorily by showing
how the system actually behaves.   A second purpose is to deduce
some relations for the CNS and thermoreceptors' control function
in this system so that the mathematical model may be completed;
that is, although the body's passive thermal properties as a controlled process have been modeled a priori, however crudely, nothing
is known a priori about the CNS, the thermoreceptor, and the
effector characteristics, so that modeling necessarily depends upon
using experimental data.   Steady-state results are presented first,
followed by some transient performance results.

(A) STEADY–STATE PERFORMANCE

Extensive experimental data on thermoregulatory heat flows is
available from the work of Benzinger et al. [3.5], from which they
postulate the control pathways illustrated in Figs. 3.3 and 3.4.   The
data concern the metabolic heat-production rates, including shivering, and the evaporative heat-loss rates, at various steady-state
conditions of core and skin temperature (Fig. 3.12).   The wide
ranges of skin and core temperatures were obtained by immersing
the subject in a water bath, and the large evaporative losses by
using a gradient calorimeter at high temperatures.   It is important
to note that the original data points, to which the illustrated curves
are "best fitted," were each obtained at a steady-state condition,
that is, after transients due to changing the conditions had settled
out.   These data points are omitted from the figure for simplicity,
but the referenced work shows there is fair confidence in the curves.

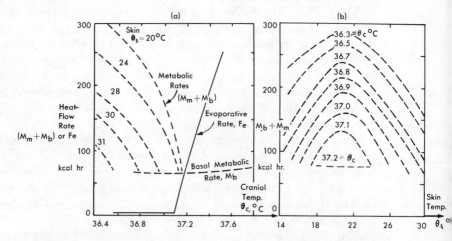

FIG. 3.12   Steady-state thermoregulatory heat-flow rates.   (*After Benzinger et al.*)

The curves of Fig. 3.12a obviously imply a thermoregulatory set point of 37.1°C for the core temperature. Evaporative heat loss F_e is essentially zero below this set point but increases steeply and almost linearly above it at a rate of about 70 kcal/hr per $\frac{1}{10}$°C of cranial temperature increase. This heat-loss characteristic was found to be essentially unchanged by varying the skin temperature at any given core temperature, at least until the skin temperature was artificially forced below 33°C [3.14]. A plausible conclusion, therefore, is that the heat-loss center of the hypothalamus controls on the basis of core temperature. There seems to be no clear functional or anatomical evidence in favor of a role for the skin-warmth receptors in sweating control.

Metabolic heat-production rates, on the other hand, were not quantitatively meaningful until the skin temperature was considered as well as the core temperature. The rate fell to the BMR M_b of approximately 70 kcal/hr for all core temperatures above the set point, implying that muscle-shivering metabolism M_m was zero. Below the set point, however, the rate was clearly a function of both θ_s and θ_c, which is the second basic control function postulated by Benzinger et al. (Sec. 3.3). The individual curves for constant skin temperature are similar to the sweating curve, with a lower average gain of about 25 kcal/hr per $\frac{1}{10}$°C (core) in the skin-temperature range 20 to 30°C. The apparent failure of the system to hold the core set point when the skin temperature is higher (e.g., $\theta_s = 31$°C, $\theta_c = 36.5$°C) is overcome by the person undertaking gross movement or exercise, so that this experimental point would not actually occur in practice.

In Fig. 3.12b the data with skin temperature as abscissa and core temperature as parameter are cross-plotted. Data for a wider range of skin temperature can be conveniently included, and reveal that there is one skin temperature, about 20°C, at which heat-production rate is maximum, independent of the core temperature. The average linear gain in the region $22 < \theta_s < 30$ is 18 kcal/hr per skin °C. In the next section this curious heat-production characteristic is shown to correlate with thermoreceptor properties, but we should now note that it implies thermoregulatory system failure if skin temperature falls to 20°C, because an unstable regime is then entered (less heat generation—skin temperature falls—even less heat produced) unless the animal can attain a more favorable environment. Is this phenomenon perhaps relevant to the happy disinterest with which persons reportedly enter the stage of freezing to death?

Crosbie et al. [3.4] have conducted considerable experimental work with the aim of synthesizing a suitable mathematical model. Their slab model was simulated on an analog computer and subjected to various input conditions for comparison with experimental data. Experiments were conducted on nude subjects in an air calorimeter. Figure 3.13 shows the steady-state responses for vari-

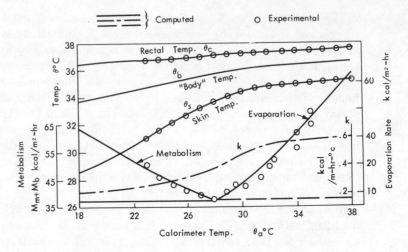

FIG. 3.13 Steady-state thermoregulation—experimental and computed. (*From Crosbie et al.*)

ous calorimeter ambient temperatures. Note that the core set point of 37.1°C in Fig. 3.12 is apparently achieved without need for control effort, by either shivering or sweating, when the ambient temperature is about 28°C. The smooth lines represent analog-computer solutions of the model, and the circles represent measured physiological data. It will be observed that agreement is generally good. The computed curve of "body" temperature relates to an averaged temperature which they postulate (see Sec. 3.8). In their slab model no direct "conduction" under vasomotor control is provided between core and skin, so that the k of both their conduction paths is under vasomotor control; in this case, k varies by a factor of 3 in about 12°C change of ambient temperature. Note that all heat flows, etc., are referred to unit area of the slab model; the total area in man is about 2 m².

(B) DYNAMIC PERFORMANCE

Benzinger et al. [3.5] noted that sudden changes in skin temperature, either up or down, produced transient overshoots in the metabolic heat rates before these settled down to suitable new steady states. Presumably this is because both cold and warm cutaneous thermoreceptors respond with overshoot in their particular modality. Crosbie et al. [3.4] have performed dynamic tests upon their subject's thermoregulatory system in the calorimeter and compared them with the computer model's response. Figure 3.14 shows the step response on moving suddenly from essentially steady state at an ambient of 32°C to one of 16°C. The most striking result is the response's long duration, which is indicated by the computer to be about 10 hr, although the corresponding physiological experiment was only conducted for 3 hr. In other computer responses with their model, the time to reach maximum sweating after stepping into a hotter environment is found to be very much less than the 10 hr predicted by Fig. 3.14 to reach maximum shivering. These predictive probes on the computer are of course part of the justification for developing a mathematical model, especially for those experiments which would be physiologically difficult or even dangerous. In addition, a computer model can map out quickly and easily the form of response to a wide variety of stimulations. Of course such responses carry less certainty as one moves out of the range for which the model is known to be valid.

The core temperature (rectal measurement, Fig. 3.14) shows an interesting "wrong-way-rising" response of about 0.2°C after entering the colder environment. This is confirmed physiologically and could be predicted by Crosbie's model when a rate-sensitive effect was incorporated in the vasomotor-control function (see Sec. 3.8). Eventually core temperature drops somewhat, although only about 1°C compared with 7°C drop in skin temperature.

We now return to modeling the control and transducer functions of this system, which can only be based on such experimental data as those just given, rather than on the a priori method which was justifiable for the controlled process itself.

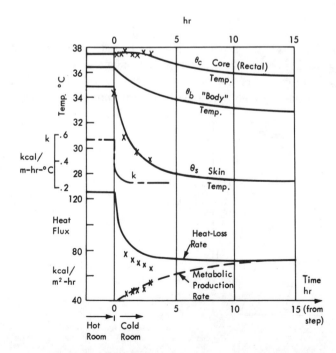

FIG. 3.14 Thermoregulatory response to "cold step." (*From Crosbie et al.*)

3.8   Mathematical Model of the Control Functions

Referring back to Fig. 3.3 we note that the model derived in Sec. 3.6 only applied to the block marked "the body." One must immediately admit that present physiological knowledge does not permit individual modeling of each of the remaining eight blocks. However, on the basis of the thermoreceptor knowledge mentioned in Sec. 3.3 and the experimental data presented in Sec. 3.7, a limited combined modeling can be attempted.

(A) THERMORECEPTOR CHARACTERISTICS

Some detailed electrophysiological information is available concerning the response characteristics of cutaneous thermoreceptors [3.8, 3.15–3.17]. The response to temperature changes has both proportional and rate-of-change components.

Steady-state Responses.   The typical steady-state proportional characteristic of a cold receptor is a humped curve showing a maximum impulse frequency at some temperature and falling to zero on either side in a total range of about 25°C. Each individual receptor peaks typically at a different temperature, however, as shown in Fig. 3.15 from data for the cat's lingual cold receptors [3.15]. These experimenters also found that when the impulse frequencies from 10 to 20 fibers were averaged, the combined characteristic peaked in the 15 to 20°C range and fell to zero by about 38°C. These characteristics are similar to the heat-production-rate curves of Fig. 3.12*b*, so that one is tempted to suggest a direct controller action from cold thermoreceptors to heat production (although the latter is of course also modulated by core temperature).

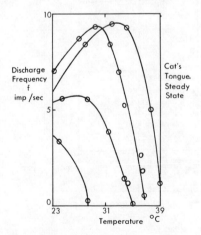

FIG. 3.15   Isolated thermoreceptor characteristics—cold.   (*From Hensel and Zotterman.*)

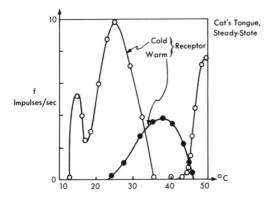

FIG. 3.16 Isolated thermoreceptor characteristics—warm and cold. (*From Zotterman.*)

The warmth receptors have rather similar individual response characteristics to steady-state temperatures, except that their region of maximum firing frequency is more closely bunched around a temperature very close to 37°C. Since the warm and cold receptors have opposite characteristic slopes in the temperature region where heat production is normally needed, they could act in push-pull fashion to achieve thermoregulation; that is, as shown in Fig. 3.16 for representative single cold and warmth receptors [3.17], a decrease of temperature in the range 25 to 37°C produces an increased *difference* between the two frequencies. If, therefore, this mode of action is utilized physiologically, the cold fiber's synaptic connections would be excitatory to heat production, whereas the warmth fiber's connections would be inhibitory. However, direct physiological evidence on this point is lacking, and indeed the exact function of the warmth receptors is not clear. The increasing impulse frequency of cold receptors with temperature above 45°C is thought to be responsible for the "paradoxical-cold" effect, which is experienced if one plunges a hand into very hot water without first knowing whether it is hot or cold.

Dynamic Responses. Both cold and warm cutaneous receptors are sensitive to rate of change of temperature, but not in an identical manner. Figure 3.17 shows the cold-receptor response following sudden reduction of tongue temperature. There is an initial jump in impulse frequency followed by a decay to the steady-state frequency which is so typical of receptor characteristics. If the decay is modeled by an exponential curve, the time constant is about 2 sec. Benzinger et al. [3.5] report that the shivering metabolism response to rapid temperature changes also overshoots its steady-state value; although this overshoot is not evident in Fig. 3.14, it is not surprising since the first measurement was not made until after about 1 hr.

Unlike the pupillary receptors, the cold receptors are bidirectionally rate-sensitive, because a rapid increase of temperature sharply

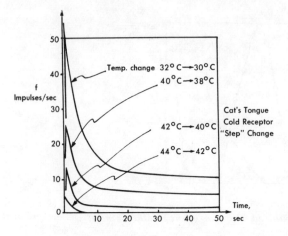

FIG. 3.17   *Isolated thermoreceptor characteristics—transient.* (*From Hensel and Zotterman.*)

reduces the existing impulse frequency initially, often to zero, in addition to the more expected increase of frequency upon temperature decrease.   On the other hand, while the warmth receptors exhibit a similar rate sensitivity to rapid temperature increases, they also have a tendency to increase their impulse frequency upon rapid reduction of temperature [3.10].   The significance of this unexpected characteristic is again not clear.

The *hypothalamic thermoreceptors* probably act through setting up slow potentials in a local neuronal region, thus effectively changing threshold levels.   It is not clear whether separate cold and warmth reception exists, or whether rate sensitivity is exhibited.

▼

(B) HEAT–FLUX CONTROL FUNCTIONS

Evaporation Rate, F_e. Figure 3.12 provides basic empirical information on the control functions.   Thus, the evaporation rate F_e in Fig. 3.3 is essentially an almost linear threshold phenomenon with core temperature θ_c as input.   Hence, the linearized transfer relation is

$$\begin{aligned} F_e &= k_1(\theta_c - \theta_c^*) \quad \text{kcal/hr} \quad \theta_c > \theta_c^* \\ F_e &= F_{eo} \quad \text{kcal/hr} \quad \theta_c < \theta_c^* \end{aligned} \tag{3.37}$$

where θ_c^* = desired core temperature $\approx 37.1°\text{C}$
$\quad\quad F_{eo}$ = basal evaporation rate ≈ 0
$\quad\quad k_1 \approx 700 \text{ kcal/(hr)(°C)}$

This is a steady-state relation, because there are insufficient data for dynamic modeling.   However, evaporation is known to increase within a few minutes after entering a hot environment and to reach its maximum rate within less than 1 hr [3.11].   This relatively rapid response probably depends in part upon thermoreception elsewhere than θ_c because, for example, sweating is apparently initiated if the ambient temperature exceeds 30°C, without change of core temperature [also 3.11]. Obviously the complete picture is more complex than that presented in Eq. (3.37).

Crosbie et al. have used a somewhat different evaporation relation, based upon a different controlling variable. They define the weighted average body temperature θ_b as

$$\theta_b = \frac{\theta_s L_s + \theta_m L_m + \theta_c L_c}{L_s + L_m + L_c} \tag{3.38}$$

where L is the thickness of the appropriate slab (Fig. 3.7a). They then define evaporation heat flux as

$$
\begin{aligned}
F_e &= F_{eo} + k_4[k_2(\theta_b - \theta_b^*) + k_3(\theta_b - \theta_b^*)^4] & \theta_b &> \theta_b^* \\
&\leq 50 F_{eo} & & \\
F_e &= F_{eo} & \theta_b &< \theta_b^*
\end{aligned} \tag{3.39}
$$

where k_2, k_3 = appropriate empirical constants
  k_4 = "exercise" constant
  θ_b^* = desired set-point body temperature

Relative humidity, air velocity, etc., are presumed to provide an upper limit to F_e, which is specified rather arbitrarily as $50 F_{eo}$.

Muscle Shivering M_m and BMR M_b. Figure 3.12b provides a two-variable nonlinear function

$$M_b + M_m = f(\theta_c, \theta_s)$$

For approximate purposes it might be adequate to linearize the relation for which average gain constants have already been quoted:

$$
\begin{aligned}
M_m + M_b &= M_b + 250(\theta_c^* - \theta_c) + 18(\theta_{so} - \theta_s) & \theta_c &< \theta_c^* \\
M_m + M_b &= M_b & \theta_c &> \theta_c^*
\end{aligned} \tag{3.40}
$$

where $\theta_c^* = 37.1°C$
  $\theta_{so} \approx 27°C$
  $M_b \approx 70 \text{ kcal/hr}$

The linearized relation in θ_s is found by drawing an average slope line in the linear-looking area of Fig. 3.12a which also passes through the M_b value at $\theta_c = 37.1°C$. This is a line for $\theta_c = 27°$ which is thus the required value of θ_{so}. When parallel straight lines are drawn for several values of θ_s, they do in fact represent the local domain reasonably well. The gain of core-temperature control is then the slope of these parallel lines, namely about 250 kcal/hr of M_b per °C of core temperature, while the gain of skin-temperature control is the ratio of change in M_m for a small change in θ_s at constant θ_c, for example 70 kcal/hr between 24 and 28°C, so that the gain equals about 18 kcal/(hr)(°C). Alternatively, this gain can be measured from the cross-plotted graph of Fig. 3.12b. Often this linearization would unnecessarily restrict the applicability of the model, and in such cases relevant nonlinear relations should be used.

Crosbie et al. defined a simple linear relation for muscle shivering based upon their averaged body temperature, but this of course automatically includes core and skin temperatures (as well as muscle temperature) in the relation:

$$M_m + M_b = M_b + k_5(\theta_b^* - \theta_b) \qquad \theta_b < \theta_b^*$$
$$M_m + M_b = M_b \qquad \theta_b > \theta_b^* \tag{3.41}$$

Vasomotor Action. The modeling of this control action depends to some extent on the form of model for the controlled process. With the slab model, Crosbie et al. found it necessary to incorporate both proportional and rate-sensitive action in order to have the overall model's dynamic responses compatible with experimental data. Thus their thermal-conductivity relation is

$$k = k_o \left[1 + k_6(\theta_b - \theta_b^*) + k_7 \frac{d}{dt} \theta_b \right] \qquad \theta_b > \theta_b^*$$
$$\leq 1.7k_o$$
$$k = k_o \left[1 + k_8(\theta_b - \theta_b^*) + k_7 \frac{d}{dt} \theta_b \right] \qquad \theta_b < \theta_b^* \tag{3.42}$$
$$\geq 0.56k_o$$

where k_6 is about twice as large as k_8. Figure 3.13 illustrates the k changes during a transient response.

In the concentric-cylinder model, vasomotor action affects only the core-skin thermal conductance. Here Stolwicjk [3.3] quotes a range of k per unit length as

$$8 < \frac{k}{L_{cs}} < 50 \text{ kcal}/(\text{hr})(\text{m}^2)(°\text{C})$$

while the causative skin blood flow ranges over 1 decade (factor of 10) either side of the normal flow of 0.2 liter/(min)(m²).

This completes our present, and admittedly grossly simplified, modeling of the thermoregulatory system. The feedback loops of Fig. 3.3 can now be modeled by approximate relations (with basal and shivering metabolisms lumped together), even though they are not fully satisfactory dynamic relations. The sluggishness of the controlled process itself emerges perhaps as the outstanding feature, so that the fast dynamics of many of the loop components can probably be safely ignored. Many phenomena of greater or lesser relevance have been ignored in this modeling—for example, the hunting or on-off type of action which in detail is observed in the sweating and shivering mechanisms, or the local operation of thermoregulation in specific areas. However, the intention has been to illustrate the mathematical approach to modeling in general. Hopefully the investigator will tackle such further complexities with confidence that a mathematical model can be built up, provided only that experimental data or relevant physical data are available. The availability of computers assures him that almost all models which he may conceive can be dynamically tested.

4 Energy Storage and Dissipation

4.1 Introduction

The thermoregulatory control system demonstrates fluctuations of temperature because the ideal control action of matching heat loss and heat production cannot be achieved instantaneously under varying conditions. The resulting temperature fluctuations involve changes in the body's stored energy. Ultimately all dynamic systems can be viewed as devices through which energy flows occur, even though the actual rate of energy flow (power) may vary tremendously in different systems; thus, communicating devices such as the brain absorb infinitesimal power per unit element (of the order of 10 watts for the order of 10^{10} neurons) in performing their vitally important biological functions, whereas the skeletal muscles can briefly produce an order of 1 kw (about 1 hp) in direct mechanical work, together with somewhat more in the degraded form of heat.

The fact that we can meaningfully describe the skeletal mechanical output in terms of electric power (kilowatts) hints at the unitary nature of all forms of energy, which is formalized in the principle of conservation of energy, often called the first law of thermodynamics: Energy can be transformed between various forms but can be neither created nor destroyed. In recent years this law has been generalized to include also the equivalence of energy and matter.

$$\Delta E = \Delta Q - \Delta W \tag{4.1}$$

where ΔE = change of internal energy storage in system

ΔQ = net flow of heat into system

ΔW = external work done by system, including such forms as mechanical, chemical, electromagnetic

Figure 4.1 illustrates some of the energy-conversion phenomena in biology and engineering, in five different energy fields, chemical, electromagnetic, mechanical, fluid, and thermal. These fields do not necessarily represent all phenomena satisfactorily, but they do permit a fairly comprehensive picture. The arrows on the diagram represent both specific feedback receptors (for example, thermoreceptors) and energy transduction "engines" (for example, photosynthetic cells). Transducers often work through several domains; for example, the heart converts chemical energy into fluid energy via mechanical work of heart muscle. Indeed the closed-loop chain of cardiac control can be traced around the diagram as follows:

Chemical → mechanical; heartbeats

Mechanical → fluid; bloodflow characteristics

Fluid → mechanical; mechanostimulation of baroreceptor

Mechanical → electromagnetic; baroreceptor afferent path

Electromagnetic → chemical; cardiac innervation

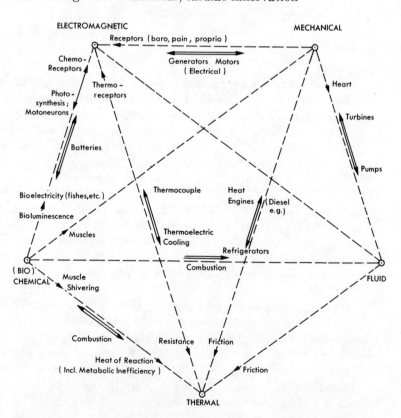

FIG. 4.1 Energy-conversion diagram.

Some common engineering transducers are also shown on this diagram, and it is interesting that they tend to occur in reversible pairs, for example, motors and generators.

Since many biological activities are basically "fueled" by the biochemical compound adenosine triphosphate (ATP), such activities are shown as one-way arrows from the chemical field. All biochemical energy, of course, ultimately derives through the single arrow of photosynthesis. However, in his capacity as engineer, man can make use of some other energy sources, including water and wind power and electromagnetic radiation. Like photosynthesis, however, these sources are ultimately derived from the sun's electromagnetic radiation.

A feature of considerable practical importance in biology and engineering is the set of arrows from all other fields to the thermal one. This describes the universal tendency in nonliving systems for energy to be degraded irreversibly into heat energy. The widespread phenomenon of resistance accomplishes this in the electromagnetic, mechanical, and fluid fields, as well as the heat-of-reaction effect in chemical reactions. This "heat-death" phenomenon is formally embodied as the second law of thermodynamics, of which three aspects are discussed below. They are equivalent to one another, but this is not immediately obvious.

1. A perpetual-motion machine which will cool some system and convert the abstracted heat into useful work without other supply of energy is, most unfortunately, impossible. If this were not so, one could, for example, build a machine which could heat a house and cool a refrigerator simultaneously by pumping cold water from a nearby river through a heat exchanger and cooling it a little to abstract heat. The colder water could then pass through the refrigerator while the abstracted heat simultaneously warmed the house—all for the expense only of pumping the water! Indeed, both functions can be performed, but at the expense of some external work for such items as compressors [see, for example, (4.1)].

    The basic snag to this sort of perpetual-motion machine is that heat in the water at say 4°C cannot be upgraded to the same quantity of heat at say 40°C without expending energy. A new state variable called entropy is needed to quantify this idea [1.3, 1.4, 4.1]. Thus, although there is almost a limitless supply of heat in the water it is unavailable for our purposes unless we first pay for it with a high-energy input. There is an analogy here to the amount one can save at a sale—one certainly cannot save anything until one first spends something.

2. Any transducing device can only convert a proportion of its energy input into another form unless the desired output form is heat, in which case 100 percent conversion is only too easy. Muscles are relatively efficient transducers from chemical to mechanical energy, comparing well with the engineer's heat engines. Nonetheless,

their efficiency is only about one-third, so that two-thirds of the chemical-energy input inevitably appears as heat. Nature's great achievement has been to package a small chemical combustion engine which is able to work essentially cold (37°C) as compared with the many hundreds of degrees centigrade typical of the engineer's combustion engines.

3. A third version of the second law states that an isolated system in equilibrium must be at a state of maximum entropy, which means in uniformly low order. The living world through its ability to deisolate itself by trapping solar energy is able to build up pockets of very high order indeed—namely living organisms—within an overall world of relatively low order. If the solar-energy source were cut off, all life would slowly disappear, and the whole world would degrade itself to uniformly low order.

In the last few paragraphs the problem of converting energy between fields has been introduced. However, this chapter is more concerned with the relations for storing or dissipating energy in any one medium, although the relations will be generalized for several media. In Chap. 3 it has been shown that heat conduction in a medium is described by a partial differential equation, and in particular that the time derivative of temperature $\partial\theta/\partial t$ at any location is proportional to the rate at which energy storage changes at that location. In another basic form of partial differential equation, the wave equation, there are two forms of energy storage, potential and kinetic (for example, pressure and velocity in the acoustic wave, capacitance and inductance in the electrical transmission cable). These give rise to such properties of wave motion as the propagation velocity and resonance. Thus, ideal modeling of energetic phenomena should allow for spatial distribution as well as time distribution. However, the simplification of lumping the phenomena within discrete spatial regions will be utilized for most modeling in this book because sufficiently accurate results can be obtained so much more easily.

The process of lumping results in a set of ordinary differential equations, which are each first order for heat conduction and second order for the wave equation. Such a set of equations corresponds to a network of dynamic elements, as shown by the block diagram of Fig. 3.6b. The two laws basic to circuit analysis of lumped networks are presented in Sec. 4.3, after which the actual energy-storage and dissipation phenomena are considered quantitatively in successive sections. As a preliminary, however, the energetic aspects of some very simple and familiar circuits are explored qualitatively in Sec. 4.2.

4.2   Energetic Aspects of Active and Passive Control Systems

The conditions under which an active control system could go into unending oscillation were developed qualitatively in Chap. 2. From an energy point of view it is clear that this oscillation results because the phasing of the input-output relations of the loop elements is such that the amplifiers pump energy into the system during each cycle, as does the child in working up an oscillation on his swing (Fig. 4.2). When the child has reached sufficient amplitude he may cease to input energy and instead sit quietly on the seat. It continues to oscillate, except that the amplitude decays more or less gradually in time. The swing is an example of an almost conservative system, that is, one in which there is little degradation of system energy into the unwanted form of heat through inevitable and ubiquitous frictional effects. The system is passive when the child sits quietly, but is actively controlled if he decides that there is some swing amplitude he desires and

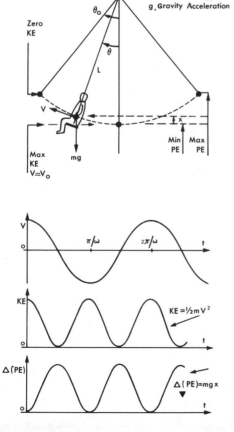

FIG. 4.2   Dynamic and energetic relations for conservative swing.

toward which he will manipulate the system. A conservative system is one in which the total energy remains constant, without external supply, although it may be converted periodically among various forms. The second law of thermodynamics asserts that in practice one will always irreversibly degrade some energy to heat during such conversions, although the rate may be made vanishingly small if the process is expensively designed and occurs at an infinitesimal rate.

In this particular example, there is rhythmic interchange between energies of two types:

1. *Potential energy* (PE), or energy of the child's position as a result of doing work against gravity.
2. *Kinetic energy* (KE), or energy of the mass velocity.

In the ideal conservative system the sustained oscillation is essentially sinusoidal both in angle and in the two energies, as sketched in Fig. 4.2.

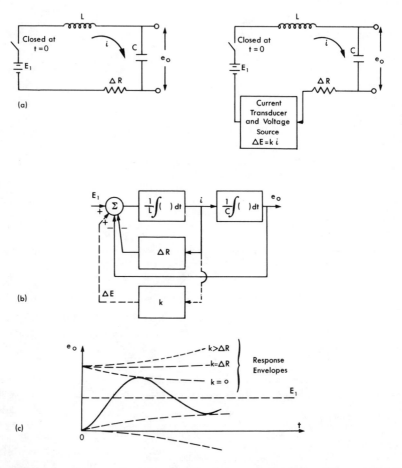

FIG. 4.3   *Active and passive feedback in resistive-inductive-capacitive system.*

Since no passive systems are conservative, their useful energy level can only decay. However, as just discussed, a sustained oscillation can be obtained when the process is actively controlled, at the expense of a continuing controlled energy input to the system. Consider now the basic inductor-capacitor (*LC*) circuit shown in Fig. 4.3. This is an oscillator except that the inevitable small resistance ΔR of components such as the battery cause decay of the oscillations which result when the switch is closed. This system could be actively controlled in several ways, for example, by inserting a source which generates a voltage proportional to a constant k multiplied by the measured current. This can be considered a negative resistance for, as Fig. 4.3 shows, this active feedback path can cancel out the undesired decaying effect of the passive feedback path containing ΔR. Topologically there is no difference between the ΔR and k paths, but from the control point of view they represent the difference between a passive system and an actively controlled one.

4.3   Generalized Networks and Kirchhoff's Laws

In this section we establish some general rules for the analysis of networks of interconnected lumped elements. Mathematical relations are developed for the operating laws of three generalized linear energetic elements, and then their energy-storage relations are given in later sections.

Kirchhoff's laws of electric circuit analysis are suitable for generalized networks including mechanical (masses, springs, dashpots), thermal (conduction, convection, thermal energy), fluid (resistances, etc.), and chemical (diffusion, etc.) [see, for example, (4.2), (4.3)]. We first define two generalized variables:

Generalized effort e
Generalized flow f

The resulting generalized *law* of a lumped energetic network element is

$$e = Zf \tag{4.2}$$

where the detailed dynamic properties of the element are defined by its

Generalized impedance Z

whose properties are subsequently examined. A second important relation is that the product of these two variables is the generalized power, or rate of change of energy,

Generalized power $P = ef$ $\hspace{3cm}$ **(4.3)**

TABLE 4.1   GENERALIZED VARIABLES IN SEVERAL FIELDS

System	Electrical	Mechanical	Fluid	Thermal	Chemical
Effort e	Voltage v	Force F	Pressure p	Temperature θ	Concentration C
Flow f	Current i	Velocity V	Volume rate of change $\dot{V} = \dfrac{dV}{dt}$	Kcal/hr q	Mole flow rate q
Power $P = ef$	Watts P	HP, ft-lb/ sec, etc.	HP, etc.	"Available" part q (see notes)	

Analogous sets of commonly used variables for different fields are suggested in Table 4.1, but they are not the only possible sets. Unfortunately, the power flow of heat cannot be meaningfully expressed except through consideration of the availability or entropy effects.   Kirchhoff's laws specify the equations for interconnecting networks of these elements.

1. *Kirchhoff's first law—the loop (e.g., voltage) law.* The algebraic sum of effort differences (voltage drops) across the three impedance elements (resistors, capacitors, inductors) and across effort or flow sources (batteries, etc.) is zero around any closed loop; that is (Fig. 4.4),

$$\Sigma_{\text{loop}}(e_j - e_k) = 0 \qquad e_j \neq e_k \tag{4.4}$$

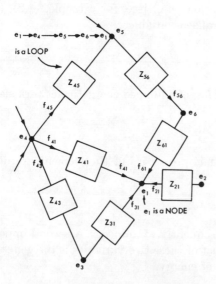

FIG. 4.4   Generalized lumped network.

2. *Kirchhoff's second law—the node (e.g., current) law.* The algebraic sum of flows (currents) into a node from all its branches is zero, that is (Fig. 4.4),

$$\Sigma_{\text{node}} f_{jk} = 0 \tag{4.5}$$

The *loop* law is true for *all* closed loops, and not just for the "minimum" closed loop in any region. Physically this is exemplified by the result that if one walks around a mountain and returns to the same spot, the net change of elevation is always zero no matter what different elevations may have been attained during the traversal. The actual solution of the problem is usually obtained by inserting the impedance laws of the elements into the equations also, so that, for example, a solution for the flows is obtained by using the loop law for efforts. Although the directions of flow arrows may be arbitrarily chosen, care must be taken with the *signs* of efforts and flows, in writing out the Kirchhoff relations.

4.4 The Dynamic Nature of the Network Equations

An electric network consisting only of resistors and some voltage-driving sources can easily be visualized, and a vascular network of interconnecting vessels in steady-state flow provides a superficially analogous example, since both network elements produce resistance to flow. In the resistor network a new steady-state distribution of current and voltage is in principle instantaneously established after any changes of driving external voltage. This inertialess type of network depends upon the fact that ideal resistors do not store energy, although they do dissipate it as heat at a known rate. Consequently, the set of network equations, either in current or in voltage variables, is purely algebraic in form. However, when energy-storage elements are introduced into the network, the behavior becomes dynamic and a transient settling time is necessary to reestablish equilibrium after any change of forcing conditions. This is exemplified by the hydraulic network since energy is stored in the fluid as well as being dissipated through flow resistance. Therefore, the hydraulic and electric resistor networks are only analogous in the steady state. In general the lumped hydraulic network and all other lumped energy-storing networks are characterized by a set of ordinary differential equations, as noted already for the thermoregulatory system. In the next three sections the mathematical nature of the three generalized impedance elements satisfying Eq. (4.2) is examined.

TABLE 4.2   THE GENERALIZED DISSIPATIVE ELEMENT

Area	Electrical Resistance	Fluid Resistance (Liquid, Gaseous—Incompressible)	Mechanical Friction	Heat Conduction	Chemical Diffusion
Linear element and symbols				Conduction	
Linear law—dissipative relation	Resistor $v = Ri$ volts amp.	$p = Rq$ pressure volume drop flow rate Viscous flow (laminar, Poiseuille)	$F = RV$ force velocity	$\theta = \dfrac{1}{k} q$ θ temp. diff. °C q kcal/(hr)(°C) (for example)	$c = \dfrac{1}{k} q$ conc. mole diff. flow rate
Power dissipation	$P = Ri^2$ $\ = \dfrac{1}{R} v^2$	$P = Rq^2$ $\ = \dfrac{1}{R} p^2$	$P = FV = RV^2$ $\ = \dfrac{1}{R} F^2$	(See text)	(See text)
Nonlinear element	Thyrite, diodes, etc.	Turbulent flow, orifices, etc.	"Stiction," etc.	Radiation	
Nonlinear dissipative relation		$p = Rq^2$	Stiction $F = F(V)$, above. Square-law friction $F = RV^2$	$q \propto k(\theta_1^4 - \theta_0^4)$	

▼

4.5 The Generalized Dissipative Element

In the physical world, flow is always opposed by a resistance. The resistance is overcome by applying a suitable effort, and the result is a continuous dissipative loss of energy, or more correctly a conversion to the usually undesired form of heat. The generalized form of Eq. (4.2) for impedance elements becomes

$$e = Rf \qquad (4.6)$$

where R is called the *resistance*, and in electricity this is known as Ohm's law. When the resistance can be presumed constant the element is linear. Linear resistive elements are perhaps more closely approximated in the electric field than in any other. In the nonlinear case we simply write

$$e = e(f) \qquad (4.7)$$

Table 4.2 shows linear and some nonlinear relations for resistive elements in the five fields already mentioned, all of which exist in biological systems. The power dissipation in the linear dissipative element follows by combining Eqs. (4.3) and (4.6).

$$P = Rf^2 = \frac{e^2}{R} \qquad (4.8)$$

The last two columns of Table 4.2 represent diffusion phenomena, for which the power term is somewhat different. For example, one cannot look upon heat-flow rate as the power since heat leaves the conduction element as fast as it enters. Rather, it is the availability of this heat for doing useful work in other domains of energy (for example, mechanical) that has been dissipated by the lowering of the forcing variable of temperature. This unavailable energy depends upon the temperature and the special form of heat capacity called *entropy*, as already mentioned.

▼

4.6 The Generalized PE-storage Element

An element stores PE when its effort level increases as a result of influx of the flow variable. Thus the voltage across a capacitor increases in proportion to the amount of electrons (coulombs) which has inflowed, while the force in a spring increases directly with its displacement. It is natural, then, to define a

Generalized displacement h

which is directly the integral of the flow variable f:

$$h = \int f \, dt \qquad \text{or} \qquad f = \frac{dh}{dt} \qquad (4.9)$$

Thus, for example, velocity V is the time derivative of displacement dx/dt. Consequently, the law of the linear PE element is

$$
\begin{array}{ll}
e = \dfrac{h}{C} & \quad e = \dfrac{1}{C} \displaystyle\int f\, dt \\[2mm]
\qquad\qquad \text{or} & \\[1mm]
h = Ce & \quad f = C\, \dfrac{de}{dt}
\end{array}
\tag{4.10}
$$

where C is the generalized *capacitance* with units of displacement per unit effort. Table 4.3 presents data for linear PE-storage elements in the electric, fluid, mechanical, and thermal fields.

For this generalized element the PE stored E_p is a more natural function to use than the power transfer P_p, which emerged for the dissipative element. However, the two are always related by the identity

$$
P \equiv \frac{dE}{dt}
\tag{4.11}
$$

The elemental work done dW when an effort e displaces an elemental distance dh in the direction of the effort is

$$
dW = e\, dh
$$

and in the ideal PE element, this is all stored as potential energy, so that

$$
E_p = \int_o^h e\, dh
\tag{4.12}
$$

By combining the law of the element equations (4.10) with Eq. (4.12) one obtains

$$
E_p = \frac{h^2}{2C} = \frac{Ce^2}{2}
\tag{4.13}
$$

For the example of the electric capacitor

$$
E_p = \frac{Cv^2}{2} \quad \text{joules (watt-sec)}
\tag{4.14}
$$

where C = capacitance, farads
  v = volts

The energy stored against gravity by the child on the swing (Fig. 4.2) is also normally called PE. Indeed, if the relevant effort variable is taken as the component of gravity force tangential to swing motion, an analogous "gravity spring" emerges in which this force component is approximately proportional to angular deflection θ, and hence the stored energy is proportional to θ^2. However, the gravity force and stored energy E_g are more conventionally expressed with regard to the vertical force F and displacement x; that is,

$$
F = mg
\tag{4.15}
$$

$$
E_g = Fx = mgx
\tag{4.16}
$$

TABLE 4.3 THE GENERALIZED POTENTIAL-ENERGY STORAGE ELEMENT

$$h = \int f\, dt$$

Area	Electrical Capacitor	Fluid Compressibility	Mechanical Spring, etc.	Thermal Mass
Linear element	$$Q = \int i\, dt$$	$$Q = \int q\, dt$$	$$x = \int v\, dt$$	$$Q = \int q\, dt$$ $$C = mVc$$
Law of the element	Volts $$v = \frac{1}{C} \int i\, dt = \frac{Q}{C}$$ Amperes Coulombs $$L = C\frac{dv}{dt} \quad Q = Cv$$	Pressure $$p = \rho\,\frac{g}{A} \int q\, dt = \rho\,\frac{g}{A}\, Q$$ $$= \rho g x$$	Force $$F = kx$$	Temp. $$\theta = \frac{1}{C} \int q\, dt = \frac{Q}{C}$$ $$Q = C\theta$$
Energy stored	$$E_p = \frac{Cv^2}{2} = \frac{Q^2}{2C} \quad \text{joules}$$	$$E_p = \frac{\rho g}{2A}\, Q^2$$	$$E_p = \frac{F^2}{2k} = \frac{kx^2}{2}$$	(See text)

where g is the gravity constant. The gravity and pressure forms of energy storage are mutually convertible in many cases; for example, in a vertical column of blood the pressure energy is maximum at the bottom of the column, whereas the gravity energy is maximum at the top. If the column is static the sum of the two energies is constant at any height.

▼

4.7   The Generalized KE-storage Element

Elements store KE when their effort level increases as a result of the flow variable increasing. In electricity an inductance exhibits this phenomenon, and in the fluid-mechanical area it is associated with the energy of motion. Naturally oscillatory circuits arise when energy interchange between KE and PE elements is possible. However, since there is no KE phenomenon in heat transfer and chemical reactions, networks of these types do not naturally oscillate.

For the KE phenomenon it is convenient to define one further generalized variable:

Generalized momentum p

which is directly the integral of the *effort* variable e:

$$p = \int e\, dt \qquad \text{or} \qquad e = \frac{dp}{dt} \tag{4.17}$$

The law of the generalized element is

$$\begin{array}{|ccc|}
\hline
 & & \\
p = Lf & & e = L\,\dfrac{df}{dt} \\
 & \text{or} & \\
f = \dfrac{p}{L} & & f = \dfrac{1}{L}\displaystyle\int e\, dt \\
 & & \\
\hline
\end{array} \tag{4.18}$$

where L is the generalized inductance (or perhaps mass) with units of effort per unit rate of change of flow. Table 4.4 shows linear examples in the electric and mechanical fields, results for fluid mass being the same as for mechanical mass. Since electric inductance is a familiar phenomenon, we derive the relevant equations for the accelerating-mass example. From Newton's second law of motion,

$$F = \frac{d}{dt}\, mV \tag{4.19}$$

Therefore,

$$p \triangleq \int F\, dt = mV \tag{4.20}$$

The quantity mV is classically defined as the momentum of m, which provides the motivation for the name here given to the generalized variable p.

TABLE 4.4   THE GENERALIZED KINETIC-ENERGY STORAGE ELEMENT

Area	Electrical Inductance	Mechanical (or Fluid) Inertia
Linear element		
Law of the element	Volts $\quad$ Momentum $v = L\dfrac{di}{dt} \quad p = Li$ Amperes $i = \dfrac{1}{L}\displaystyle\int v\,dt = \dfrac{p}{L}$	Force $\quad$ Momentum $F = m\dfrac{dV}{dt} \quad p = mV$ Velocity $V = \dfrac{1}{m}\displaystyle\int F\,dt = \dfrac{p}{m}$
Energy stored	$E_k = \tfrac{1}{2}Li^2$	$E_k = \tfrac{1}{2}mV^2$

The KE E_k stored by this generalized linear element can be obtained by

$$dE_k = p\,df \quad \text{or} \quad E_k = \int_0^f p\,df$$

Therefore, using Eqs. (4.18),

$$E_k = \frac{Lf^2}{2} = \frac{p^2}{2L} \tag{4.21}$$

Thus, for the electric inductance,

$$E_k = \tfrac{1}{2}Li^2 \tag{4.22}$$

and for a moving mass,

$$E_k = \tfrac{1}{2}mV^2 \tag{4.23}$$

4.8   State-variable Diagram Representing Energetic Components

The relations among the three generalized energetic elements, dissipative, PE-storage, and KE-storage, can be usefully represented on a diagram of the four state variables which have now been defined. These variables are

Effort	e	
Flow	f	
Momentum	$p = \int e\,dt$	(4.24)
Displacement	$h = \int f\,dt$	

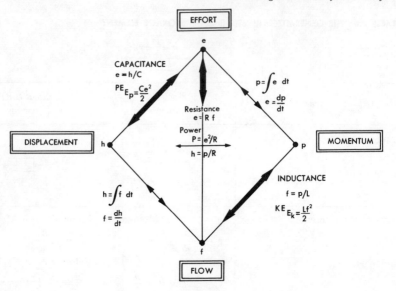

FIG. 4.5   State diagram for energetic components.   (*After Paynter.*)

The associated power and energy functions are

Power $\qquad P = ef = \dfrac{dE}{dt}$

PE $\qquad\quad E_p = \int e\,dh$ $\qquad\qquad\qquad\qquad\qquad\qquad$ **(4.25)**

KE $\qquad\quad E_k = \int p\,df$

The generalized elements are defined from the basic impedance relation

$e = Zf$ $\qquad\qquad\qquad\qquad\qquad\qquad\qquad\qquad\qquad\qquad$ **(4.26)**

and are

Resistance $\qquad R = e/f \qquad P = Rf^2 = e^2/R$

Capacitance $\qquad C = h/e \qquad E_p = \tfrac{1}{2}Ce^2$ $\qquad\qquad$ **(4.27)**

Inductance $\qquad L = p/f \qquad E_k = \tfrac{1}{2}Lf^2$

The relations of these elements are shown on the state diagram of Fig. 4.5.   Two sides of the diamond are involved in the definitions of the extra two state variables, while the other two sides define the energy-storage elements.   Finally, the dissipation element connects across either diagonal.   Although this generalized presentation may be confusing initially, the reader should appreciate that this standard scheme can incorporate a wide range of energetic phenomena from the physical world.   Such generalizations are of conceptual value.   All the phenomena discussed occur in biological systems.

The scheme is not fundamentally limited to linear systems, but on the other hand, spatially distributed effects cannot be included directly.   The process of generalizing has emphasized that the thermal and chemical diffusion phenomena differ from the electrical, fluid, and mechanical ones in that power cannot be so simply defined, and in that there is only one form of energy storage.

5 The First-order Linear Lumped Model

The Operational Transfer Function

5.1 Introduction

In this chapter a number of biological phenomena are modeled as first-order linear lumped systems. These often represent rather simplified models but the major inaccuracies are pointed out. Furthermore, the first-order model represents only one part of the overall system in many cases, and indeed may be one lump of a repeated process.

On the subject of analogy, it has become fashionable to represent many diverse systems by electrical models, presumably due to widespread knowledge of the basic electric elements and to the facility with which electric elements can be built up into analogous circuits. In principle there is nothing more universal in electrical analogies than in mechanical or fluid ones, but admittedly it is only in the electric domain that one can build elements which are quite so close to pure lumped dissipative and energy-storage elements (resistors, capacitors, and inductors) with such cheapness, linearity, smallness, cleanliness, and safety. The electronic analog computer itself provides an illustration of this point.

With regard to energy, the various models developed in this chapter are shown to be characterized by one energy-storage element (either potential or kinetic) and one form of dissipating element:

there are various possible forms of input to such systems. This generalization results in a unique form of the describing differential equation, and consequently of the corresponding block diagram. This standardized or *canonical* form is first order, and therefore there is only one overall mathematical problem to be solved. At this stage it is worthwhile to introduce a rather powerful mathematical tool: *operational mathematics* in general and the *Laplace transform* in particular. Initially we only use it to recast the differential equation into a convenient *transfer function*, and the reader should look upon the transform only as a convenient symbol. Some of its advantages are demonstrated in later sections when dynamic behavior is being analyzed, and indeed its importance for the reader should grow throughout the book with increasing familiarity and comprehension. It should be emphasized that the Laplace transform can be used immediately for dynamic analysis work by the reader, even if he remains for a period somewhat unsure of its exact nature and physical meaning.

The transfer-function approach in analyzing dynamic systems has both conceptual and practical value. First, the Laplace transformation of variables bears some analogy to the use of logarithms; that is, manipulation is simplified and useful tabulations are available. Second, the transfer function provides a convenient shorthand symbolism for expressing the responses and interactions of dynamic systems, as will become more obvious with some practice.

5.2   The Thermal Lumped Element; the Time Constant

In the analysis of the body's thermoregulatory system in Chap. 3, each of the three lumps (core, muscles, and skin) is respectively specified by a single time-variable temperature. Consequently, each lump is described by a first-order differential equation [Eqs. (3.30) to (3.32)], so that for our prototype first-order example we now consider the muscles' lump in more detail. The relevant equation (3.31) is here repeated:

$$m_m c_m \frac{d\theta_m}{dt} = M + q_{cm} - q_{ms} \tag{5.1}$$

where $M = M_m + M_x$ is the total muscle metabolic rate. For simplicity but without affecting principles, let us presume that the heat flow between core and muscles q_{cm} is zero. Then by including Eq. (3.34) for muscle-skin conduction we obtain

$$m_m c_m \frac{d\theta_m}{dt} = M - \frac{kA_{ms}}{L_{ms}} (\theta_m - \theta_s) \tag{5.2}$$

or, by collecting and manipulating terms in θ_m,

$$\frac{m_m c_m L_{ms}}{kA_{ms}} \frac{d\theta_m}{dt} + \theta_m = \frac{L_{ms}}{kA_{ms}} M + \theta_s \tag{5.3}$$

For generality we now define new variables and constants, and hence obtain

$$T\frac{dy(t)}{dt} + y(t) = k_1 x_1(t) + k_2 x_2(t) \qquad (5.4)$$

where

$$T = \frac{m_m c_m L_{ms}}{k A_{ms}} \qquad k_1 = \frac{L_{ms}}{k A_{ms}}$$

and

$$k_2 = 1$$

This equation describes one standard form of first-order system. The variables $x_1(t)$ and $x_2(t)$ on the right-hand side are input-forcing and disturbance variables, in this case, muscle metabolism M and skin temperature θ_s, respectively. In order to reemphasize that they are functions of time, they are so written explicitly. If a core temperature different from muscle temperature is allowed, there is one more input to the system and the constant T changes in value. The variable $y(t)$ is the output variable of the system, in this case, the muscle temperature θ_m.

(A) TIME CONSTANT

The term T is the most important characteristic parameter of the first-order system because the k's are essentially only scaling constants. T is defined as the *time constant* of this system and defines its dynamic characteristics as explained in Chaps. 6 and 7. T arises as the ratio of mc and kA/L and is therefore the ratio of the thermal-energy storage constants and the thermal resistance, or dissipative constant. More generally, the time constant of the first-order system is always the ratio of the system's energy-storage and dissipative parameters.

The form of Eq. (5.2) shows the details of the physical processes involved, as illustrated by Fig. 5.1a. On the other hand, Eq. (5.4) represents a standard mathematical form for the first-order system, for which the slightly simpler block-diagram form is given in Fig. 5.1b. The unity gain of the negative-feedback path from the output variable results directly from incorporating the time constant in the forward path.

To evaluate the thermal time constant, the following approximate values of parameters for the slab model of an average man are used:

$$m_m = 30 \text{ kg}$$
$$c_m \approx 1 \text{ cal}/(g)(°C)$$
$$k = 0.0011 \text{ cal}/(cm)(°C)(sec)$$
$$A_{ms} = 2 \text{ m}^2$$
$$L_{ms} = 1.6 \text{ cm}$$

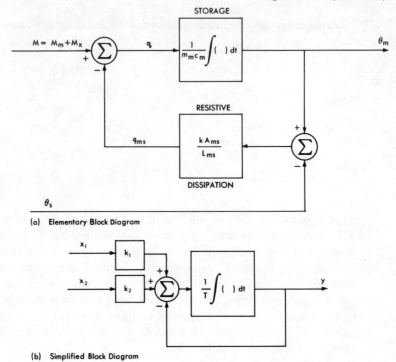

FIG. 5.1   Thermal lumped element.

Therefore,

$$T = 2{,}180 \text{ sec} \approx 36 \text{ min} \tag{5.5}$$

This large value of time constant immediately tells us that this system is sluggish; that is, the individual settling time after a transient is of the order of 1 hr (Fig. 5.3 and Chap. 6). It is also interesting to reconsider the first-order model derived from lumping the homogeneous conducting slab, Eq. (3.8) and Fig. 3.6, by rewriting it in the standard form to obtain

$$2\left[\frac{\rho c(\Delta x)^2}{2k}\frac{d\theta_n}{dt} + \theta_n\right] = \theta_{n-1} + \theta_{n+1} \tag{5.6}$$

This is an iterative unit lump because it is repeated N times, and its time constant varies as the square of the path length Δx, that is, inversely as the square of the number of lumps N.

5.3   *Compartment Models for Storage and Excretion Flows*

Cells individually, and collectively as organs, are continually engaged in transferring materials across membranes. The cells must metabolize both for maintenance of their own life processes and to perform the tasks necessary in such specialized cells as neurons, liver cells, bioluminescent cells, and thermoreceptors. This metabolism involves a constant inflow of nutrition and outflow of waste and other products through the cell membrane. At first glance a passive diffusion-type process should suffice for membrane transfer. For example, the plasma concentration (extracellular) of nutrition molecules should remain higher than that intracellularly because the nutrition molecules are consumed within the cell, which is therefore a sink for such material. However, there are many materials circulating within the body, and the cells must be selectively permeable to the ones that are needed. Also, some cells exhibit a transmembrane potential which is actively sustained by a sodium-potassium pump [5.1]. Transfer of ionized material across a charged membrane involves both electric and chemical potentials, and the equations are complex and nonlinear [(1.4), (1.5)]. Instead, we therefore concentrate here upon some basic chemical-diffusion examples, although in detail nonmetabolically active processes such as facilitated diffusion and solvation may also be included [5.2].

The model to be developed represents two lumped compartments between which a particular material flows at a rate proportional to its concentration difference. The restriction to linearity is not essential except for present didactic purposes. There are many applications of this approach in the literature, of which a few examples are: water and electrolyte flows in extracellular and intracellular fluids [5.3]; extravascular storage in the liver [5.4]; flows and storage of CO_2 and O_2 in the lungs and arterial-venous systems [5.5]; the effects on renal function of sodium and water-retaining hormones [5.6]. The last two studies involve nonlinearities in at least some aspects.

We now consider an idealized linear lumped model in which a *supply* compartment interchanges material with a *consuming* compartment (Fig. 5.2). The concentration C_1 of material in the supply compartment's fluid is assumed to be constant, to be under the control of the investigator, or to be the known output of another system; in other words, to be the input variable to the present system. The concentration C_2 in the consuming compartment is an

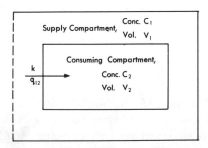

FIG. 5.2   Two-compartment flow model.

obvious choice of output variable.   The combined storage and flow equation is

$$V_2 \frac{dC_2}{dt} + M = q_{12} = k(C_1 - C_2) \tag{5.7}$$

where k = diffusion constant, cm^3/sec
$\quad\quad C$ = concentration, moles/cm^3
$\quad\quad V$ = volume, cm^3
$\quad\quad q_{12}$ = net flow rate from compartment 1 into compartment 2, moles/sec
$\quad\quad M$ = rate of consumption inside V_2, moles/sec

The concentrations are assumed to be sufficiently small that the flow does not produce volume changes.   Equation (5.7) is rearranged to yield a first-order equation in the standard form of Eq. (5.4):

$$T \frac{dC_2}{dt} + C_2 = C_1 + \frac{M}{k} \tag{5.8}$$

where the time constant $T = V_2/k$.   Some obvious steady-state results are:

Without consumption in V_2 $\quad\quad M = 0 \quad\quad C_1 = C_2$
With consumption in V_2 $\quad\quad\quad C_1 = C_2 - M/k$

Different permeabilities in each direction can readily be incorporated by rewriting the net flux as

$$\begin{aligned} q_{12} &= k_{12}(C_1 - C_2) - k_{21}(C_2 - C_1) \\ &= (k_{12} + k_{21})(C_1 - C_2) \end{aligned} \tag{5.9}$$

Thus, the net permeability is merely the sum of the two.   In order to represent active transfer outward from the consuming compartment, k_{21} would be negative, as indeed could also be the sum of $k_{12} + k_{21}$ if k_{21} were sufficiently negative.   However, because of membrane potential effects, this is not a realistic model.

Pace [5.3] postulates a slightly different linear model of the form

$$q_{12} = k_{12}C_1 - k_{21}C_2 \tag{5.10}$$

This particular model allows a steady-state concentration difference to exist, a known condition of cells, even when the consumption rate M is zero since Eq. (5.8) is now modified to

$$T \frac{dC_2}{dt} + C_2 = \frac{k_{12}}{k_{21}} C_1 - \frac{M}{k_{21}} \tag{5.11}$$

There is still little physical justification for the flow relation of Eq. (5.10), however, and a somewhat better postulate is that the "pumped" outward flow is governed actively by some such law as

$$q_{21} = k_{21}[(C_2 - C_1) + E] \tag{5.12}$$

where E is the active metabolic potential driving the pump. With the usual linear inward flow from Eq. (5.7), the net flux becomes

$$q_{12} = (k_{12} + k_{21})(C_1 - C_2) - k_{21}E \qquad (5.13)$$

If $k_{21} \gg k_{12}$, this system has the steady-state solution

$$C_2 = C_1 - E \qquad (5.14)$$

The assumption of constant pumping rate $k_{21}E$ is gross. However, although the necessary energy relations for pumping across charged membranes are known, the method and dynamics of the sodium-potassium pump's flow rates are not. In particular one asks whether there is a negative-feedback controller operating to regulate such a variable as internal sodium concentration in the cell. Such a question is especially relevant for the neuron where the average flow rates of sodium and potassium ions depend in part upon the frequency of action potentials.

5.4 Nanoplankton Respiration

A model of nanoplankton respiration has been proposed [5.7] which accounts reasonably for gross community photosynthesis, storage, and respiration of nanoplankton in tropical waters. This model has been presented in terms of direct analog elements in an electric circuit.

The sunlight is the input to the system and is represented by a battery of voltage e_b (Fig. 5.3a). The production rate f of material by photosynthesis is proportional to the difference between e_b and a "backup" potential e, of material in the system, with the constant of proportionality being looked upon as the battery conductance $1/R_b$. The community respiration rate f_R is assumed proportional to the potential e, and the storage rate f_C proportional to the rate of change of potential in the community cellular storage capacity C. Finally, the total production rate f must equal the sum of respiration and storage rates. The equations therefore are

$$f = \frac{1}{R_b}(e_b - e) \qquad f_R = \frac{1}{R}e$$

$$f_C = C\frac{de}{dt} \qquad \text{and} \qquad f = f_R + f_C$$

with the resulting block diagram of Fig. 5.3b.

These equations can usefully be solved for both f and f_R as outputs, with e_b as the input:

$$\frac{C}{1/R + 1/R_b}\frac{df}{dt} + f = \frac{1}{R + R_b}\left(\frac{C}{1/R}\frac{de_b}{dt} + e_b\right) \qquad (5.15)$$

$$\frac{C}{1/R + 1/R_b}\frac{df_R}{dt} + f_R = \frac{1}{R + R_b}e_b \qquad (5.16)$$

The dependency of respiration rate f_R upon the sunlight as input provides another example of the canonical first-order differential

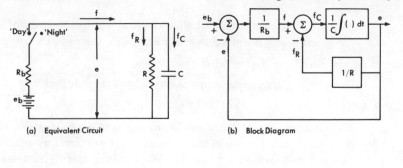

(a) Equivalent Circuit   (b) Block Diagram

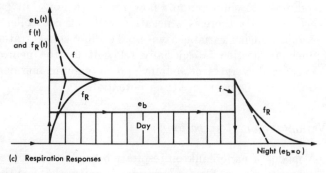

(c) Respiration Responses

FIG. 5.3   *Nanoplankton respiration model.*

equation (5.4). Since the sunrise of the tropics can reasonably be
approximated by a step input, the normal respiration-rate response
will be approximated by an exponential-type curve to a new steady
state (Fig. 5.3c), which is studied in Chap. 6.   However, the differential equation relating total photosynthesis rate f to sunlight has
an extra term, the derivative de_b/dt. In consequence, the step
response is different, as shown in Fig. 5.3c. Physically the initial
jump and subsequent exponential decay to a steady state of photosynthesis rate represent the initial short-circuit effect when a voltage is placed across an uncharged capacitor.   The two exponential
curves are governed by the same time constant.   Mathematically
the system is still first order since no derivatives higher than the
first appear in Eq. (5.15).   In control engineering, circuits satisfying Eq. (5.15) are widely used in the controller component, and are
called *lead-lag elements*, for reasons which are explained in Chap. 7.
In biology the step responses of many receptors are similar, as we
have noted.

5.5   *The Laplace Transform and the Transfer Function*

In the preceding three sections we modeled system components in
three different energy domains. They were of course carefully
chosen to produce first-order differential equations of the linear

kind because this is our present purpose. In each case, the basic first-order differential equation arose as the mathematical model for input $(x_1, x_2, \ldots)$–output (y) relations, namely

$$T\frac{dy}{dt} + y = k_1 x_1 + k_2 x_2 + \cdots \tag{5.17}$$

and in addition the last model also produced a slightly more complex form

$$T\frac{dy}{dt} + y = k\left(\alpha T\frac{dx}{dt} + x\right) \tag{5.18}$$

where α was a constant. Actually Eq. (5.17) is a special case of Eq. (5.18) (for one input variable) for which $\alpha = 0$, and Eq. (5.18) represents the most complex form possible for a linear constant-coefficient first-order system. However, the form of Eq. (5.17) is met more often than that of Eq. (5.18), with the result that both are termed *canonical*, or standardized, forms.

Such systems can be represented by elementary block-diagram operations involving feedback, as already exemplified in Fig. 5.1b for the thermal lump, but this method becomes clumsy in complex systems. We therefore seek a more compact symbolism by which to represent any given dynamic system as a single transfer block between inputs and outputs (Fig. 5.7). Several such symbolisms are available deriving from the area of operational mathematics; the differential operator D, the Heaviside operator p, the Fourier operator $j\omega$, and the Laplace operator s are all possibilities which may be met by the student in his readings. They are in many respects similar, but the Laplace transform and its operational variable s have gained wide acceptance in mathematics, physics, and engineering and so will henceforth be used in this book. At this stage it is not necessary for the reader to understand thoroughly the mathematical background of Laplace transformation, but rather he should accept it as a useful and powerful shorthand tool for the representation and solution of linear dynamic systems.

In this chapter the Laplace transform is defined, together with a few of its properties. However, only the simplest of manipulations are then made in order that transfer functions can be abstracted from ordinary differential equations to provide us with the simple symbolism we seek. Furthermore, the reader who is not mathematically inclined, but who wishes to utilize the symbolism, may jump now to Eq. (5.26). It is important to realize that the use of Laplace transforms is ultimately simplifying because it converts difficult integration procedures in the time domain into multiplication in the s domain of the Laplace operator or variable (complex frequency).

(A) DEFINITION OF LAPLACE TRANSFORM

A time-domain variable $f(t)$, such as the various flows, potentials, etc., of this chapter, can be expressed as a new variable $F(s)$ in the complex frequency domain s defined by the Laplace transform relation

$$\pounds[f(t)] \triangleq F(s) \triangleq \int_0^\infty f(t)e^{-st}\, dt \qquad (5.19)$$

where $\pounds[\ \ ]$ means "Laplace transformation of [ ]." There are certain restrictions on the form of $f(t)$, and there is also a corresponding backward relation from $F(s)$ to $f(t)$ (Sec. 6.8).

▼

(B) TRANSFORM OF DERIVATIVE OF A FUNCTION

Since ordinary differential equations comprise variables and their time derivatives, or integrals, we need to obtain transform relations for these functions. Thus, for example, the form of $F(s)$ corresponding to df/dt is obtained after integrating by parts the defining relation (5.19), as follows:

$$\pounds\left[\frac{df(t)}{dt}\right] = \left[f(t)e^{-st}\right]_0^\infty + s\int_0^\infty f(t)e^{-st}\, dt$$
$$= sF(s) - f(t = 0+) \qquad (5.20)$$

This is a very significant relation, since it tells us that if a time function $f(t)$ transforms into a complex domain function $F(s)$, its time derivative df/dt transforms essentially into the original complex domain function multiplied by s. There is, however, the complication that the initial condition of the function $f(t = 0)$ must be included; actually it is shown as $t = 0+$, that is, the value of the function an infinitesimal period after zero time. The reason for this is that in transient analyses discontinuous input functions are often applied at $t = 0$ precisely, and we do not want to be confused about what value should be taken by $f(t = 0+)$ [or more simply $f(0+)$] if there is a discontinuity in the response at $t = 0$.

▼

(C) ON INITIAL CONDITIONS

In order to compute the dynamic behavior of a system defined by a differential equation, the initial condition of the system must be specified at the beginning of the time interval in which the analyst is interested. In studying transient responses it is normal to set $t = 0$ at the beginning of the interval. Thus, Eq. (5.17) becomes

$$T\frac{dy}{dt} + y = k_1x_1 + k_2x_2 \qquad t > 0 \qquad (5.21)$$

TABLE 5.1 INTEGRAL–DERIVATIVE TRANSFORMS

Time Function $t \geq 0$	Laplace Transformed Function
$f(t)$	$\rightleftharpoons$ $F(s)$
$\dfrac{d}{dt} f(t) = \dot{f}(t)$	$\rightleftharpoons$ $sF(s) - f(0+)$
$\int f(t)\, dt$	$\rightleftharpoons$ $\dfrac{1}{s} F(s) + \dfrac{[\int f(t)\, dt]_{t=0}}{s}$

with the initial condition $t = 0$, $y = y_0$, and of course the input functions $x_1(t)$ and $x_2(t)$ must also be specified for the time interval after the initial condition. In general a system defined by an nth-order differential equation must have n initial conditions specified.

The inclusion of initial conditions in the Laplace-transformed variables actually represents a simplification, therefore, since they need not be separately specified. However, initial conditions do not determine the functional nature of a system's dynamic response, but only its precise phasing. Therefore, the analyst finds it convenient to ignore initial conditions in developing transfer relations; in particular, if they are all set to zero, the transform relations such as Eq. (5.20) are considerably simplified. In nonlinear systems the form of dynamic response is no longer necessarily independent of the initial conditions, and the transfer-function technique then becomes invalid.

(D) INTEGRAL–DERIVATIVE TRANSFORMS

In Table 5.1 the Laplace transforms are given of a function $f(t)$, its first derivative, and its first integration, where the symbol $\rightleftharpoons$ means "transforms reversibly into." The initial-condition expression for the integration is evaluated as $t \rightarrow 0+$ from positive time values; it is usually zero. Note that the integrator operator, which has so far been shown in block diagrams as $\int(\ )\, dt$, is now seen to be replaceable by $1/s$ (for example, in Figs. 2.10 and 3.11).

▼ (E) TRANSFORM OF THE STEP FUNCTION

The step input function has already been utilized, and it provides a simple example for direct calculation of a Laplace transform from the defining relation. The unit step function (Fig. 5.4) is defined as a function which equals zero before some convenient time origin

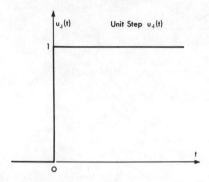

FIG. 5.4   Unit step function $u_{-1}t$.

($t = 0$) and which jumps to unity value thereafter.   For reasons which become clear later, it is usually symbolized by $u_{-1}(t)$.

$$
\begin{aligned}
u_{-1}(t) &= 0 \qquad t < 0 \\
u_{-1}(t) &= 1 \qquad t \geq 0
\end{aligned}
\tag{5.22}
$$

Its Laplace transform therefore follows as

$$
\mathcal{L}[u_{-1}(t)] \triangleq u_{-1}(s) = \int_0^\infty 1e^{-st}\, dt = \frac{1}{s}
\tag{5.23}
$$

That is, the transform pair is

$$
u_{-1}(t) \rightleftharpoons \frac{1}{s}
\tag{5.24}
$$

(F) TRANSFER FUNCTION OF FIRST-ORDER SYSTEM

Some of these rules are now applied to obtain the transfer functions of the two canonical forms of the first-order system [Eqs. (5.17), (5.18)].   Different symbols should formally be used to distinguish an s-domain function from its corresponding time-domain function and in the literature either capital or Greek letters are often used for this purpose.   However, in transfer-function analysis the work is rather needlessly complicated thereby, for there should be no doubt when one is in the transformed s domain, since the s symbol is then present rather than the t symbol.   Therefore, in this book no distinction will normally be made between the variables in s and t domains except to add these latter in parentheses until the reader becomes acclimatized and whenever there seems to be any ambiguity.

   If initial conditions are assumed to be zero, the Laplace-transformed version of Eq. (5.17) is

$$
Tsy(s) + y(s) = kx(s)
\tag{5.25}
$$

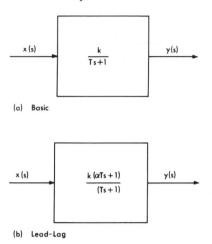

FIG. 5.5   *Transfer functions of canonical first-order systems.*

where, for simplicity, only one input variable x is included.   Now the complex variable can be treated in all normal respects as an algebraic variable so that Eq. (5.25) can be manipulated to obtain the standard transfer-function form

$$\frac{y}{x}(s) = \frac{k}{Ts + 1} \qquad (5.26)$$

This is the desired transfer-block symbolism we have sought, as shown in Fig. 5.5a.   It provides a complete description of the corresponding physical system, on the understanding that initial conditions must also be included when needed.   The symbolism has the advantage of ready manipulation in block diagrams because transfer functions can be multiplied together algebraically when they appear in series (see Sec. 5.6).   The transfer function of Eq. (5.18) for the nanoplankton example becomes (Fig. 5.5b)

$$\frac{y}{x}(s) = k\frac{\alpha Ts + 1}{Ts + 1} \qquad (5.27)$$

The transfer function defines the output-input ratio in the s domain, and as such expresses the dynamic dependence of y upon x. However, it does not express time-domain functions except after converting back through the Laplace transformation.   It is worth noting that since 1 and Ts in the denominator must have the same dimensions, the dimensions of s must be inverse time.   Indeed, one can also work through the basic definition to show this, or alternatively note the working equivalence $s = d(\ )/dt$.

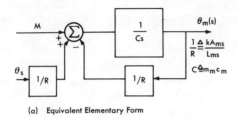

(a)   Equivalent Elementary Form

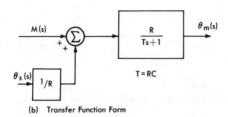

(b)   Transfer Function Form

FIG. 5.6   *Block diagrams for thermal lump.*

5.6   *Block-diagram Manipulation*

Elementary block diagrams have been drawn directly from the describing differential equations for the thermal lump and nano-plankton examples (Figs. 5.1 and 5.3), and they involve feedback paths.   On the other hand, Fig. 5.5 asserts that a first-order system can be represented in an open-loop single-block form, incorporating an overall transfer function.   Clearly the two forms are equivalent,

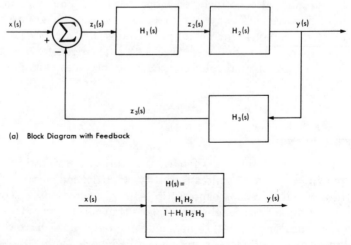

(a)   Block Diagram with Feedback

(b)   Equivalent Transfer Function

FIG. 5.7   *Block-diagram manipulation of transfer functions.*

and hence simple rules for manipulation of the various block-diagram forms would be useful.

Figure 5.6a shows an equivalent elementary block diagram for the thermal lump of Fig. 5.1a. The elementary block diagram with feedback loop has now been prepared for collapsing into a single open-loop transfer function. For this collapsing a general rule can be derived quite simply, which is illustrated in Fig. 5.7. By the definition of transfer function, we can write the following relations:

$$\frac{z_2}{z_1}(s) = H_1(s) \qquad \frac{y}{z_2}(s) = H_2(s)$$

$$\frac{z_3}{y}(s) = H_3(s) \qquad \text{and} \qquad z_1(s) = x(s) - z_3(s)$$

Solving these equations yields

$$\boxed{\frac{y}{x}(s) = H(s) \triangleq \frac{H_1 H_2}{1 + H_1 H_2 H_3}} \tag{5.28}$$

This is an especially important formula for feedback-control systems, as explored in Chap. 11. It should also be noted that collapsing to an overall transfer function is formally equivalent to elimination of intermediate variables in the describing set of differential equations, but that it is far less tedious.

Returning now to the thermal lump (Fig. 5.6), it is easily seen that, in Eq. (5.28),

$$H_1 = \frac{1}{Cs} \qquad H_2 = 1 \qquad H_3 = \frac{1}{R}$$

so that

$$\frac{y}{x}(s) = H(s) = \frac{R}{Ts + 1}$$

as noted in Fig. 5.6b.

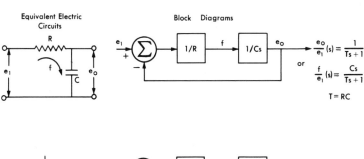

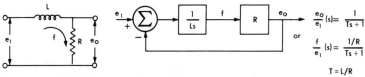

FIG. 5.8   *Some general passive first-order configurations.*

The equivalence of the elementary block diagram and the overall transfer function has now been demonstrated, and ease of manipulation already appears as one of the advantages of the transfer-function approach. The basic relation for collapsing the transfer functions within the loop is frequently needed.

5.7 Generalized First-order Configurations

Passive first-order systems usually have one element of energy storage and one element of dissipation, but in actively controlled systems more elements may be present. Some useful configurations are shown in Fig. 5.8, which utilizes the generalized symbolism of Chap. 4 for PE storage C, KE storage L, and resistive dissipation R. Although they are suggestive of electric circuits as shown, they are widely applicable, as Chap. 4 should have made clear.

6 Transient Response Characteristics

Example of First-order System; Laplace-transform Techniques

6.1 Introduction

In earlier chapters the ideas of information flow and energy flow have been built up for different models of biological systems, although many gross assumptions have had to be made to present a simple picture, as is inevitable in any formal discipline. The concept has been established that a system, described by differential equations, or by their linear equivalent, the transfer function, has output variables which are causally produced by inputs, and in detail modified by the system's initial conditions; more shortly, inputs to the system cause the outputs.

Previously we have concentrated upon the nature and modeling of the system, that is, the block between the input and output variables. In this and the next chapter we concentrate upon the actual output behavior in response to two classes of input forcing, namely the transient and frequency response characteristics, but in order not to confuse the input-output ideas by too much mathematical complexity at this stage, only first-order systems are considered.

Inputs can be rather arbitrary in general biological systems, but a basic feature of the experimental approach is to control the input variables significant to the particular investigation. Thus, part of the trouble that the social sciences have found in applying formal scientific methods is due to the impossibility of controlling even the important variables. Instead, it has often been necessary to measure input-output relations during normal operations and try to infer something of the system's structure from analysis of these records. This procedure can be called the *identification* or *transfer function discovery problem* and is of increasing interest also to the biomedical and control engineering analyst (Secs. 10.4, 15.4).

Given the ability to control the inputs, the analyst naturally chooses ones which give the most conceptual insight into his problem. In general this means that the signals should be suitable for analytical operations as well as approximating to realistic signals. There are, in fact, three main classes of input functions:

Transient inputs—especially the impulse, step, and ramp functions
Periodic inputs—especially the sinusoidal, square, and triangular waves
Statistical inputs—especially random and pseudorandom signals

In this chapter transient inputs are specified mathematically, and the resulting transient responses of first-order systems are investigated. For such simple systems these responses can be calculated by intuitive methods, but with high-order systems either the Laplace-transform technique or numerical solution by computer is desirable. In principle the transient response of a given system completely characterizes its dynamic nature; transient inputs are therefore widely used in biological, physical, and engineering experimentation.

Responses to the periodic inputs are investigated in Chap. 7. The sinusoidal frequency response is a widely used analytical and experimental tool in engineering and has important meaning for biological systems also. In principle the frequency response information of a given system is not independent of the transient response information, but in practice both may play important roles by illuminating different aspects of the system's dynamic characteristics. Statistical signals occur widely in biological systems. Hence the study of the response of dynamic systems to these signals is important. However, their study requires some further theoretical background and is not tackled until Chap. 14.

6.2   The Transient Input Functions

The step change of an input condition has already been introduced, for example, the walking of a thermoregulated nude man from one room at 32°C into another at 16°C. The change of input condition can be completed before the system has started to respond appreciably, so that for all experimental and analytical purposes this input can be regarded as equivalent to the idealized mathematical step input, although the change is not in fact completed in an infinitesimally small time. Similarly, the rapid "hypodermic" type of input can be considered an impulse input provided the input is completed before the resultant diffusive or other transfer responses of the system become appreciable. Thus we can investigate the responses to transient inputs of our mathematically modeled processes with confidence that they can have real practical significance.

(A) THE UNIT STEP FUNCTION $u_{-1}(t)$

The unit step function is defined mathematically as a sudden change, at some particular time, of unit magnitude in the value of a variable from a previously constant value. Usually the previously constant value is chosen to be zero, and the time origin to coincide with the step change. It is therefore defined as

$$u_{-1}(t) = 0 \qquad t < 0$$
$$u_{-1}(t) = 1 \qquad t \geq 0 \tag{6.1}$$

as illustrated in Fig. 6.1*b*. (The significance of the subscript becomes clear shortly.) Figure 6.1*b* shows that an infinite rate of change is needed at $t = 0$, which means that an idealized step input cannot occur physically, although we have shown that it can often be approached as closely as needed for all practical purposes.

A more subtle point of caution arises, however, particularly for biological systems. In many natural situations the environment does not produce many step inputs; instead, the inputs are smoothly varying. Consequently the step response may not represent an important evolutionary performance criterion of the system, and if the investigator finds a saturating or other nonlinear step response it does not necessarily mean that the system's normal operation cannot be linear, and vice versa. Another important point arises if the system is nonlinear, because then superposition is no longer valid; that is, the system response to two separate inputs cannot be calculated by adding the response of each input considered as if the other were not present. One striking example already met is the pupillary-light reflex where the negative step response is quite different from the positive step response (Fig. 2.22). In this case, positive and negative steps applied in rapid sequence do not cancel out as they would if the system were linear and the interval between steps were small.

The Laplace transform of the unit step input was considered in Sec. 5.5 as an elementary transformation exercise:

$$u_{-1}(t) \rightleftharpoons \frac{1}{s} \tag{6.2}$$

A step function of any desired strength can be obtained by multiplying the unit step function by k. Owing to linearity of the transformation, the following applies:

$$ku_{-1}(t) \rightleftharpoons \frac{k}{s} \tag{6.3}$$

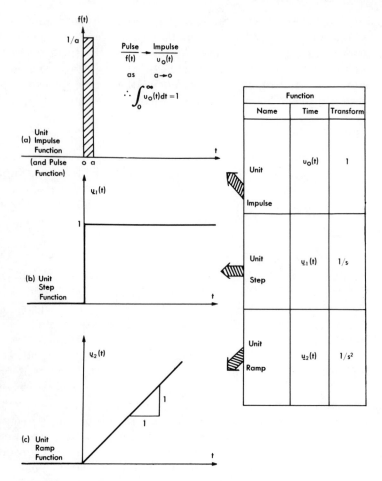

FIG. 6.1   The basic transient input functions.

(B) THE UNIT IMPULSE FUNCTION $u_0(t)$

Mathematically the unit impulse function is synonymous with the Dirac delta function. It may be defined by considering a function which has the value $1/a$ in the region between $t = 0$ and $t = a$, and then allowing the value of a to tend toward zero. Under this condition the time integral of the function equals the product $(1/a)a$, which clearly remains constant at unity whatever the value of a (Fig. 6.1a).

$$\int_{-\infty}^{+\infty} u_0(t)\, dt = 1 \tag{6.4}$$

$$\text{where } u_0(t) = 0 \quad t < 0 \quad t > 0$$
$$= \frac{1}{a} \quad 0 < t < a$$
$$\to \infty \quad 0 < t < a \quad \text{as } a \to 0$$

In practical systems a short-pulse input is equivalent to the idealized impulse if no significant response occurs during the pulse. Thus, when a pile driver hits a pile the impulsive force becomes very high for a very brief period. In detail, however, high-speed electronic-flash photographs reveal that physical systems are deflected during such impacts, which in turn usually increases impact time. For example, a golf ball may become almost flattened momentarily on being struck, and the boxer's anatomy suffers similarly.

The Laplace transform of the impulse function may be found directly from the defining relation, by noting that the necessary limits of integration are only $t = 0$ and $0+$ $(= a)$, and that during this interval $e^{-st} = e^{-0} = 1$. Thus the transform pair is

$$u_0(t) \rightleftharpoons 1 \tag{6.5}$$

The step and impulse functions have a direct relationship. We notice in Fig. 6.1 that the step input has a nonzero value of derivative only at $t = 0$, and here the value must be infinite since the step change is a sudden jump. On the other hand, the integral of the unit impulse is defined to be unity and to occur between $t = 0$ and $0+$ only, so that the value of the integral changes by a unit amount around $t = 0$. Clearly, therefore, the step is the integral of the impulse, and the impulse is the derivative of the step. Having first established this relation physically, we also note that it follows by definition from the two Laplace transforms, since s corresponds to time differentiation and $1/s$ to time integration; that is,

$$u_0(t) = \frac{d}{dt} u_{-1}(t) \tag{6.6}$$

(C) THE UNIT RAMP FUNCTION $u_{-2}(t)$

The unit ramp function is defined as one which increases linearly with time after $t = 0$ (Fig. 6.1c):

$$\begin{aligned} u_{-2}(t) &= t \qquad t > 0 \\ u_{-2}(t) &= 0 \qquad t < 0 \end{aligned} \tag{6.7}$$

and the Laplace-transform pair may be shown to be

$$u_{-2}(t) \rightleftharpoons \frac{1}{s^2} \tag{6.8}$$

This function is the integral of the unit step function.

Ramp inputs are particularly useful in studying the velocity-tracking response of servomechanisms; for example, in studying compensatory eye movements when the skull and therefore the semicircular canals (Sec. 8.3) are being rotated, a revealing input is that of steadily increasing angular velocity.

The reader should study the table in Fig. 6.1 showing the three transform pairs and their relations to each other. It should be clear now that the symbolism $u_{-n}(t)$ indicates a transient time function whose Laplace transform involves $1/s^n$.

(D) OTHER TRANSIENT INPUTS

Higher-order analytical inputs may be useful in some cases. For example, some investigators of retinal ganglion cell responses have used exponentially increasing light stimuli. Computation becomes generally more complex in such cases, although the exponential input itself is very convenient.

6.3   The Step Response of the Basic First-order System

The step response is now considered in more detail, and in particular for the basic first-order system, whose describing differential equation is, after Eq. (5.17),

$$T \frac{d}{dt} y(t) + y(t) = k x(t) \tag{6.9}$$

and whose transfer function (Fig. 6.2a) is

$$\frac{y}{x}(s) = H(s) = \frac{k}{Ts + 1} \tag{6.10}$$

The step response can be found by using the Laplace-transform method, and this saves labor when the systems become higher than about second order. However, for the first-order system such a powerful technique is not particularly advantageous, and an intuitive approach is first presented here to provide some useful physical insights. There is also the "classical" approach, which we only mention briefly since it is not necessary for the system in question and since it becomes involved when the system is complex (see Bibliography on Differential Equations).

(A) SOLUTION BY INTUITIVE APPROACH

If the system is presumed to be stable and hence to reach a new steady-state value y_∞ in response to a step input as $t \to \infty$, the derivative term $T(dy/dt)$ of Eq. (6.9) can be set to zero in evaluating y_∞. Therefore, if x_a is the magnitude of the step input, that is, the number of unit step functions,

$$y_\infty = k x_a \tag{6.11}$$

In general, the system can be in any state x_0 before the onset of the step; for simplicity here, however, x_0 is set equal to zero, and furthermore the step input is presumed to occur at $t = 0$. Immediately after the step input occurs there can be no change in $y(t)$ from 0 because there has been no chance yet for energy transfer. However, Eq. (6.9) must be satisfied, so that there must be a finite rate

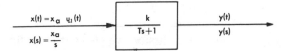

(a)　The System and the Step Input

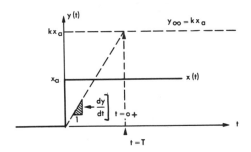

(b)　The Initial and Final Responses

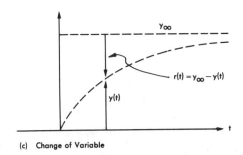

(c)　Change of Variable

FIG. 6.2 Constructing the step response of the basic first-order system.

of change of output as soon as the step is applied, namely,

$$\left[\frac{dy}{dt}\right]_{t=0+} \triangleq \dot{y}(0+) = \frac{kx_a}{T} \qquad (6.12)$$

and using Eq. (6.11)

$$\left[\frac{dy}{dt}\right]_{t=0+} = \frac{y_\infty}{T} \qquad (6.13)$$

Equation (6.13) is an important relation which specifies that the initial slope crosses the steady-state response after one time constant (Fig. 6.2b). This initial slope is not maintained, however, and from the population example we might predict an exponentially decaying continuation to the steady-state response. In general terms, the initial slope arises from a changing amount of energy stored in this first-order system. The final steady-state condition corresponds to either zero or maximum energy storage in which any energy flow escapes through the dissipative element (Fig. 5.8).

So far, the initial slope and final steady state have been determined, and we now need to know the shape of the curve connecting these values. It turns out that a change of variable reduces Eq. (6.9) to the simpler form already met, so that we define

$$r(t) \triangleq y_\infty - y(t)$$

Therefore

$$\frac{dr}{dt} = -\frac{dy}{dt} \tag{6.14}$$

Substituting Eq. (6.14) in Eq. (6.9),

$$T\frac{dr}{dt} + r = 0 \tag{6.15}$$

This equation is identical with Eq. (2.8) for the population-dynamics model, where $T = -1/k$, so that the response is

$$r(t) = y_\infty e^{-t/T} \qquad t > 0 \tag{6.16}$$

and by using Eq. (6.14)

$$y(t) = y_\infty(1 - e^{-t/T}) \qquad t > 0 \tag{6.17}$$

These two responses are plotted in Figs. 6.2c and 6.3a, respectively. As far as $r(t)$ is concerned, the output decays exponentially from some initial condition to a zero steady-state value, whereas $y(t)$ rises monotonically with exponentially decaying slope to reach the new steadystate $y_\infty = kx_a$.

(B) CHARACTERISTICS OF THE STEP RESPONSE

The step response of the first-order system [Eq. (6.17)] is shown on Fig. 6.3a, and the initial-slope condition has already been noted. In addition, a useful property of such an exponentially decaying curve (that is, decaying to y_∞) is that the tangent from any point, say at time t_1, intersects the steady-state condition one time constant later ($t_1 + T$). The response has completed 63 percent of its total change after a period equal to one time constant, or more generally, a proportion $1/e^n$ of the total change remains to be completed at time $t = nT$, that is, after n time constants. Although the plot of Eq. (6.17) is not a straight line on semilogarithmic coordinates, as was true for the population problem (Fig. 2.9), that of Eq. (6.16) is. Unfortunately, in any given experiment it may be difficult to wait long enough to establish the steady-state value y_∞ needed to test an exponential-curve hypothesis by plotting data as $r(t) = y_\infty - y(t)$. In this case, one must guess likely values of y_∞

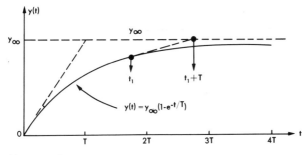

(a) The Basic Response

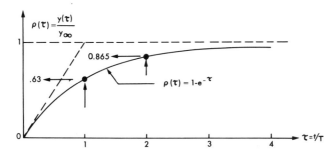

(b) Nondimensional Response

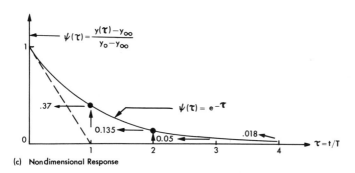

(c) Nondimensional Response

FIG. 6.3 Characteristics of the step response of the basic first-order system.

and plot the resulting curves; if no reasonable y_∞ value yields a straight line, the curve is not a single exponential.

A nondimensional form of presentation produces a unique, monotonically rising step response for all first-order systems. Namely, define

$$\rho(\tau) \triangleq \frac{y(\tau)}{y_\infty} \qquad \tau \triangleq \frac{t}{T}$$

whence Eq. (6.17) becomes

$$\rho(\tau) = 1 - e^{-\tau} \qquad \tau > 0 \qquad (6.18)$$

which is illustrated in Fig. 6.3*b*. A more complete nondimensional form eliminates the unity term on the right-hand side and also incorporates any initial condition y_0. Thus by defining

$$\psi(\tau) \triangleq \frac{y - y_\infty}{y_0 - y_\infty} \tag{6.19}$$

there results

$$\psi(\tau) = e^{-\tau} \qquad \tau > 0 \tag{6.20}$$

which is the simple and unique exponential decay already encountered (Fig. 6.3*c*). Readers who are not currently interested in other techniques for finding the step response may now jump to Sec. 6.4.

▼

(C) CLASSICAL SOLUTION METHOD

The classical method of mathematics has well-developed techniques which provide shortcuts in special cases, but we do not explore these here. An important principle is that the total solution may be viewed as the sum of two parts: the *complementary* or *homogeneous solution* $y_H(t)$ which results from the system's internal structure—that is, its transfer function—and found by placing the forcing function (input) to zero; and the *particular* or *forced solution* $y_P(t)$, which results from the external forcing function or input. Then

$$y(t) = y_H(t) + y_P(t) \tag{6.21}$$

1. To find y_H we write the homogeneous equation

$$T\frac{dy}{dt} + y = 0 \tag{6.22}$$

We solve this by a more general method than previously by assuming a solution of the form

$$y_H = Ce^{\alpha t} \tag{6.23}$$

where C and α are arbitrary constants. (In the general case, n such additive solutions must be postulated for an nth-order equation.) Therefore

$$\frac{dy_H}{dt} = \alpha Ce^{\alpha t} \tag{6.24}$$

Hence, by inserting Eqs. (6.23) and (6.24) in Eq. (6.22),

$$(\alpha T + 1)Ce^{\alpha t} = 0$$

whence

$$\alpha = -\frac{1}{T} \tag{6.25}$$

which checks the result of Eq. (6.16). The constant C is evaluated in finding the total solution, as discussed in (3).

2. In the general case of y_P, there are a number of possible methods of evaluation, but here it is simply obtained again by recognizing that, in the steady state, $dy/dt = 0$. Hence, from Eq. (6.9),

$$y_P = kx_a \qquad (6.26)$$

3. The complete solution of y is obtained by adding y_H and y_P:

$$y(t) = y_H(t) + y_P(t) = Ce^{-t/T} + kx_a$$

and C is evaluated by incorporating the initial condition $y_0 = 0$; that is,

$$C = -kx_a = -y_\infty$$

The complete solution becomes, as before,

$$y(t) = y_\infty(1 - e^{-t/T}) \qquad t > 0 \qquad (6.17)$$

▼

(D) OPERATIONAL METHOD OF LAPLACE TRANSFORMATION

The most practical pencil-and-paper method for finding transient responses of moderately complex linear systems is certainly the operational one, although computer solutions are still often necessary because of nonlinearity, etc. The Laplace-transformation technique provides a simple and direct methodology for working out the required solutions.

In the s domain the output is the product of input and transfer function, the order of multiplication being unimportant.

$$y(s) = x(s)H(s) = H(s)x(s) \qquad (6.27)$$

Solutions are obtained by two methods, as follows:

Table Look-up. Frequently encountered transform pairs are widely tabulated. In consequence the ... ne-domain response $y(t)$ may often be found by consulting such tables when $y(s)$ is known. Some useful transform pairs are tabulated in Sec. 6.8, but the reader is referred to the Bibliography for extensive tables. In the present example of the first-order system's step response, the transform of the system output is defined by Eq. (6.27):

$$y(s) = \frac{x_a}{s}\frac{k}{Ts+1} = y_\infty \frac{1}{s(Ts+1)} \qquad (6.28)$$

The time-domain solution follows directly from consulting the tables, as illustrated for this first-order system in Fig. 6.4. Thus table look-up can be successfully used without understanding the underlying theory of Laplace transformation, rather as one uses a table of logarithms to perform multiplications.

Method of Partial Fractions. The method of partial fractions is a powerful but somewhat tedious technique which ensures an answer, even if the particular transform pair is not tabulated. Fundamentally it breaks the transformed system down into first-order (or perhaps second-order) subsystems, which are certainly tabulated. The following procedure for the first-order system is very simple.

In Eq. (6.28), $1/s(Ts + 1)$ can be expressed by an alternative but identical expression, involving so-called *partial fractions*.

$$\frac{1}{s(Ts + 1)} \equiv \frac{A}{s} + \frac{B}{Ts + 1} \qquad (6.29)$$

where A and B are constants to be determined. Each side can be rearranged to give

$$\frac{0s + 1}{s(Ts + 1)} \equiv \frac{(B + AT)s + A}{s(Ts + 1)} \qquad (6.30)$$

where the 0 on the left-hand side indicates a zero coefficient for the s term. Since Eq. (6.30) is an identity, a set of equations is obtained to solve for the coefficients A, B, etc. Here we have

$$
\begin{aligned}
1 &= A & &\text{for } s^0 \text{ term} \\
0 &= B + AT & &\text{for } s^1 \text{ term}
\end{aligned}
\qquad (6.31)
$$

so that $\quad B = -T$

In consequence, the right-hand side of Eq. (6.29) can readily be evaluated because the inverse transforms for the first-order terms in the resulting expression

$$y(s) = y_\infty \left(\frac{1}{s} - \frac{T}{Ts + 1} \right) \qquad (6.32)$$

are basic and easily remembered. $1/s$ is the unit step function [Eq. (6.2)], and $\dfrac{T}{Ts + 1}$ the basic first-order expression with time-domain equivalent $e^{-t/T}$ (see table of Fig. 6.4). Therefore, Eq. (6.32) inverse-transforms into

$$y(t) = y_\infty(u_{-1}(t) - e^{-t/T}) = y_\infty(1 - e^{-t/T}) \qquad t > 0 \qquad (6.17)$$

This example presents the basis of the partial-fraction method, but a number of developments are necessary to cope with more complicated expressions (Sec. 6.8). The power of the method lies in its ability to resolve each expression into easily soluble components.

6.4   The Impulse Response of the Basic First-order System

The Laplace-transform technique produces the impulse response of the basic first-order system [Eqs. (6.9), (6.10)] in a particularly simple manner. The impulse response is the response of the system to a unit impulse input, and since the transform of the unit impulse

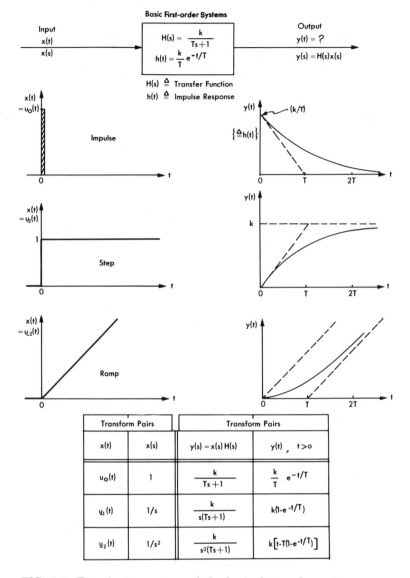

FIG. 6.4 Transient responses of the basic first-order system.

function in $x(t)$ is 1 [see Eq. (6.5), where for compactness we now work with unit-strength input functions],

$$y(s) = 1\,\frac{k}{Ts+1} = \frac{k}{Ts+1}$$

This result shows that the transform of the impulse response is identically the transfer function of the system itself. As a consequence, its impulse-response function is (Fig. 6.4)

$$y(t) = \frac{k}{T}\,e^{-t/T} \qquad\qquad (6.33)$$

This is of course the exponential decay to which we have become accustomed. However, a fresh and important aspect of the dynamic behavior of systems is revealed by this result. Since the impulse input is completed by $t = 0+$, all the exponentially decaying response after the initial jump is unforced. In general, an impulse input can be considered to produce initial conditions in the system, and the system's dynamic characteristics are then revealed by the manner in which the decay ensues. Systems which are more complex than first order typically do not respond with an initial jump, although some derivative of the output then does so. In energy terms the impulse input instantaneously transfers a quantity of energy to the system, and the subsequent response characterizes the way in which the energy is dissipated.

We now formally note that the time function which constitutes a transform pair with the transfer function $H(s)$ is called the *impulse response* $h(t)$ of that system:

$$h(t) \rightleftharpoons H(s)$$

Thus, when the input forcing $x(t)$ is an impulse, the output function $y(t)$ identically equals $h(t)$. This result is true for linear systems of whatever complexity.

▼

We now also establish the impulse response in the time domain using the differential equation (6.9) because this procedure aids understanding of the impulse function. We integrate each side of Eq. (6.9) between $t = 0$ and $0+$ $(= a)$, noting that $x(t)$ is the unit impulse function.

$$T \int_0^{0+} \frac{dy}{dt}\, dt + \int_0^{0+} y\, dt = k \int_0^{0+} x\, dt = k \qquad (6.34)$$

Integration of the first term on the left-hand side yields

$$T \int_0^{0+} \frac{dy}{dt}\, dt = T[y(0+) - y(0)]$$

The interval a from $0 \to 0+$ spans the impulse and we expect therefore $y(0+)$ to differ from $y(0)$, the latter being defined equal to zero. Integration of the second term on the left-hand side can at most yield a value ya (Fig. 6.5) which is one order lower than the first integral and tends to zero in the limit as $a \to 0$. Thus Eq. (6.34) yields

$$y(0+) = \frac{k}{T} \qquad (6.35)$$

The problem is indeed now changed to an initial-condition problem identical with those for the $r(t)$-variable problem [Eq. (6.15)] and

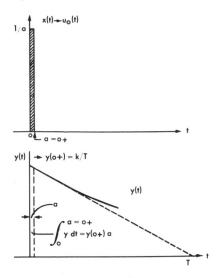

FIG. 6.5 *Initial phase of basic first-order impulse response.*

the population problem [Eq. (2.8)], so that the complete impulse response is

$$y(t) = y(0+)e^{-t/T} = \frac{k}{T} e^{-t/T} \tag{6.36}$$

6.5 The Ramp Response of the Basic First-order System

In the s domain the ramp response is readily written as

$$y(s) = \frac{k}{s^2(Ts + 1)} \tag{6.37}$$

by use of the transform pair [Eq. (6.8)]. From appropriate tables of transform pairs (Fig. 6.4) or by the partial-fraction technique (Sec. 6.8), the ramp response may be evaluated as

$$y(t) = k[t - T(1 - e^{-t/T})] \qquad t > 0 \tag{6.38}$$

It can also be found by classical techniques.

As shown in Fig. 6.4 the response increasingly lags behind the input, approaching an asymptotic value (for $k = 1$). This is evaluated by letting $e^{-t/T} \to 0$ as $t \to \infty$:

$$y(t \to \infty) = t - T \tag{6.39}$$

whereas $x(t) = t$

Thus $y(t)$ is delayed a constant time T behind $x(t)$.

6.6   Transient Responses of the "Lead-Lag" First-order System

The nanoplankton example of Sec. 5.4 produced the most general form of the linear first-order system,

$$T\frac{dy}{dt} + y = k\left(\alpha T\frac{dx}{dt} + x\right) \tag{6.40}$$

$$\text{or} \quad H(s) = \frac{y(s)}{x(s)} = k\frac{\alpha Ts + 1}{Ts + 1} \tag{6.41}$$

where α is the ratio of the numerator and denominator time constants. The steady-state response of y to a step input in x of magnitude x_a can be written by placing derivatives equal to zero as previously:

$$y = kx_a$$

Hence k can still be called the *steady-state gain* of this system.

(A) STEP RESPONSE

The unit step response can be found directly by the Laplace-transform method:

$$y(t) = k[1 - (1 - \alpha)e^{-t/T}] \qquad t > 0 \tag{6.42}$$

and the resulting form is sketched in Fig. 6.6a. Four different subclasses of this system arise according to the numerical value of the parameter α.

$\alpha > 1$—*The Lead-Lag System.*   A typical passive RC circuit for this system is shown in Fig. 6.6b. In principle this circuit responds initially to an ideal step input with a finite jump, subsequently decaying back to a new steady-state level.

Previously the rate-sensitive property of biological receptors has been remarked, as well as the proportional property. The simplest transfer function for a proportional-plus-rate-sensitive element would be

$$H(s) = k_p + k_R s \tag{6.43}$$

However, such a system is unrealizable if rapidly changing inputs, such as steps, are applied, because the rate of change of the output becomes very large. A more reasonable model would consist of an approximately rate-sensitive element plus proportionality, viz.,

$$H(s) = k_p + \frac{k_R s}{Ts + 1} \tag{6.44}$$

which after rearrangement equals Eq. (6.41) in basic form.

$\alpha = 1$—*An Element without Dynamics*—*A Static Gain* k.   If $\alpha = 1$, the numerator and denominator dynamics exactly cancel and we are left with a *static* gain; that is, if a step input occurs, the output is also an immediate step (Fig. 6.6a).   Modern electronic amplifiers do a creditably good job of achieving this in many different parts of the frequency spectrum; for example, an amplifier of neuron signals may respond in about $\frac{1}{10}$ msec (10^{-4} sec) and a digital-computer amplifier in about 10^{-9} sec.

$\alpha < 1$—*The Lag-Lead System.*   A typical passive RC circuit for this system is shown in Fig. 6.6b.   The word *lag* now precedes *lead* in the title, and the relevant frequency response characteristics in Sec. 7.8 clarify the use of these terms.   The step response consists of a small initial jump plus an exponentially decaying rise with the same time constant T as for the decay of the lead-lag system's response.

$\alpha = 0$—*The Basic First-order System.*   When $\alpha = 0$, the lag-lead system degenerates into the basic first-order system examined in previous sections (Fig. 6.6a).

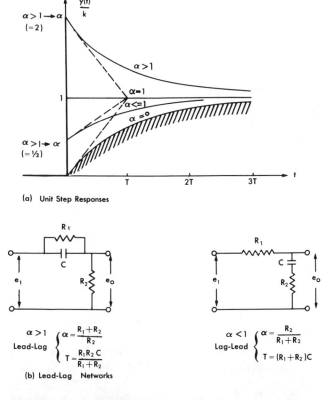

(a) Unit Step Responses

(b) Lead-Lag Networks

FIG. 6.6 First-order lead-lag systems.

▼

6.7   On Superposition, Initial Conditions, Convolution, and First-order Cascades

In the preceding sections some techniques have been presented for finding transient responses, and these have been applied to generalized first-order systems. In this section some further topics are introduced which relate this work to wider considerations.

(A) SUPERPOSITION AND INITIAL CONDITIONS

Linear systems provide for the analyst the advantage of superposition. Thus, if multiple inputs occur either together or apart their effects can be calculated separately and the results combined to give the overall response. We shall consider this quantitatively with the example of the basic thermoregulatory lump [Eq. (5.3)], in which both metabolic rate M and skin temperature θ_s are to be input variables. We have

$$T \frac{d\theta_m}{dt} (t) + \theta_m(t) = k_1 M(t) + \theta_s(t) \tag{6.45}$$

With two inputs there are two transfer functions θ_m/M and $\theta_m{}^s/\theta_s$, although they can both be incorporated in a single equation

$$\theta_m(s) = \frac{k_1}{Ts + 1} M(s) + \frac{1}{Ts + 1} \theta_s(s)$$

or more simply

$$\theta_m(s) = \frac{k_1 M(s) + \theta_s(s)}{Ts + 1} \tag{6.46}$$

This equation clearly shows that the two inputs are processed by the same dynamic system.

In order to be more general, let us include the initial conditions

$$
\begin{aligned}
\theta_m &= \theta_{m0} \\
\theta_s &= \theta_{s0} \qquad t = 0 \\
M &= M_0
\end{aligned}
\tag{6.47}
$$

Returning therefore to Eq. (6.45) and performing the Laplace transformation again to include initial conditions, we obtain

$$T[s\theta_m(s) - \theta_{m0}] + \theta_m(s) = k_1 M(s) + \theta_s(s)$$

Thus, only one of the three stated initial conditions is necessary to this problem, that is, the value θ_{m0} defining the condition of the system when the inputs are applied. Hence, the transfer-function form of Eq. (6.46) can be modified to include the initial condition

$$\theta_m(s) = \frac{k_1 M(s) + \theta_s(s) + T\theta_{m0}}{Ts + 1} \tag{6.48}$$

where θ_{m0} is just a constant, not a Laplace-domain variable of the system.

The validity of superposition for this linear system allows three separate transfer functions to be written, with the understanding that the complete response of θ_m comprises the sum of three components.

$$\frac{\theta_m}{\theta_s}(s) = \frac{1}{Ts+1} \qquad \frac{\theta_m}{M}(s) = \frac{k_1}{Ts+1}$$

and

$$\theta_m(s) = \frac{T\theta_{m0}}{Ts+1}$$

The system described by the last equation has no input except the initial condition. It is identical to the population-decay problem, with k negative, and its time solution is

$$\theta_m(t) = \theta_{m0}e^{-t/T} \qquad t > 0$$

It specifies reasonably enough that $\theta_m(t)$ will decay to zero after $t = 0$ unless there is some $\theta_s(t)$ or $M(t)$ to maintain it (Fig. 6.7a). However, if the physical situation is that the initial values of Eqs. (6.47) continue unchanged after $t = 0$ and furthermore satisfy the condition

$$\theta_{m0} = \theta_{s0} + k_1 M_0 \tag{6.49}$$

then θ_m will also remain unchanged (Fig. 6.7b). This static output can be validly considered a result of three exactly canceling transient responses, as shown in Fig. 6.7b. That is, the exponential decay of the initial condition is canceled by rising exponentials due to the presence of "steps" of M_0 and θ_{m0} after $t = 0$. The latter are considered steps because by specifying the initial condition of the system θ_{m0}, the values of M and θ_m for $t < 0$ are automatically ignored. Now let us consider the case when Eq. (6.49) is satisfied for $t < 0$, but when the skin temperature $\theta_s(t)$ is suddenly lowered as a step function at $t = 0$. Furthermore, we assume the person feels cold after time T and starts to warm himself with a step in $M(t)$. The temperature history $\theta_m(t)$ then follows some such combined curve as that sketched in Fig. 6.7c, the overcompensation shown being arbitrary.

With regard to initial conditions, it is often easier to work with variables which represent changes from their initial values. These are usually called $\Delta\theta_m(t)$, etc., so that, in the present example,

$$\Delta\theta_m(t) = \theta_m(t) - \theta_{m0}$$
$$\Delta\theta_s(t) = \theta_s(t) - \theta_{s0}$$
$$\Delta M(t) = M(t) - M_0$$

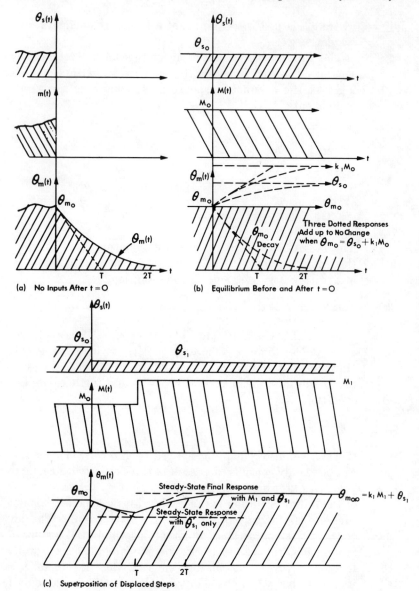

FIG. 6.7   Superposition and initial conditions.

The form of Eq. (6.46) would then automatically result *without the initial condition* since, by definition, $\Delta\theta_m(0) = 0$. It should be noted that Δ refers to a finite change in the variable and is not involved in the differentiating operation; that is,

$$\frac{d}{dt}\,\Delta\theta_m(t) = \frac{d}{dt}\,\theta_m(t) - \frac{d}{dt}\,\theta_{m0} = \frac{d}{dt}\,\theta_m(t)$$

Stated alternatively, the rate of change of a variable does not depend upon the arbitrary definition of its zero level.

(B) CONVOLUTION

In developing the transfer function $H(s)$ by applying Laplace transformation to the describing differential equation for a system, it emerged that the time-domain curve constituting the impulse response has great significance because it characterizes the system completely and also is the inverse transform of the transfer function. The output of a system is by definition the product of the input function $x(s)$ and the transfer function $H(s)$:

$$y(s) = H(s)x(s) \tag{6.50}$$

and it has been mentioned that one advantage of the Laplace-transform technique is the resulting ability to deal only in algebraic operations. What, then, is the corresponding relation in the time domain among $y(t)$, $x(t)$, and $h(t)$? Referring back to a typical describing differential equation (6.45), which is the preliminary to obtaining the transfer-function equation (6.46), we see that $h(t)$ does not appear at all. However, we know that $h(t)$ is called the impulse response, and further that it turns out to be related directly to the solution of the unforced homogeneous equation for the system $y_H(t)$ [Eq. (6.23)].

The inverse transformation of Eq. (6.50) into a time-domain relation yields the *convolution integral*

$$y(t) = \int_{-\infty}^{t} h(t - \sigma)x(\sigma)\, d\sigma = \int_{-\infty}^{t} h(\sigma)x(t - \sigma)\, d\sigma \tag{6.51}$$

These equations look difficult but they do have simple graphical representations which facilitate an understanding of the concept of impulse response (Fig. 6.8). The variable σ is merely ordinary time t, but it is given the new symbol because it is a "dummy" variable which disappears after the integration is complete, and because we still want to reserve the symbol t for the time variable of the output $y(t)$. A dynamic system may be looked on qualitatively as filtering the past history of its inputs, and in general this takes the form of weighting, or remembering, the recent past more heavily than the distant past, and perhaps more heavily than the immediate present. This wording suggests that the process is qualitatively what the CNS must do in learning, forgetting, and/or conditioning.

Convolving means folding back and this is precisely what has been done to the basic first order's impulse response (Fig. 6.8) in weighting or filtering the history of the $x(t)$ input, which is shown in this case as a step function. Each point on the response $y(t)$ results from an integrated product such as suggested by the shaded areas. The same process is illustrated for a second impulse response which remembers the middle past most heavily (shown chain-dotted). Of course, the convolution process also applies to arbitrary time histories of the input.

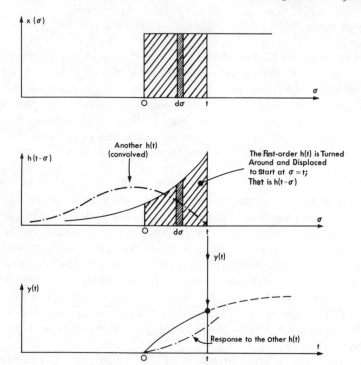

FIG. 6.8   The convolution integral.

In digital-computer programming of convolution, both the input function and the impulse response are broken into sets of samples at discrete intervals of time $\Delta\sigma$. The integral product is then replaced by a summation of products. It is this procedure which is followed in *time series prediction* such as when the economist tries to find the memory characteristic of the stock market (impulse response) from daily, weekly, or other data in order to predict its future behavior from past data. This sort of problem raises many interesting aspects, which are discussed as the problem of identification in Sec. 15.4.

In summary, the convolution integral expresses mathematically the physical process by which a dynamic system weights or remembers its past history of inputs in generating its current output. This weighting function is identically the impulse response $h(t)$ which we use more frequently for manipulative purposes in its Laplace-transform version as the transfer function $H(s)$.

(C) FIRST–ORDER CASCADES

So far only the standard transient inputs of impulse, step, and ramp functions have been considered in finding the responses of the first-order system; for example, the unit impulse response of the basic first-order system

$$H_1(s) = \frac{k_1}{T_1 s + 1}$$

is given by

$$y(t) = \frac{k_1}{T_1} e^{-t/T_1}$$

It is relevant to ask what would be the output $z(t)$ if another first-order system of transfer function

$$H_2(s) = \frac{k_2}{T_2 s + 1}$$

received this $y(t)$ as its input. $z(t)$ can be found by the Laplace-transform method as follows:

$$z(s) = H_2(s) y(s) = H_1(s) H_2(s)$$

Therefore

$$z(s) = \frac{k_1 k_2}{(T_1 s + 1)(T_2 s + 1)}$$

Either transform tables or the method of partial fractions yields the time-domain response

$$z(t) = \frac{k_1 k_2}{T_1 - T_2} (e^{-t/T_1} - e^{-t/T_2}) \tag{6.52}$$

Two equally valid interpretations of this response are possible, dependent upon the physical conditions of the problem. These are illustrated in Fig. 6.9.

1. The response of a first-order system $H_2(s)$ to an exponentially decaying input function has been found, that is, a problem concerning a first-order system.
2. The impulse response has been found of a second-order system defined by two first-order systems in cascade.

The last view can be generalized for an nth-order cascade, such as occasionally occurs physically, for example, a series of hydro-reservoirs down a river which are uncoupled because the flow through the turbines at a given reservoir is not affected by the level of the downstream one (that is, there is no feedback from one stage to an earlier stage). However, the lumps of the thermoregulatory example or of general heat conduction (Fig. 3.6) are strongly coupled. In general, the complexity of a system's response depends upon the complexity both of its input function and of the system itself.

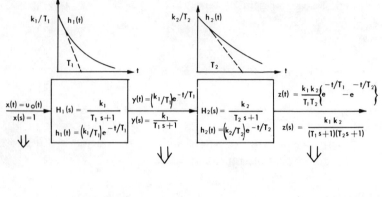

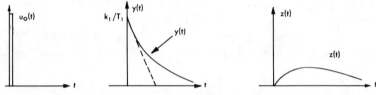

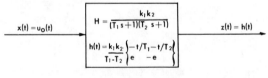

(a)   Cascade of Two First-order Systems

(b)   Equivalent Single Block

FIG. 6.9   *First-order cascades.*

6.8   *Summary of Laplace Transformation*

(A) DEFINITIONS

A function of time $f(t)$, defined for $t > 0$ and generally single-valued, can be transformed into a function $F(s)$ of the Laplace or complex variable s. This is called the forward transformation $\mathcal{L}$, whereas the inverse transformation $\mathcal{L}^{-1}$ proceeds from $F(s)$ to $f(t)$. For problems of convergence in the integrals, the Bibliography should be consulted.

Transform pair $f(t) \rightleftharpoons F(s)$ [or $f(s)$ usually in this book]

Transform symbolism $F(s) \triangleq \mathcal{L}[f(t)]$

$\qquad\qquad\qquad\qquad f(t) \triangleq \mathcal{L}^{-1}[F(s)]$

Direct transformation $F(s) \triangleq \displaystyle\int_0^\infty f(t)e^{-st}\,dt$

Inverse transformation $f(t) = \dfrac{1}{2\pi} \displaystyle\int_{c-j\infty}^{c+j\infty} F(s)e^{ts}\,ds \qquad t \geq 0$

where c must be defined to ensure convergence in this contour integration.

(B) PROPERTIES

Linearity $kf(t) \rightleftharpoons kF(s)$
$f_1(t) + f_2(t) \rightleftharpoons F_1(s) + F_2(s)$

where k is constant.

Real differentiation $\dfrac{df(t)}{dt} \rightleftharpoons sF(s) - f(0+)$

$\dfrac{d^n f(t)}{dt^n} \rightleftharpoons s^n F(s) - s^{(n-1)} f(0+)$

$- s^{(n-2)} \dfrac{df}{dt}(0+) - \cdots - \dfrac{d^n f(0+)}{dt^n}$

Real integration $\displaystyle\int f(t)\, dt \rightleftharpoons \dfrac{F(s)}{s} + \dfrac{[\int f(t)\, dt]_{0+}}{s}$

where the integral on the right-hand side is evaluated at $t = 0+$, as approached from $t > 0$, and can usually be taken as equal to zero unless impulse-type functions are being considered.

Time translation $f(t - T) \rightleftharpoons e^{-Ts} F(s)$

Real convolution $\displaystyle\int_0^t f_1(t - \sigma) f_2(\sigma)\, d\sigma \rightleftharpoons F_1(s) F_2(s)$

Initial-value theorem $\displaystyle\lim_{t \to 0} [f(t)] \equiv \lim_{s \to \infty} [sF(s)]$

Final-value theorem $\displaystyle\lim_{t \to \infty} [f(t)] \equiv \lim_{s \to 0} [sF(s)]$

▼

(C) METHOD OF PARTIAL FRACTIONS

This procedure is useful to reduce a *rational algebraic fraction*, consisting of numerator and denominator polynomials in s, into first-order components which are then inverse-transformable by inspection into time-domain exponentials. The first-order components each have as their denominator one root of the denominator polynomial. The roots may in general be real and separate, real and repeated, or complex conjugate; modified techniques are needed in the last two cases. The numerator polynomial must be one order lower than the denominator if the time function is to remain finite. In the usual procedure the inverse transformation is needed for a function $y(s)$ which is the output of a system defined by a transfer function $H(s)$ when the input is $x(s)$. $y(s)$ is then a *rational algebraic fraction* in s.

$$y(s) = \frac{A(s)}{B(s)} = \frac{a_0 + a_1 s + \cdots + a_m s^m}{b_0 + b_1 s + \cdots + b_n s^n}$$

$$\equiv \frac{A(s)/b_n}{(s + p_1)(s + p_2) \cdots (s + p_i) \cdots (s + p_n)} \qquad (6.53)$$

where the a's and b's are constants and the $-p$'s are the n roots of the denominator.

1. *Roots real and separated:*

$$\frac{A(s)/b_n}{(s+p_1)(s+p_2)\cdots(s+p_i)\cdots(s+p_n)} \equiv \frac{C_1}{s+p_1} + \frac{C_2}{s+p_2}$$
$$+ \cdots + \frac{C_i}{s+p_i} + \cdots + \frac{C_n}{s+p_n} \quad (6.54)$$

where

$$C_i = \left[(s+p_i)\frac{A(s)}{B(s)} \right]_{s=-p_i}$$

Each component then forms a Laplace-transform pair

$$\frac{C_i}{s+p_i} \rightleftharpoons C_i e^{-p_i t} \quad (6.55)$$

2. *Roots separated, but including complex-conjugate pairs:* In this case, two roots form a complex-conjugate pair

$$\begin{aligned} p_i &= +\sigma_i + j\omega_i \\ p_{i+1} &= +\sigma_i - j\omega_i \end{aligned} \quad (6.56)$$

The same inverse transformation is used as in (1), and the identity

$$e^{(-\sigma_i \pm j\omega_i)t} \equiv e^{-\sigma_i t}(\cos \omega_i t \pm j \sin \omega_i t) \quad (6.57)$$

is then used to obtain real time functions.   See Chap. 8 for examples.

3. *Roots including one or more repeated values:*

$$\frac{A(s)/b_n}{(s+p_1)(s+p_2)\cdots(s+p_i)^r \cdots (s+p_{n-r})} \equiv \frac{C_1}{s+p_1} + \frac{C_2}{s+p_2}$$
$$+ \cdots + \left[\frac{C_{i1}}{s+p_i} + \cdots + \frac{C_{ik}}{(s+p_i)^k} + \cdots + \frac{C_{ir}}{(s+p_i)^r} \right]$$
$$+ \cdots + \frac{C_{n-r}}{s+p_{n-r}} \quad (6.58)$$

Special treatment is needed for the repeated roots p_i.

$$\begin{aligned} C_{ir} &= \left[(s+p_i)^r \frac{A(s)}{B(s)} \right]_{s=-p_i} \\ C_{i(r-1)} &= \frac{1}{1!} \left\{ \frac{d}{ds}\left[(s+p_i)^r \frac{A(s)}{B(s)} \right] \right\}_{s=-p_i} \\ C_{ik} &= \frac{1}{(r-k)!} \left\{ \frac{d^{(r-k)}}{ds^{(r-k)}}\left[(s+p_i)^r \frac{A(s)}{B(s)} \right] \right\}_{s=-p_i} \\ C_{i1} &= \frac{1}{(r-1)!} \left\{ \frac{d^{(r-1)}}{ds^{(r-1)}}\left[(s+p_i)^r \frac{A(s)}{B(s)} \right] \right\}_{s=-p_i} \end{aligned} \quad (6.59)$$

Each of the repeated root's components then forms a transform pair

$$\frac{C_{ik}}{(s+p_i)^k} \rightleftharpoons \frac{C_{ik}}{(k-1)!} t^{(k-1)} e^{-p_i t} \quad (6.60)$$

Equations (6.37) and (6.38) exemplify this procedure, as the reader should check.

TABLE 6.1   SOME USEFUL TRANSFORM PAIRS

Transformed Function $F(s)\ [\equiv f(s)]$	Time Function $f(t) \quad t > 0$	Location in Book
1 $1/s$ $1/s^2$	Unit impulse, $u_0(t)$ Unit step, $u_{-1}(t)$ Unit ramp, $u_{-2}(t)$	Figs. 6.1, 6.4
$\dfrac{1}{Ts + 1}$	$\dfrac{1}{T}e^{-t/T}$	Fig. 6.4
$\dfrac{1}{s(Ts + \ )}$	$1 - e^{-t/T}$	Fig. 6.4
$\dfrac{1}{s^2(Ts + 1)}$	$t - T(1 - e^{-t/T})$	Fig. 6.4
$\dfrac{1}{s}\dfrac{\alpha Ts + 1}{Ts + 1}$	$1 - (1 - \alpha)e^{-t/T}$	Fig. 6.6
$\dfrac{1}{(Ts + 1)^2}$	$\dfrac{1}{T}\left(\dfrac{t}{T}\right)e^{-t/T}$	Table 8.1
$\dfrac{1}{s(Ts + 1)^2}$	$1 - \left(1 + \dfrac{t}{T}\right)e^{-t/T}$	Table 8.2
$\dfrac{1}{(Ts + 1)^n}$	$\dfrac{1}{(n-1)!T}\left(\dfrac{t}{T}\right)^{n-1}e^{-t/T}$	Eq. (9.23)
$\dfrac{1}{(\alpha Ts + 1)\left(\dfrac{T}{\alpha}s + 1\right)}$	$\dfrac{1}{T}\dfrac{\alpha}{\alpha^2 - 1}(e^{-t/\alpha T} - e^{-\alpha t/T})$	Table 8.1, Fig. 6.9
$\dfrac{1}{s(\alpha Ts + 1)\left(\dfrac{T}{\alpha}s + 1\right)}$	$1 + \dfrac{1}{\alpha^2 - 1}(e^{-\alpha t/T} - \alpha^2 e^{-t/\alpha T})$	Table 8.2
$\dfrac{1}{\left(\dfrac{s}{\omega_n}\right)^2 + 2\zeta\dfrac{s}{\omega_n} + 1}$	$\dfrac{\omega_n}{\sqrt{1 - \zeta^2}}[e^{-\zeta\omega_n t}\sin(\omega_n\sqrt{1 - \zeta^2}\,t)]$	Table 8.1
$\dfrac{1}{s\left[\left(\dfrac{s}{\omega_n}\right)^2 + 2\zeta\dfrac{s}{\omega_n} + 1\right]}$	$1 - \dfrac{e^{-\zeta\omega_n t}}{\sqrt{1 - \zeta^2}}\sin(\omega_n\sqrt{1 - \zeta^2}\,t + \theta)$ where $\theta = \arctan\dfrac{\sqrt{1 - \zeta^2}}{\zeta}$	Table 8.2
$\dfrac{1}{\left(\dfrac{s}{\omega_n}\right)^2 + 1}$	$\omega_n \sin \omega_n t$	Tables 8.1, 7.2
$\dfrac{\dfrac{s}{\omega_n}}{\left(\dfrac{s}{\omega_n}\right)^2 + 1}$	$\omega_n \cos \omega_n t$	Table 7.2
$\dfrac{1}{s\left[\left(\dfrac{s}{\omega_n}\right)^2 + 1\right]}$	$1 - \cos \omega_n t$	Table 8.2
$e^{\pm Ts}$	$u_0(t \pm T)$	Eqs. (9.7), (9.10)

7 Frequency Response Characteristics

Example of First-order System

7.1 Introduction

Frequency response is a well-established tool of electrical and control engineers. Initially it applied only to linear systems, but it has since been extended to include some nonlinearities. The ideas behind frequency response also provide a conceptual base for generalized harmonic analysis, which is the basic tool for dealing with statistical signals of continuous waveform in dynamic systems (Sec. 14.3).

Frequency response has a firm mathematical foundation which is demonstrated in this chapter, but one of its important advantages is that any system's frequency response can be obtained by direct experimental measurements, and from these data useful predictions can be made almost without any knowledge of the underlying theory. For these reasons the frequency response technique is also proving useful in the investigation of biological systems, and it is therefore presented here reasonably carefully.

The basic ideas behind frequency response are extremely simple and perhaps better appreciated by our muscles than by our brains. Thus, in the swing example already mentioned and in others such as rocking a car out of an icy hole, one has learned without mathematical calculation that a satisfactory oscillation can be built up with little physical effort if this effort is applied in an oscillatory manner, at a certain frequency, and with a certain phase relation to the object's motion. The resulting oscillation is at the system's *natural frequency*, and although oscillations can be built up at other frequencies it is readily found that much greater effort is required to do so. The frequency response technique enables one to present such dynamic characteristics both graphically and analytically for all types of linear dynamic systems.

The first basic result underlying frequency response is that if a sinusoidal wave $x(t)$ of some angular frequency ω rad/sec and some constant maximum amplitude A is inputted into a linear system, then the steady-state output $y(t)$ is also a sinusoidal wave of the same frequency but of generally different amplitude B and phase ϕ relative to the input (Fig. 7.1). The definitions of relevant terms are:

1. The *angular frequency* ω defines a rate of rotation in radians per second. One radian equals $360°/2\pi$, approximately $57°$, and therefore the radian frequency ω is proportional to the conventional frequency f in cycles per second; furthermore, the *period* of the cycle, T in seconds, is inversely related to these measures. Thus

$$f = \frac{\omega}{2\pi} = \frac{1}{T} \qquad (7.1)$$

These relations are obvious enough for a rotating wheel, but in fact are relevant for the sinusoidal wave also since this repeats itself every 2π rad. This point is clarified by the rotating-vector concept, which is introduced shortly.

2. The *amplitude* of a sinusoidal wave defines its maximum excursion from the mean value, which latter is usually taken as zero.

3. The *phase* ϕ is the angle by which instantaneous values of the input and output sinusoidal waves differ; usually the phase is that of the output relative to the input and is *lagging* or *leading* according to whether a point on the output lags (is behind; occurs at a later time) or leads (is in advance of) a similar point on the input wave. We defer the problem of how to measure the phase quantitatively until advantage can be taken of the rotating-vector concept, but Fig. 7.1 illustrates the idea qualitatively.

If the input is applied to a system at rest, steady oscillatory conditions clearly cannot apply until after some suitable period during which a transient buildup occurs. This buildup involves increasing the total amount of energy stored in the system, but when the steady state is ultimately established the rate of change of energy is zero over each cycle; that is, the amount accepted from the input exactly balances that lost by dissipation. The transient part of the response can be calculated by the methods of Chap. 6, being the homogeneous solution y_H, although it is not of interest to do so here. On the other hand, the steady-state frequency

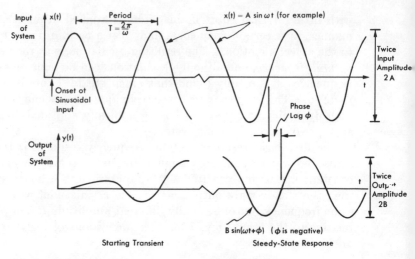

FIG. 7.1   Nature of frequency response curves.

response is the particular or forced solution y_P and can be obtained in a particularly simple way from the transfer function. Finally, the reader may be wondering how a sinusoidal output wave can be considered a steady state. It is steady in the sense that the amplitude ratio and relative phase do not vary in time.

The *amplitude ratio* (AR) is defined as the ratio of output and input amplitudes, and is often called the *gain*. Both it and ϕ are functions of the sinusoidal frequency ω and are the sole frequency response measures applying to any given linear system. In general these measures are different for different systems and therefore are largely able to characterize them. Although it is important that the characteristics are easily calculated from a knowledge of the transfer function, the reverse problem of identification provides another of the important uses of frequency response techniques— that is, given the ability to measure these characteristics experimentally, to identify the form of transfer function that applies.

In principle these frequency response measures add no new information to that obtainable from a transient such as the step response, but in practice the high-frequency end of the system's behavior is not determined with any resolution by transient responses. The analyst is well advised, therefore, always to use both time-domain and frequency-domain techniques.

In order to orient the reader regarding the biological significance of frequency response, a number of topics are considered below.

(A) THE SEMICIRCULAR CANALS

The semicircular canals of the inner ear are the basic transducers for a vestibulo-ocular reflex which compensates the eyeball for the inevitable rotational movements of the body and skull encountered in normal life. A simple linear transfer function can be derived

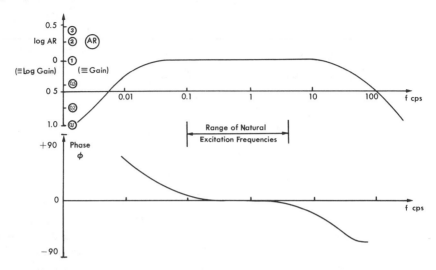

FIG. 7.2 Frequency response plots of semicircular canals. (*From Jones and Milsum.*)

from the hydrodynamic and other characteristics of this transducer, and the resulting transient responses do provide clues to what the important input variable is. However, the AR and phase characteristics make it very clear that the element must normally be "intended" to act as a transducer of the skull's angular velocity. Thus, in the normal range of the skull's stimulation frequency, say $\frac{1}{10}$ to 5 cps, there is a constant AR and only small phase when the angular velocity is regarded as input (Fig. 7.2). This would not be the result if either the skull angle or the skull acceleration were considered as input. With regard to the choice of logarithmic scales on Fig.7.2 for log AR and f the reader should only note for now that these provide the standard axes for the so-called *Bode plots*, and that the advantage of this particular mode of presentation is demonstrated in Sec. 7.6. The AR in this case is that of the skull's angular velocity as input and the deflection of a component in the semicircular canal called the *cupula* as output.

It is perhaps not yet entirely clear how the skull's angular velocity as input is distinguished from the angular frequency f of this same input's sinusoidal motion. When an animal walks or runs steadily, there is a periodic stimulus which usually causes some head motion and the input frequency f is concerned with the repetition rate of this motion. On the other hand, the magnitude of the skull's angular velocity which results depends on the strength of the stimulus and the use of head muscles, among other factors.

(B) HARMONIC CONTENT

The periodic excitation of the head, just discussed, cannot in practice be the exact sinusoidal form of Fig. 7.1. Instead, the waveform is typically irregular within the T between successive footfalls, but

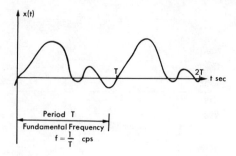

FIG. 7.3 *A repetitive irregular waveform.*

is then repeated in further T intervals (for example, Fig. 7.3). Almost all repetitive waveforms, except those from the specially designed class of oscillators, exhibit this characteristic to some degree, one familiar example being the waveform of a musical instrument. Normal musical instruments cannot generate a "pure" musical tone because the generating process always adds *harmonics*, that is, oscillations at multiples of the fundamental frequency $1/T$. Since all musical instruments, including biological ones, have differing harmonic content, middle C (262 cps) on the violin can be distinguished from a singer's similar note.

(C) FOURIER SERIES

There is a very powerful mathematical theorem due to Fourier which enables any irregular but repetitive waveform (Fig. 7.3) to be broken down into a set of sine and cosine waves of various amplitudes, comprising in general the fundamental and all its harmonics. This is exemplified in Fig. 7.4 for the waveform of Fig. 7.3. The Fourier series furthermore defines how to evaluate A_n and relative ϕ_n's of the harmonics (see Bibliography on Linear Analysis).

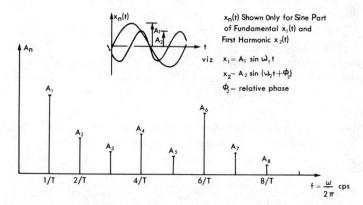

FIG. 7.4 *Harmonic content—amplitude spectrum.* (*Not to scale.*)

(D) SUPERPOSITION

The principle of superposition states that when there are multiple inputs, each is filtered by a linear system as if the others were not present, and the total output is the sum of the component outputs (Sec. 6.7). In this regard *filtering* means the action of the dynamic system in processing the input to yield the output. Thus the frequency response has relevance in such tasks as calculating the output to complex repeated waveforms, since each of the output components can be separately computed by using the system's frequency response data.

(E) HEARING

We hear many sounds, and their transduction and perception are of obvious interest. Since a repetitive waveform can be broken down into the simpler form of separate harmonics, and since waveform analysis is usually performed in this way, it is pertinent to ask whether the cochlea operates the same way. Although the auditory mechanism is not yet completely understood quantitatively, there seems little doubt that the answer is no; rather, auditory perception seems to involve some kind of pattern recognition of a time-frequency field. Indeed, the ear seldom encounters sustained noises in a natural habitat, so that a purely steady-state frequency analysis could be led astray by the transient buildup, such as shown in Fig. 7.1. Furthermore, it would be an inefficient way of decoding the rather limited amount of information, in the mathematical sense, normally transmitted by nature's sounds and human speech.

(F) BANDWIDTH AND FIDELITY

Another important aspect of human speech is its communication by such means as telephones. Now a direct cost can be placed upon the *bandwidth* of a channel—that is, upon the range of fre-

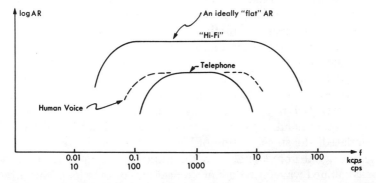

FIG. 7.5 Frequency response characteristics of communication channels. (*All approximate.*)

quencies transmitted by the channel without significant attenuation of the AR and without distortion in the phase relations.   Communications engineers find that acceptable intelligibility results if frequencies from about 300 to 3,000 cps only are transmitted (Fig. 7.5), although voice harmonics extend considerably beyond these limits.

More rigorous demands are placed upon the fidelity of reproduction for such communication equipment as radios, phonographs, and tape recorders.   In particular, music has significant overtones and undertones which require the bandwidth to extend perhaps from 50 cps up to 15,000 cps (Fig. 7.5).   However, even here it is not so critically necessary that the AR of the frequency response be absolutely flat in the bandpass region as that any variations be smooth.

(G) FREQUENCY RESPONSE OF VISUAL TRACKING

The use of frequency response is further illustrated by work on the visual-tracking servomechanism (see the references mentioned in Sec. 12.2).   In this work the phenomenon of pattern recognition has been quantitatively established, which perhaps we already accept qualitatively as part of the general adaptive behavior pattern of living organisms with appreciable "brains."   In the case of visual tracking with the head still, the relevant frequency response data quantify this phenomenon very clearly (Fig. 7.6).   Thus, if the subject is asked to follow a light-spot target moving to and fro horizontally with a sinusoidal motion, he rapidly "locks on" to the target because it is predictable.   In fact, he usually responds with a small phase lead and with slightly larger amplitude of eye movement than of target movement, that is, gain greater than unity or $\log AR > 0$. This condition applies at all frequencies up to about 1 cps, above which the gain and phase both fall off rapidly.   This combination of gain and phase characteristics turns out to be very difficult to achieve by any realistic physical system which does not have some such element as that of pure prediction.   This provides a hint about the brain's operation which is confirmed if the target motion is made unpredictable to the brain's "pattern recognizer," because falling gain and phase characteristics, such as the control engineer would expect of a typical servomechanism, then result.

The nature of the unpredictable signal used in this test is of interest.   Recall that an irregular but repeatable wave can be constructed from a set of harmonically related sine waves.   On the other hand, if the sine waves are selected to be nonharmonically related, the superposition of between three and five of them makes the signal unpredictable to the human's "computer," because the combined waveform does not repeat except over long time intervals. Therefore, the servomechanism presumably reverts to a simple watch-and-follow mode.   For analysis the output wave must be

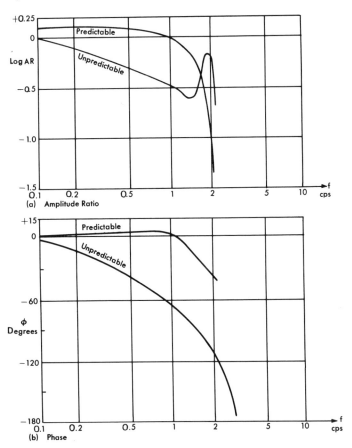

FIG. 7.6 *Frequency response of visual tracking.* (*From Young.*)

"unscrambled" into the corresponding nonharmonic set of components in order to evaluate gain and phase values at the several frequencies.

(H) PREVIEW OF THE CHAPTER

So far these introductory notes have explained something of the motivation and utility of frequency response analysis. Several examples of biological importance have been given, with results taken from work which will be developed in detail later. The idea of steady-state AR and phase characteristics as a function of frequency is thereby illustrated. The ideas of harmonic content and bandwidth have also been introduced qualitatively.

In succeeding sections the necessary mathematical tools to handle frequency response are built up, and the results are illustrated for the basic and lead-lag first-order systems. The generalized rotating-vector representation is developed in Sec. 7.2, and the filtering of a sinusoidal wave by a linear system is examined in Sec. 7.3. For the reader who wishes to avoid these mathematical details, however, at least temporarily, a summary of the results and operating rules developed in these two sections is given in Sec. 7.3D. Specific results for the first-order system follow in Sec. 7.4, and the generalized methods of presentation called *polar* and *Bode plots* are given in Secs. 7.5 and 7.6, respectively. Section 7.7 shows how the Bode plots of first-order cascades are combined, and Sec. 7.8 analyzes the lead-lag's frequency response. Section 7.9 correlates frequency and transient response characteristics, and finally in Sec. 7.10 two other periodic waveforms, the square and triangular, are introduced, leading rather naturally to the element called the *pure time delay.*

▼

7.2   *The Generalized Sinusoidal Wave*

(A) GENERAL

The basis of the frequency response method is the sinusoidal wave, of which the sine function itself is an example. By trigonometric definition the value of sin α is the ratio of the side opposite to the angle α and the hypotenuse in a right-angled triangle (Fig. 7.7). As α increases from zero, with the counterclockwise direction taken positive by convention, sin α first rises to a maximum value of unity at $\alpha = 90° = \pi/2$ rad, reduces again to zero at 180°, and then mirrors this characteristic with negative sign during the next 180° before it is back where it started at $\alpha = 360° \equiv 0°$. Clearly any further increase in α only repeats the cycle just outlined.

In frequency response work we are concerned with oscillations in time of variables such as pressure, temperature, and position. In fact, if we look upon the hypotenuse mentioned above as an arm rotating at a constant rate, we can define the angle α at any time t by

$$\alpha = \omega t \qquad \text{rad} \tag{7.2}$$

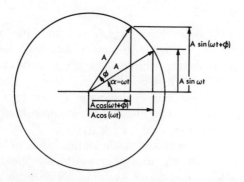

FIG. 7.7   Rotating arm.

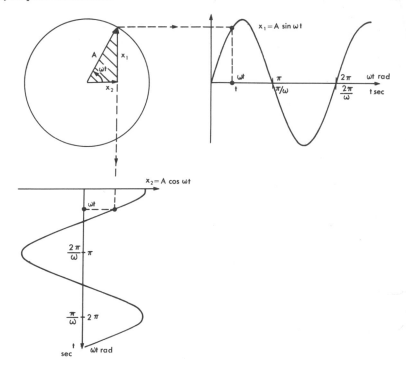

FIG. 7.8   *Generation of sinusoidal waves from rotating vector.*

where ω is a constant of proportionality which is called the *angular velocity* or *angular frequency* in radians per second. On the other hand, f in cycles per second is always called *frequency*. The definition of a variable $x_1(t)$ varying as a sine wave in time with angular frequency ω is then

$$x_1(t) = A \sin \omega t = A \sin 2\pi ft \tag{7.3}$$

where A is the maximum magnitude of the wave (Fig. 7.7), called the *amplitude*. Since one cannot ensure that the phasing of the wave is always such that $x = 0$ at $t = 0$, a general translation of angle is useful:

$$x_1(t) = A \sin (\omega t + \phi) \tag{7.4}$$

The angular relations are shown in Fig. 7.7 while Fig. 7.8 demonstrates the generation of both sine and cosine waves from the rotating arm of length A.

(B) DIFFERENTIATION AND INTEGRATION OF SINE WAVES

In frequency response the differential equation, or transfer function, of the system must operate upon the sinusoidal input, so that differentiation and integration operations must be quantified.

Given a sine wave (Eq. 7.3), the first derivative is

$$\frac{d}{dt} x_1(t) = A\omega \cos \omega t \tag{7.5}$$

This is a cosine wave of the same frequency but of amplitude equal to A multiplied by ω. The wave can of course be evaluated from this formula, but for manipulative convenience it is better to replace this cosine function with a corresponding sine function advanced or leading in phase by 90° or $\pi/2$ rad (Fig. 7.9), that is,

$$\frac{dx_1}{dt} = A\omega \sin\left(\omega t + \frac{\pi}{2}\right) \tag{7.6}$$

A phase lag of 270° instead of the 90° lead would also be applicable, but we take the positive value in differentiating. A plausible justification is that one expects a differentiated sine wave to be advanced in phasing since there must be a rate of change before any change occurs. On the other hand, the process of integration involves phase lag and the above argument in reverse.

Further differentiations produce extra successive phase leads of 90° and magnitude multiplications by ω (Fig. 7.9):

$$\frac{d^2x_1}{dt^2} = A\omega^2 \sin(\omega t + \pi) \tag{7.7}$$

On the other hand, integrations produce successive phase lags of 90° and magnitude divisions by ω (Fig. 7.9).

$$\int x_1\, dt = \frac{A}{\omega} \sin\left(\omega t - \frac{\pi}{2}\right) \tag{7.8}$$

$$\iint (x_1\, dt)\, dt = \frac{A}{\omega^2} \sin(\omega t - \pi) \tag{7.9}$$

We should note that although the second derivative and the second integral fall on the same radius line of Fig. 7.9, they are not in fact of the same phase because the second derivative term leads the second integral term by 360° or 2π rad; also, their magnitudes are different.

Our present purpose is to obtain an operational form for the input-output relation when sine waves are processed by linear dynamic systems. This could be done directly through the Laplace-transform symbolism, but we shall defer this development until the results can first be demonstrated by some mathematical arguments which may be more familiar to the student and in any case are essential to later developments. So far, then, the rotating arm has been shown to generate sine waves and the basic diagram also shows the corresponding arm's phase and magnitude when differentiation and integral operations are performed. However, the analytic form of such relations [Eqs. (7.5) to (7.9)] is not convenient for the manipulations needed in determining the input-output relations of frequency response. Convenience is readily achieved, however, by

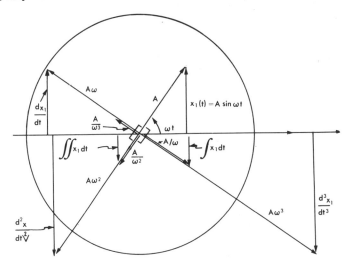

FIG. 7.9 *Derivatives and integrals of* $x_1(t) = A \sin \omega t$.

modifying the rotating-arm model slightly so that it is a formal rotating vector in the complex plane. The complex-plane representation is also important elsewhere in control work, and this is a convenient time for its introduction.

(C) COMPLEX PLANE AND ROTATING VECTOR

The development of the complex-plane relations follows logically from the above groundwork, but the reader unacquainted with the complex plane would benefit by reading from the Bibliography. The key idea is that a complex number is comprised of real (Re) and imaginary (Im) parts, and in the complex plane the abscissa is the Re axis and the ordinate the Im axis. As shown in Fig. 7.10, the complex number z is then the hypotenuse of a right-angled triangle whose other sides are the real part u and the imaginary part v. Therefore, the magnitude of z, conventionally defined as r, is

$$|z| = r = (u^2 + v^2)^{1/2} \qquad (7.10)$$

The complex number z is a *vector*, which must be specified regarding both magnitude and direction. In Fig. 7.10 an analogy is drawn with Fig. 7.7 so that the direction equals the phase, or angle α, and is defined by

$$\alpha = \arctan \frac{v}{u} \qquad (7.11)$$

that is,

$$\tan \alpha = \frac{v}{u}$$

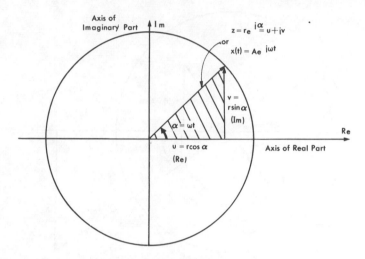

FIG. 7.10   The complex variable plane.

It remains now to make the formal definition of the complex number.   In cartesian space this is done in terms of the rectangular coordinates u and v of the real and imaginary axes, respectively, as

$$z = u + jv \tag{7.12}$$

where $j^2 = -1$. j is needed because u and v are merely quantities, but at the same time they must be distinguished from each other since they are always mutually perpendicular.   It is shown shortly that j specifies the action of a 90° rotation.   Equation (7.12) is thus a complete description of a vector in the complex plane, which in our present problem can be identified with the rotating arm of Fig. 7.7.

z may be specified equally well by two polar coordinates, that is, by the length, modulus, or magnitude r of the vector z, and by its phase α.   These polar coordinates have already been specified in Eqs. (7.10) and (7.11) as functions of u and v, and the definition of z in polar coordinates follows after a little manipulation.   In Fig. 7.10

$$\begin{aligned}\text{\textit{The real part}} &\qquad u = r \cos \alpha \\ \text{\textit{The imaginary part}} &\qquad v = r \sin \alpha \end{aligned} \tag{7.13}$$

so that Eq. (7.12) becomes

$$z = r(\cos \alpha + j \sin \alpha) \tag{7.14}$$

From the identity known as Euler's theorem,

$$\cos \alpha + j \sin \alpha \equiv e^{j\alpha} \tag{7.15}$$

The definition of a complex number in polar coordinates follows from Eq. (7.14) as

$$z = re^{j\alpha} \tag{7.16}$$

The formal definitions of a complex number are now complete for both rectangular and polar coordinates [Eqs. (7.12), (7.16)], and we utilize these results for the rotating-arm definition of Fig. 7.7. Thus, by making the following equivalences,

$x(t) \equiv z$ = rotating vector
$A \equiv r$ = magnitude of vector
$\omega t \equiv \alpha$ = angle of vector from real axis (counterclockwise positive)

Eqs. (7.16) and (7.12) become, respectively,

$$x(t) = Ae^{j\omega t} \tag{7.17}$$
$$x(t) = A(\cos \omega t + j \sin \omega t) \tag{7.18}$$

We shall utilize the polar form of Eq. (7.17) because of the manipulative advantages of exponential functions which have been frequently demonstrated already. As shown by Eq. (7.18), the $x(t)$ so defined has both cosine and sine components. However, because these components are the real and imaginary parts of a complex number, they are only additive if the appropriate rules developed above are used. This new $x(t)$ is not, therefore, quite the same as the $x_1(t)$ of Eq. (7.3) used in Sec. *B*.

It should also be noted that because Eq. (7.17) defines a vector in the complex plane, the notation $x(j\omega)$ is often used rather than $x(t)$. In this case x is regarded as a function of frequency rather than of t.

(D) DERIVATIVES AND INTEGRALS OF ROTATING VECTOR $Ae^{j\omega t}$

The general rotating vector [Eq. (7.17)] is directly differentiable, yielding

$$\frac{dx}{dt} = j\omega Ae^{j\omega t} = j\omega x(t) \tag{7.19}$$

We have already noted that differentiation of a sine wave produces 90° phase advance and magnitude multiplication by ω (Fig. 7.9). Consequently, one can reasonably expect that the j multiplier in Eq. (7.19) corresponds to 90° phase advance. Indeed, this can be formally shown as a particular result of DeMoivre's theorem, but also can be readily demonstrated in a special case. Thus a particular vector which lies along the abscissa has only a real part (say β) and is written in rectangular coordinates as $z_1 = \beta + j0 = \beta$. On the other hand, a vector of the same magnitude lying along the ordinate has only an imaginary part $z_2 = 0 + j\beta = j\beta$. Consequently, the result z_2 of multiplying a vector z_1 by j is to rotate it through 90° counterclockwise.

The magnitude and phase relations for both successive differentiations and integrations may readily be found by extending the result of Eq. (7.19), and these are tabulated in Table 7.1. Figure 7.9 illustrates the magnitude and phase relations still, except that

TABLE 7.1   ROTATING-VECTOR DERIVATIVES AND INTEGRALS

Vector $x(t) \equiv x(j\omega)$	Magnitude, $\|x(t)\|$	Phase (in radians)
$x(t) = A e^{j\omega t}$	A	ωt
Derivatives		
$\dfrac{dx}{dt} = j\omega x(t)$	$A\omega$	$\omega t + \dfrac{\pi}{2}$
$\dfrac{d^2 x}{dt^2} = (j\omega)^2 x(t) = -\omega^2 x(t)$	$A\omega^2$	$\omega t + \pi$
$\dfrac{d^n x}{dt^n} = (j\omega)^n x(t)$	$A\omega^n$	$\omega t + \dfrac{n\pi}{2}$
Integrals		
$\displaystyle\int x(t)\, dt = \dfrac{x(t)}{j\omega}$	$\dfrac{A}{\omega}$	$\omega t - \dfrac{\pi}{2}$
nth integral $= (j\omega)^{-n} x(t)$	$\dfrac{A}{\omega^n}$	$\omega t - \dfrac{n\pi}{2}$

the $x(t)$ now being considered is the complete rotating vector, marked A on the figure, rather than merely its vertical component $x_1(t)$, and similarly for the successive integrals and derivatives. This table provides the key to the filtering of sinusoidal waves by linear dynamic systems.   Before proceeding to this problem, however, the equivalent Laplace-transform result can be quickly stated.

(E) LAPLACE TRANSFORM OF ROTATING VECTOR $A e^{j\omega t}$

The rotating vector, which provides the generalized sinusoidal wave [Eq. (7.17)], may be Laplace-transformed, but in order to keep certain points clear in the following development it is preferable to label the frequency of this generalized form ω_n rather than ω. However, ω_n can take on any value in the permitted range of ω; therefore

$$
\begin{aligned}
x(s) &= \int_0^\infty A e^{(j\omega_n - s)t}\, dt \\
&= \frac{A}{s - j\omega_n} \equiv A\, \frac{s + j\omega_n}{s^2 + \omega_n{}^2}
\end{aligned}
\tag{7.20}
$$

The last expression was obtained by multiplying the term $1/(s - j\omega_n)$ by its *complex conjugate* $1/(s + j\omega_n)$ in both numerator and denominator.   The complex-conjugate expression has the same real part as its original, but the imaginary part has the opposite sign.   Complex-conjugate expressions therefore have equal magnitudes, as the reader should check.   The second form of Eq. (7.20) is convenient because it can immediately be separated into its real and imaginary

TABLE 7.2 SOME LAPLACE-TRANSFORM PAIRS

$x(t)$	$\rightleftharpoons$	$x(s)$
$e^{j\omega_n t}$	$\rightleftharpoons$	$\dfrac{1}{s - j\omega_n} \equiv \dfrac{s + j\omega_n}{s^2 + \omega_n^2}$
$\cos \omega_n t$	$\rightleftharpoons$	$\dfrac{s}{s^2 + \omega_n^2}$
$\sin \omega_n t$	$\rightleftharpoons$	$\dfrac{\omega_n}{s^2 + \omega_n^2}$

parts and the Laplace-transform pair becomes

$$A e^{j\omega_n t} \rightleftharpoons A \frac{s + j\omega_n}{s^2 + \omega_n^2} = A \frac{1}{s - j\omega_n} \tag{7.21}$$

By use of Eq. (7.18) on the left-hand side of expression (7.21), the transform pairs of Table 7.2 result.

7.3 Filtering of Sinusoidal Waves by Linear Systems; Self-reproduction

▼

(A) GENERALIZED TRANSFER FUNCTION

We now demonstrate the result presumed in Fig. 7.1, that when an arbitrary linear dynamic system is subject to a sinusoidal input, in the steady state the output is also sinusoidal and of the same frequency, though generally of different AR and phase. The lead-lag system's differential equation

$$T \frac{dy}{dt} + y = \alpha T \frac{dx}{dt} + x$$

and its transfer function $\hspace{4cm}$ (7.22)

$$\frac{y}{x}(s) = \frac{\alpha T s + 1}{T s + 1}$$

are merely first-order examples of the more general forms for complex linear systems.

$$b_n \frac{d^n y}{dt^n} + \cdots + b_1 \frac{dy}{dt} + b_0 y = a_m \frac{d^m x}{dt^m} + \cdots + a_1 \frac{dx}{dt} + a_0 x \qquad m < n \tag{7.23}$$

and

$$\frac{y}{x}(s) = H(s) = \frac{a_m s^m + \cdots + a_1 s + a_0}{b_n s^n + \cdots + b_1 s + b_0} \tag{7.24}$$

▼

(B) ROTATING–VECTOR TECHNIQUE

Consider as the input the generalized rotating vector, and apply it to the right-hand side of Eq. (7.23), using Table 7.1.   Therefore

$$\text{Right-hand side} = Ae^{j\omega t}[a_m(j\omega)^m + \cdots + a_1(j\omega) + a_0] \tag{7.25}$$

We have no a priori reason to assume that the output $y(t)$ will indeed be sinusoidal and of the same frequency.   However, unless this assumption is made there will be no term of the form $e^{j\omega t}$ on the left-hand side, in which case the required equality between the right-hand and left-hand sides cannot hold for arbitrary time values t. Therefore, we presume that the output is of the form

$$y(t) = Be^{j(\omega t + \phi)} \tag{7.26}$$

and hence obtain for the left-hand side of Eq. (7.23)

$$\text{Left-hand side} = Be^{j(\omega t + \phi)}[b_n(j\omega)^n + \cdots + b_1(j\omega) + b_0] \tag{7.27}$$

The arbitrary unknowns introduced by this procedure are the AR B/A and ϕ, and to solve the problem they must therefore be related to the a and b constants of the system, as well as to the sinusoidal frequency ω.   If now the definitions of $x(t)$ [Eq. (7.17)] and $y(t)$ [Eq. (7.26)] are incorporated, the combination of Eqs. (7.25) and (7.27) yields

$$\frac{y(t)}{x(t)} = \frac{B}{A}\,e^{j\phi} = \frac{a_m(j\omega)^m + \cdots + a_1(j\omega) + a_0}{b_n(j\omega)^n + \cdots + b_1(j\omega) + b_0} \tag{7.28}$$

Several important conclusions result from this expression:

1. Although both x and y are functions of time, Eq. (7.28) shows that their ratio is independent of time, but is a function of angular frequency ω.   Thus steady-state frequency response conditions do exist which are independent of time.
2. The middle expression of Eq. (7.28) shows that the y/x ratio is a vector whose magnitude is the AR B/A of the system's frequency response and whose direction is the relative ϕ.   Furthermore, the last expression of Eq. (7.28) is a ratio of complex numbers, which in turn can be expressed as a single complex number.   This complex number is uniquely determined by the system's constants and by the frequency.   Hence the AR and phase are both determined; see Eqs. (7.32) and (7.33), respectively.
3. The last expression of Eq. (7.28) is identically the transfer function $H(s)$ of the given system except that s has been replaced by $j\omega$, and hence can pertinently be reexpressed as

$$H(j\omega) = \frac{y}{x}\,(j\omega) = \frac{a_m(j\omega)^m + \cdots + a_1(j\omega) + a_0}{b_n(j\omega)^n + \cdots + b_1(j\omega) + b_0} \tag{7.29}$$

▼

(C) LAPLACE–TRANSFORM TECHNIQUE

The same result may be established by use of the Laplace-transform technique. The transformed expression for the rotating-vector input $x(t)$ of some arbitrary frequency ω_0 is (Table 7.2)

$$x(s) = \frac{A}{s - j\omega_0}$$

so that the output expression becomes

$$y(s) = x(s)H(s) = \frac{A}{s - j\omega_0} \frac{a_m s^m + \cdots + a_1 s + a_0}{b_n s^n + \cdots + b_1 s + b_0} \tag{7.30}$$

By the partial-fraction technique, it may be shown that the steady-state solution for $y(s)$ is

$$y(s) = \frac{A \left[\dfrac{a_m s^m + \cdots + a_0}{b_n s^n + \cdots + b_0} \right]_{s=j\omega_0}}{s - j\omega_0} \tag{7.31}$$

Then, by using Table 7.2, Eq. (7.28) is reobtained.

(D) SUMMARY OF BASIC RESULTS IN FREQUENCY RESPONSE

The analytical results of Secs. 7.3B and C can be summarized to provide working rules:

1. When a linear dynamic system is subject to a sinusoidal waveform input, in the steady state the output is sinusoidal and of the same frequency, but generally differing in AR and phase.
2. The relevant frequency response characteristics may be obtained by making the substitution

$$s = j\omega$$

in the transfer function $H(s)$. The transfer function is then symbolized $H(j\omega)$ and is a *complex number* in general, having Re and Im parts.
3. The AR is the magnitude of $H(j\omega)$ and is numerically determined as

$$|H(j\omega)| = \{[\text{Re } H(j\omega)]^2 + [\text{Im } H(j\omega)]^2\}^{1/2} \tag{7.32}$$

where Re $H(j\omega)$ and Im $H(j\omega)$ denote respectively the real and imaginary parts of $H(j\omega)$.
4. ϕ is numerically determined by the relation

$$\tan \phi = \frac{\text{Im } H(j\omega)}{\text{Re } H(j\omega)} \quad \text{or} \quad \phi = \arctan \frac{\text{Im } H(j\omega)}{\text{Re } H(j\omega)} \tag{7.33}$$

5. Both AR and phase are functions of the forcing frequency ω.

These points become clarified by their application to the first-order system in the next section. Various graphical techniques are then developed which simplify matters further.

7.4 Frequency Response of the Basic First-order System

In this section the treatment of Sec. 7.3 is applied in simplified form to the basic first-order system. However, for the reader who is only interested in applying the rules to obtain specific solutions, the summarized instructions of Sec. 7.3D are first followed.

(A) SUMMARY PROCEDURE

The transfer function is

$$H(s) = \frac{k}{Ts + 1}$$

so that by substituting $s = j\omega$ we obtain

$$H(j\omega) = \frac{k}{j\omega T + 1} \tag{7.34}$$

The denominator has real and imaginary parts 1 and ωT, respectively, so that by applying Eqs. (7.32) and (7.33) the AR and phase

AR	$\|H(j\omega)\| = \dfrac{k}{[1 + (\omega T)^2]^{\frac{1}{2}}}$	(7.35)
$Phase$	$\underline{/H(j\omega)} \triangleq \phi = -\arctan \omega T$	(7.36)

are found. The negative sign in the expression for phase arises because the real and imaginary parts are in the denominator. Gain and phase can clearly be evaluated numerically at any given frequency from these equations provided that the system parameters k and T are specified. Two different presentations of these characteristics are given in Secs. 7.5 and 7.6.

▼

(B) ROTATING–VECTOR ANALYSIS

The differential equation is

$$T\frac{dy}{dt} + y(t) = kx(t)$$

and with a generalized sinusoidal input

$$x(t) = Ae^{j\omega t}$$

the output is assumed to be

$$y(t) = Be^{j(\omega t + \phi)} \equiv Be^{j\phi}e^{j\omega t}$$

so that

$$\frac{dy}{dt} = (j\omega)y(t) \tag{7.37}$$

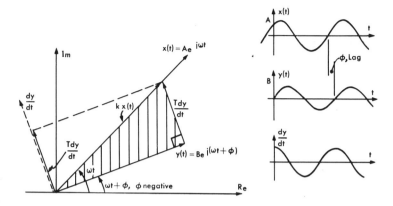

FIG. 7.11   Rotating vectors for basic first-order system.

These expressions combine to yield

$$(j\omega T + 1)y(t) = kx(t) \tag{7.38}$$

or alternatively

$$(j\omega T + 1)Be^{j(\omega t + \phi)} = kAe^{j\omega t} \tag{7.39}$$

Figure 7.11 shows the rotating-vector relations specified in the complex plane, where it should be noted that a vector is completely specified by magnitude and direction alone, without regard to actual spatial location.   Equation (7.39) can be simplified if the expression $e^{j(\omega t + \phi)}$ is divided out, leaving

$$(j\omega T + 1)B = kAe^{-j\phi} \tag{7.40}$$

As shown in Fig. 7.12 this triangle is stationary, because no time-varying terms remain.   Furthermore, because of the phase cancellation chosen the vector $y(t)$ lies along the real axis and its derivative dy/dt along the imaginary axis.   Since $T(dy/dt)$ is always positive, the triangle is always in the first quadrant and ϕ must be negative because $y(t)$ therefore lags behind $x(t)$.   Either from rearrangement of Eq. (7.40) into the form of Eq. (7.34),

$$\frac{B}{A} e^{j\phi} = \frac{k}{j\omega T + 1} \equiv H(j\omega) \tag{7.41}$$

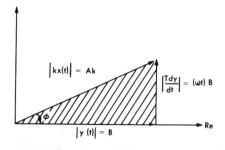

FIG. 7.12   The stationary vector triangle for Fig. 7.11.

or from the property of a right-angled triangle

$$(Ak)^2 = B^2 + (B\omega T)^2 \tag{7.42}$$

the gain and phase relations of Eqs. (7.35) and (7.36) follow directly.

▼

(C) COMPLEX NUMERATOR AND DENOMINATOR

In Eq. (7.41) the equality between the first expression's polar representation of a complex number and the second expression's complex denominator assures us that the latter is directly a complex number. The evaluation of its magnitude and phase relations is clarified by the following discussion: We first clear the denominator of imaginary terms by multiplying both numerator and denominator by the complex conjugate of the denominator:

$$\frac{1}{1 + j\omega T} \equiv \frac{1}{1 + j\omega T} \frac{1 - j\omega T}{1 - j\omega T} \equiv \frac{1 - j\omega T}{1 + (\omega T)^2} \tag{7.43}$$

The complex-plane representation of these relations is shown in Fig. 7.13. Clearly the phase is negative and is given by Eq. (7.36), because the denominator is now real and therefore has zero phase. The magnitude of the vector $1 - j\omega T$ is

$$|1 - j\omega T| = [1 + (\omega T)^2]^{1/2} \tag{7.44}$$

so that

$$\left| \frac{1}{1 + j\omega T} \right| \equiv \left| \frac{1 - j\omega T}{1 + (\omega T)^2} \right| = \frac{1}{[1 + (\omega T)^2]^{1/2}} \tag{7.45}$$

which of course confirms Eq. (7.35). However, this development additionally shows the conversion from a complex-denominator vector to a complex-numerator equivalent.

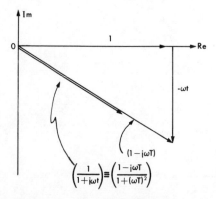

FIG. 7.13 Vectors with complex numerator and/or denominator.

7.5   The Polar, or Nyquist, Plot

The polar plot of the frequency response is a direct representation of the gain and phase data on the complex plane.   As the frequency is varied from zero to infinity, a locus of points is swept out by the tip of the $H(j\omega)$ vector, whose shape is characteristic of the system transfer function.   The particular version called the Nyquist plot includes both positive and negative frequencies, and provides the basis of a stability criterion.   In a logarithmic version it provides the basis of the Nichols chart which aids the design of closed-loop systems on the basis of their open-loop data (Chap. 11).

For the basic first-order system the frequency response relation is [Eq. (7.34)].

$$H(j\omega) = \frac{k}{j\omega T + 1}$$

Equation (7.45) is useful for plotting this function, and indeed Fig. 7.13 represents the essential first step.   Note first that k can affect the gain but not the phase, so that it is ignored here as being a trivial scaling factor.   Now, as

$$\omega \to 0 \qquad H(j\omega) \to 1 \qquad\qquad\qquad (7.46)$$

that is, unity magnitude and zero phase lag; but as

$$\omega \to \infty \qquad H(j\omega) \to 0 \qquad\qquad\qquad (7.47)$$

However, this last trend occurs in a certain definite pattern as revealed by Fig. 7.13; that is, the imaginary part ωT of $1 - j\omega T$ grows ever larger with respect to the real part 1 as ω tends to ∞, so that the phase

$$\phi \to -90°$$

while the gain $\qquad\qquad\qquad\qquad\qquad\qquad\qquad\qquad (7.48)$

$$\text{AR} = |H(j\omega)| \to 0$$

Another point of particular interest is the so-called *break frequency*, defined as

$$\omega_b = \frac{1}{T} \qquad\qquad\qquad\qquad\qquad\qquad\qquad\qquad (7.49)$$

Substituting this in Eqs. (7.35) and (7.36) shows that

$$\text{AR}_{\omega b} = \frac{1}{\sqrt{2}} \qquad \phi_{\omega b} = -45° \qquad\qquad\qquad (7.50)$$

The term ωT is a dimensionless frequency and is very useful throughout this work.   Furthermore, its value 1, corresponding to the break frequency, is particularly significant.   The resulting polar plot is shown in Fig. 7.14 and is in fact a unit semicircle.   For comparison purposes the polar plot of its inverse transfer function

$$H(j\omega) = 1 + j\omega T \qquad\qquad\qquad\qquad\qquad (7.51)$$

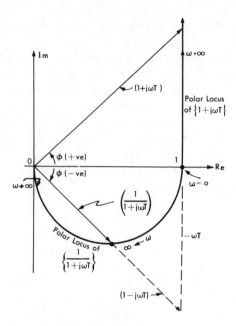

FIG. 7.14   Polar plot of basic first-order system.

is also shown on the figure.   Note its increasingly large phase lead
→ 90° and magnitude → ∞ as ω → ∞.   This element defines the
idealized proportional-plus-rate-sensitive transducer, and its increasingly large response, as frequency continues to increase, provides
another practical guarantee that it cannot be physiologically realized.

7.6   The Logarithmic, or Bode, Plots

The logarithmic plots involve logarithmic measures of AR and frequency, but linear phase.   In consequence they provide straight-line
simplifications somewhat similar to those obtained by semilogarithmic plotting of exponential time response curves.   Bode plots
are widely used in control and communications engineering practice,
and examples have already been given in Figs. 7.2, 7.5, and 7.6.

(A) LOGARITHMIC AR–LOGARITHMIC ω PLOT

By taking common (base 10) logarithms of Eq. (7.35) for the basic
first-order system, there results

$$\log_{10} \text{AR} \triangleq \log_{10} |H(j\omega)| = \log_{10} k - \tfrac{1}{2}\{\log_{10} [1 + (\omega T)^2]\} \qquad (7.52)$$

One immediately useful result is that the gain constant k is now
merely an additive constant.   However, two further useful simplifications are also available, applying to the low- and high-frequency
behavior of this system.

(B) LOW–FREQUENCY ASYMPTOTE

As ω tends toward zero the term $1 + (\omega T)^2$ in Eq. (7.52) tends toward zero. Thus the low-frequency asymptote is defined as

$$\log_{10} \text{AR} \approx \log_{10} k \qquad \omega T \ll 1 \tag{7.53}$$

which is a constant line on the Bode plot (Fig. 7.15a).

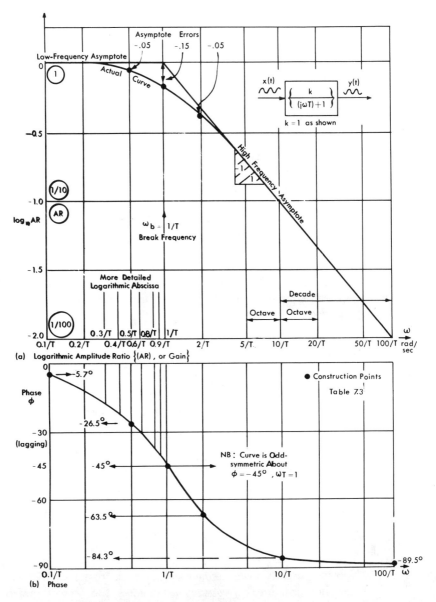

FIG. 7.15 *Bode frequency response plots—basic first-order system.*

(C) HIGH–FREQUENCY ASYMPTOTE

A similar simplification applies when ωT becomes very large, and the high-frequency asymptote is defined as

$$\log_{10} \text{AR} \approx \log_{10} k - \log_{10} (\omega T) \qquad \omega T \gg 1 \tag{7.54}$$

On a log-log plot, Eq. (7.54) is a straight line with a slope of -1, irrespective of k, which merely defines the value when $\log \omega T = 0$ (Fig. 7.15a). Furthermore, the slope is still -1 when the abscissa is $\log_{10} \omega$ rather than $\log_{10} (\omega T)$ since

$$\log_{10} (\omega T) = \log_{10} \omega + \log_{10} T \tag{7.55}$$

and $\log_{10} T$ therefore merely provides another constant term.

(D) LOGARITHMIC UNITS

The important practical meaning of this -1 slope is that the gain falls off by 1 log unit, that is, by a factor of 10, in each decade of frequency (or by a factor of 2 in 1 octave of frequency). Until fairly recently the literature has largely defined such slopes in decibels per octave or decibels per decade, but the decibel is an unnecessary unit which we shall not use. Numerically there are 20 db in each decade so that the equivalent slope to that of -1 log/log is -6 db/octave or -20 db/decade. If, however, one wishes to use a smaller unit than 1 log, then a one-tenth log or decilog seems more sensible.

$$1 \log \equiv 10 \text{ decilogs} \equiv 20 \text{ db} \tag{7.56}$$

Octaves conventionally only refer to frequency, although since they represent simple ratios, they are in principle equally applicable to gains also.

1 octave = ratio of 2, in frequency

Therefore $\tag{7.57}$

1 log (frequency) = 3.34 octaves (frequency)

(E) INTERSECTION OF LOW– AND HIGH–FREQUENCY ASYMPTOTES; BREAK FREQUENCY

Obviously it is interesting to know where the low- and high-frequency asymptotes intersect. Equations (7.53) and (7.54) are equal only when

$$\log_{10} \omega T = 0$$

that is, $\tag{7.58}$

$$\omega T = 1$$

which has already been defined as the break frequency [Eq. (7.49)]. Thus the term *break frequency* ω_b denotes the transition from a constant low-frequency asymptote to a high-frequency asymptote with -1 slope. Equations (7.53), (7.54), and (7.58) permit the asymptotic plot to be sketched (Fig. 7.15a). The only effects of varying k and T are to shift the asymptotes either up or down and either left or right. It should be noted that a semilogarithmic grid is conventionally used for this plot, and that this is because the log AR values of cascaded systems can then be directly added (Sec. 7.7).

(F) ERRORS OF THE ASYMPTOTES

The asymptotes are usually drawn for all frequencies although they are obviously inaccurate near the break frequency. The exact curve can be graphically constructed on the basis of an offset of 0.15 log from the asymptote at the break frequency [Eqs. (7.50)] and another offset of 0.05 log at octave points of frequency on either side (Table 7.3 and Fig. 7.15a). For all practical purposes there is little error in using the asymptotic construction except in the octave region on either side of the break frequency, $\omega_b T = 1$. The true curve can readily be approximated to the needed accuracy by adding to the plot the three "error" values just quoted.

(G) PHASE–LOGARITHMIC FREQUENCY PLOT

The Bode plot of phase uses a linear ordinate for phase, but the same logarithmic abscissa for frequency as in the log AR plot. For the basic first-order system, phase increases in magnitude from 0 at zero frequency to $-90°$ (phase lag) as $\omega \to \infty$ (Fig. 7.15b). On these axes the plot has odd symmetry about the break frequency, following the arctan relationship [Eq. (7.36)]; angles therefore appear in complementary pairs (Table 7.3). k has no effect whatsoever on this curve.

(H) PRACTICAL PLOTTING DETAILS

As already noted, semilogarithmic graph paper is conventionally used for Bode plots to provide a logarithmic scale for ω. However, the basic octave and decade frequency points are usually sufficient for construction purposes so that this logarithmic scale is not essen-

TABLE 7.3 LOG AR AND ϕ VALUES FOR BASIC FIRST-ORDER SYSTEM

ωT	*Actual* $\log AR - \log k$	*Asymptotic* $\log AR - \log k$	*Error of Asymptotic Data*	*Phase* $\phi°$
0.1	$-0.002 \approx 0$	0 ⎫	≈ 0	-5.7
0.5	$-0.048 \approx -0.05$	0 ⎬ LF	≈ 0.05	-26.5
1.0	-0.150	0 ⎭	≈ 0.15	-45.0
2.0	$-0.349 \approx -0.35$	-0.3 ⎫ HF	≈ 0.05	-63.5
10	$-1.002 \approx -1.0$	-1.0 ⎭	≈ 0	-84.3

tial if one remembers the values $\log_{10}(2\omega/\omega) \approx 0.3$ and $\log_{10}(5\omega/10\omega) \approx -0.3$, because a logarithmic scale can then be constructed by hand. Thus these numbers indicate that on a logarithmic scale the frequency octave points lie $\frac{3}{10}$ along the interval of a frequency decade. In general, one will seldom need to consider more than 5 decades of frequency, and often 3 will suffice. This is illustrated on Fig. 7.15 for the two higher decades, and on the lowest decade some other frequencies have also been marked, all obtained from knowing $\log_{10} 2$ and $\log_{10} 3$.

7.7 Frequency Response of First-order Cascades

Often one is concerned with a system or subsystem comprising an open chain or cascade of dynamic elements. In this section the particular case of a cascade of basic first-order elements only is considered, but generalization to all cascades is obvious.

For illustrative purposes, consider the cascade illustrated in Fig. 7.16 with three such units. The rotating-vector technique allows one to draw the three relevant output vectors as a consequence of an input vector. Each individual vector results from a first-order filtering operation (Fig. 7.11), and each first-order system contributes an appropriate AR and phase component at any given frequency. An important result is that in the multiplication of

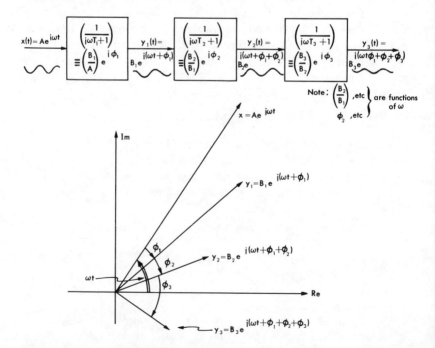

FIG. 7.16 Frequency response relations for first-order cascade.

transfer functions which results when they are in series, the individual AR's are multiplied to produce the overall AR, while the individual ϕ's are added to produce the overall phase (Fig. 7.16). If interest centers only in the cascade's input-output characteristics, the individual transfer functions may be telescoped to yield an overall function

$$H(s) \triangleq \frac{y_3}{x} (s) = \frac{y_3}{y_2} \frac{y_2}{y_1} \frac{y_1}{x} \tag{7.59}$$

and

$$H(j\omega) = H_1 H_2 H_3(j\omega) \tag{7.60}$$

The logarithmic relation is then

$$\log H(j\omega) = \log H_1(j\omega) + \log H_2(j\omega) + \log H_3(j\omega) \tag{7.61}$$

(A) LOG AR

The log AR value in this particular case is

$$\log \text{AR} = \sum_{n=1}^{3} [\log |H_n(j\omega)|] \tag{7.62}$$

$$\log \text{AR} = \sum_{n=1}^{3} \left(\log k_n - \frac{1}{2} \sum_{n=1}^{3} \{\log [1 + (\omega T_n)^2]\} \right) \tag{7.63}$$

Equation (7.62) shows that the log AR terms of the cascade elements must be added, which confirms the multiplication rule stated above. Equation (7.63) shows the breakdown of the log AR into k and ωT components, and provides basic data for constructing the Bode plot. In Fig. 7.16 all k values have been made unity for simplicity. It should be emphasized that there is a basic difference between the *steady-state* type of gain k and the *dynamic gain* term used as a synonym for the AR, which is frequency-dependent.

(B) PHASE ϕ

Figure 7.16 shows that the ratio of output and input vectors is given by

$$\frac{y_3}{x} = \frac{B_3}{A} e^{j(\phi_1 + \phi_2 + \phi_3)} \tag{7.64}$$

which, by recalling Eq. (7.41), clearly means that the phase of the cascade is the sum of the individual phases,

$$\phi = \phi_1 + \phi_2 + \phi_3 = - \sum_{n=1}^{3} \arctan \omega T_n \tag{7.65}$$

(C) EXAMPLE

The actual construction of the Bode plots for a cascaded system is best illustrated for a particular numerical example. Here we take as time constants

$$T_1 = 1 \text{ sec} \qquad T_2 = \tfrac{1}{5} \text{ sec} \qquad T_3 = \tfrac{1}{10} \text{ sec} \qquad\qquad (7.66)$$

and steady-state gains

$$k_1 = 2 \qquad k_2 = 1 = k_3$$

so that

$$H(j\omega) = \frac{2}{(j\omega + 1)(j\omega/5 + 1)(j\omega/10 + 1)} \qquad\qquad (7.67)$$

The Bode plots are then easily constructed because of the additive properties for logarithmic gains and because of the asymptotic approximations. As indicated by Eq. (7.61), no actual calculations are necessary because the plots can be directly constructed by superposing three individual plots, each of which is like that of Fig. 7.15.

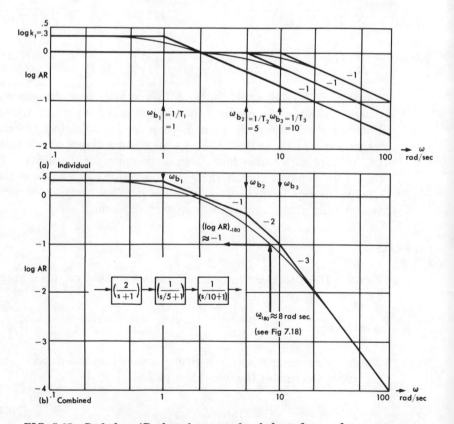

FIG. 7.17   *Bode log AR plots for cascade of three first-order systems.*

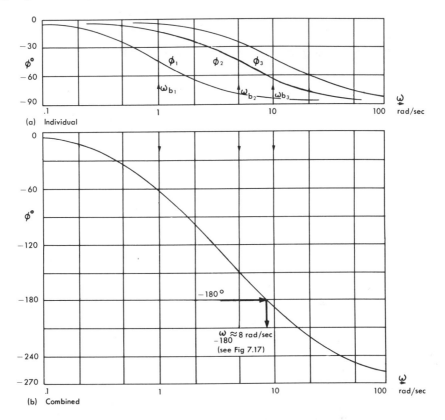

FIG. 7.18 *Bode phase plots for cascade of three first-order systems.*

However, since the time constants are different and known, there is no particular point in using a dimensionless frequency ωT. Figure 7.17 shows the resulting log AR plots, both individually in Fig. 7.17a and superposed in Fig. 7.17b. The construction essentially involves only the asymptotes, with the exact curve added as a later refinement on this approximation. The phase plots (Fig. 7.18) are also constructed by graphical addition of the individual phases at an appropriate number of frequencies. It should be emphasized that the construction is entirely graphical and based upon the few fixed values quoted. By this means an accuracy of 5 percent or better is obtained, which usually suffices.

(D) BODE PLOTS AND CLOSED–LOOP STABILITY

There are many possible uses of these Bode plots, particularly in predicting the behavior of the system when the loop is closed around it. In order to increase the reader's motivation, therefore, we now briefly consider the stability of this cascade when the loop is closed. Equation (2.23) states that instability occurs in the closed loop when $H(s) = -1$, that is, at any frequency, ω_{-180}, for which

$$AR = 1 \qquad \log AR = 0 \qquad \text{and} \qquad \phi = -180° \qquad (7.68)$$

Instability is impossible in the case of a single first-order system because phase lag never exceeds $-90°$ at any frequency. With three in cascade, however, the combined phase tends to $-270°$ as $\omega \to \infty$, and thus must pass through $-180°$ at some finite frequency, which is readily found from Fig. 7.18b to be

$$\omega_{-180} \approx 8 \text{ rad/sec} \qquad (7.69)$$

Now although this condition is necessary for instability it is not sufficient, for Eq. (7.68) also requires that the AR equal 1 at this frequency. Thus we consult Fig. 7.17b and find

$$\log AR_{-180} \approx -1 \qquad \text{or} \qquad AR_{-180} \approx \frac{1}{10} \qquad (7.70)$$

Hence, closing the loop directly would result in a stable system, but it could then be driven unstable by increasing the loop gain by a factor of 10. This subject is pursued in Chap. 11.

(E) POLAR PLOT

The polar plot can be constructed for this cascaded system by combining the data of Figs. 7.17 and 7.18 as shown in Fig. 7.19, with ω as the parameter of the polar locus. Although superficially the plot seems significant only in the low-frequency end, the area of vital interest concerning stability of the closed-loop configuration is usually in the medium-frequency region where the phase passes through $-180°$.

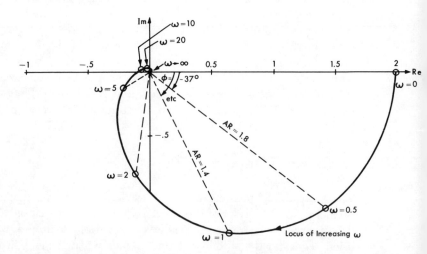

FIG. 7.19   Polar plot for cascade of three first-order systems.

7.8 Frequency Response of the Lead-Lag First-order System

The canonical transfer function for the lead-lag first-order system [Eq. (5.27)] is now written directly in the $j\omega$ form relevant for frequency response:

$$H(j\omega) = k\,\frac{j\alpha\omega T + 1}{j\omega T + 1} \tag{7.71}$$

This transfer function can be looked upon as the result of two cascaded first-order systems, in which one transfer function has terms only in the numerator, namely

$$H_1(j\omega) = j\alpha\omega T + 1 \tag{7.72}$$

and also

$$H_2(j\omega) = \frac{k}{j\omega T + 1} \tag{7.73}$$

as illustrated in Fig. 7.20. The cascade results of the previous section apply, but a new feature is that the frequency response characteristics of a transfer function having only numerator terms are now required. However, the polar plot of the vector $j\omega T + 1$ has already been presented in Fig. 7.14, and the Bode plots are presented in Fig. 7.21, together with those for its reciprocal $1/(j\omega T + 1)$. In the lead-lag system the extra constant α in the numerator produces nontrivial AR and phase values for the combined system (Fig. 7.20), as investigated below: The equivalent lead-lag expressions for Eqs. (7.63) and (7.65) are

$$\log \mathrm{AR} = \log k + \tfrac{1}{2}\log[1 + (\alpha\omega T)^2] - \tfrac{1}{2}\log[1 + (\omega T)^2] \tag{7.74}$$
$$\phi = \arctan \alpha\omega T - \arctan \omega T \tag{7.75}$$

The resulting Bode plots, with their construction from the component curves, are shown in Figs. 7.22 and 7.23, respectively, for

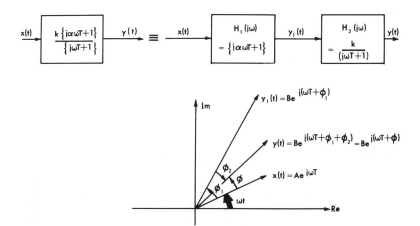

FIG. 7.20 Rotating vectors for lead-lag system.

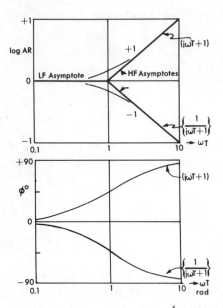

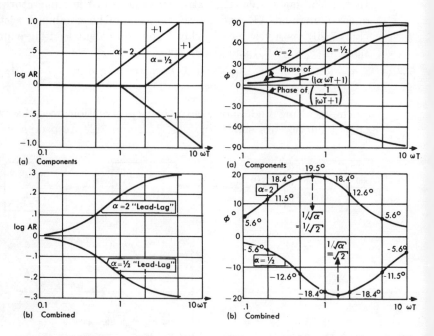

FIG. 7.21   *Bode plots for* $\dfrac{1}{j\omega T + 1}$ *and* $j\omega T + 1$.

FIG. 7.22   **Bode log AR plots for lead-lag system.**

FIG. 7.23   **Bode phase plots for lead-lag system.**

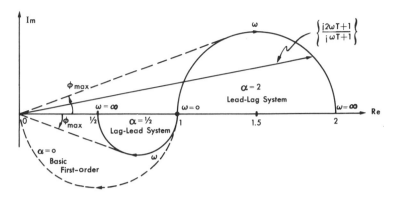

FIG. 7.24 Polar plots for lead-lag system. $k = 1$.

exemplary α values of 2 and $\frac{1}{2}$. When $\alpha > 1$ the phase is leading during a certain frequency range and is zero elsewhere; in fact, it is a symmetric curve on a log ω abscissa with a maximum value at

$$\omega = \frac{1}{\sqrt{\alpha}\, T} \tag{7.76}$$

as may be shown analytically by differentiating Eq. (7.75) and equating the result to zero for an extremum. When $\alpha < 1$ the phase curve is the same except that the phase always lags. At high frequencies, $\omega T \gg 1$, the log AR becomes constant at

$$\log \mathrm{AR} = \log \alpha \tag{7.77}$$

(where $k = 1$ has been assumed for simplicity), which represents attenuation in the *lag-lead* case $\alpha < 1$ and amplification in the *lead-lag* case $\alpha > 1$. The corresponding polar plots are illustrated in Fig. 7.24. Both types yield semicircles starting at $1 + j0$ when $\omega = 0$ and tending to α as $\omega \to \infty$. The maximum phase lead and lag are directly measurable as the tangents to the semicircles.

7.9 Correlation of Transient and Frequency Responses

The mathematical developments of Chaps. 6 and 7 have shown that both transient and frequency responses are characteristic of the system's transfer function, so that these responses cannot be independent measures of the system. It should therefore be conceptually advantageous to correlate these two response types for various systems, but at present this correlation is restricted to the basic and lead-lag first-order systems. For these types the canonical form of first-order transfer function suffices:

$$H(s) = k\frac{1 + \alpha T s}{1 + T s}$$

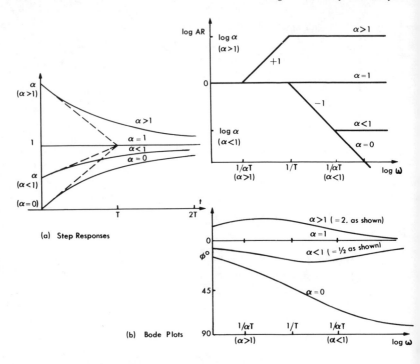

FIG. 7.25   Correlation of transient and frequency responses of first-order systems.

(A) $\alpha = 0$

The basic first-order system results when $\alpha = 0$.   The corresponding step response and Bode plots of frequency response are shown in Fig. 7.25.   (Also, the polar plot has been added to Fig. 7.24.)   The step response starts from zero, assuming zero initial conditions, and rises at an exponentially decaying rate to the steady state.   The steady-state response corresponds to the zero-frequency end of the Bode plots, showing unity gain (when $k = 1$) and zero phase lag. However, the gain and phase both fall off above the break frequency so that the response becomes negligible at high frequencies.   This negligible response at high frequencies correlates with the slow initial time response of the transient.   Thus, the frequency and time responses bear qualitatively a reciprocal relation to each other, as perhaps one should expect from the connection between the variables themselves: $\omega \sim 1/t$.

(B) $\alpha < 1$

For the lag-lead system of $\alpha < 1$, the initial jump in the step response is possible in principle because of the finite AR and zero phase lag of the frequency response as $\omega \rightarrow \infty$.

(C) $\alpha = 1$

The ideal static element is characterized by the condition $\alpha = 1$; that is, any input signal of whatever frequency content is outputted undistorted and with unchanged amplitude. This can be alternatively stated in the time domain; namely, all transient inputs are outputted undistorted.

(D) $\alpha > 1$

For $\alpha > 1$ the system is lead-lag, and leading phase now appears which correlates with the initial jump of the step response to a value higher than that of the steady state. It should be emphasized that the finite AR and zero phase response as $\omega \to \infty$ for all $\alpha > 0$ can only represent mathematical ideals, since all real physical systems must eventually fail to respond at a sufficiently high frequency. Correspondingly, no initial jumps in the step response are realizable, although with sufficient bandwidth the measurable difference may be negligible. Again it is a matter of accepting mathematical models just so long as they are sufficiently accurate for the dynamic conditions to be studied.

Bandwidth is a frequently used term in communications and control engineering. It is usually defined as the frequency above which the power falls below one-half the low-frequency value. In turn this corresponds to an AR of $1/\sqrt{2}$ or log AR of -0.15. The bandwidth of the basic first-order system is therefore equal to the break frequency, but in (idealized) lead-lag systems it is infinite.

7.10 Other Periodic Inputs

The conventional, commercial electronic function-generating unit provides the sinusoidal waveform discussed in this chapter, but in addition provides the square and triangular (sawtooth) waveforms shown in Fig. 7.26. The square wave looks like a sequence of superposed steps with alternating sign, and superposition does allow it so to be considered for linear systems. The triangular wave is just as clearly the integral of this square wave, or alternatively (but compatibly) it is a sequence of superposed ramps with alternating sign. Further simple integrations would never produce the sinusoidal form, however, and indeed this latter form is rather special. In particular, it is the only waveform having the self-reproducing property when passed through linear systems. Therefore, nonsinusoidal but periodic waveforms cannot be directly analyzed by the frequency response technique which has been presented in this chapter. However, it has been shown in Sec. 7.1 that any periodic waveform may be broken down by a Fourier series representation into a funda-

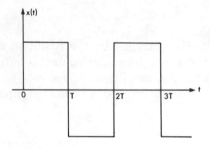

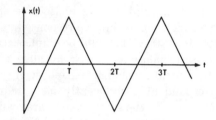

FIG. 7.26   Square and triangular waveforms.

mental sinusoidal component and a set of harmonics.   Since superposition is valid, the frequency response technique is then applicable
in this modified form.   The alternative approach is to deal with a
sequence of transient input functions delayed by successive intervals
in time.   This phenomenon gives rise to the pure-time-delay function
e^{-Ts} (Chap. 9), which is the same as the latency phenomenon of
physiology.

8 The Second-order Linear Lumped Model and Its Response Characteristics

The Pole-Zero Plot of Roots

8.1 Introduction

The first-order system arises ubiquitously as the natural result of one form of lumped energy storage combining with one form of energy dissipation. The second-order system correspondingly requires two energy-storage lumps, with an energy-dissipation element normally being unavoidably present also. Insofar as energy-storage elements are always actually distributed in space, the two-lump model for a single type of energy storage is usually just a first improvement of the single-lump model; thus in Chap. 3 the thermo-regulated body was modeled by three lumps.

There is another second-order lumped model, however, which is physically different because it involves two different types of energy storage, the generalized potential energy (PE) and kinetic energy (KE). Such systems have the ability to interchange their total stored energy between these two types and hence to exhibit oscillatory behavior. Again it should be emphasized that a lumped second-order model represents only a single-degree-of-freedom lumping of what is inevitably also a distributed system. Thus, partial differential equations such as the wave equation are needed to describe ideally the distribution of two-energy systems in space as well as in time.

Linear lumped second-order systems can be described by a canonical form of transfer function. In this transfer function, however, there is at least one more parameter than in the basic first-order system of Chap. 5, that is, three parameters instead of just the gain k and time constant T. This necessitates a family of characteristic transient and frequency responses instead of the unique response curve of the basic first-order system, and therefore is more difficult to show on universal plots.

In Sec. 8.2 the free-swinging human limb is modeled, and in Sec. 8.3 the semicircular canal is analyzed to produce a second-order model also. The first of these two models is almost conservative, while the second is highly dissipative.

The general form of the second-order transfer function and its roots is examined in Secs. 8.4 and 8.5, and the powerful method of representing the roots as a pole-zero plot in the complex plane is introduced in Sec. 8.6. The impulse, step, and frequency responses are presented in Secs. 8.7 to 8.9, and their correlation in Sec. 8.10. The responses of nonbasic second-order models are briefly discussed in Sec. 8.11.

8.2   The Free-swinging Human Limb

Walking and other modes of leg locomotion constitute a life-sustaining facility for many animals, and therefore it might be suspected that an energetic performance criterion would be relevant whenever speed is not an important aspect of the activity. Thus, since the energy input can be minimized if the pace frequency is adjusted to that resulting from free swinging of the leg as a pendulum under the effect of gravity, it is interesting that experimental evidence does show that the speed of relaxed walking is automatically adjusted to that requiring minimum respiration rate [8.1]. Of

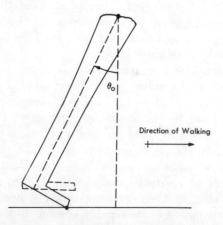

FIG. 8.1   Idealized stiff-leg walking.

course, the energy requirement never becomes ? ...o because of frictional effects and because muscle forces are necessary to achieve control of the articulated leg system and of posture during walking.

Specifically we now consider the free swing under gravity of a human leg, after release from ground contact in its rearward position by raising the toe (Fig. 8.1). However, the detailed practical problem of combining leg shortening and hip raising to enable the heel to clear the ground is not considered until Chap. 15, since it does not significantly affect the present development. Furthermore, we ignore initially the neuromuscular feedback loops of the antagonistic pairs of effector muscles, which means that we restrict attention to the passive *controlled process* itself, just as was done in Chap. 3 for the thermoregulatory control system. In due course, muscle forces are reintroduced.

(A) NEWTON'S SECOND LAW

In this problem the obvious coordinate system is that of angular motion, and therefore Newton's second law of motion must be modified from the translational form

$$F = m\ddot{x} \equiv m\dot{V} \qquad (8.1)$$

to the corresponding rotational form

$$T = J\ddot{\theta} \equiv J\dot{\omega} \qquad (8.2)$$

where T = net torque, or turning moment about axis of rotation
J = moment of inertia of accelerating body about this axis
θ = angle of rotation
ω = angular velocity of rotation

and where each superscript dot represents one differentiation of the variable with respect to time.

(B) THIN-CYLINDER MODEL OF LEG

For present computational sim, icity the leg is assumed to be a stiff uniform cylinder (Fig. 8.2) with

Length	$L \approx 90$ cm	
Diameter	$d \approx 12$ cm	
Uniform density	$\rho \approx 1.1$ g/cm³	(8.3)

Therefore

Volume $\quad V = \dfrac{\pi}{4} d^2 L = 10{,}300$ cm³

Mass $\quad m = \rho V = 11.5$ kg

which is approximately correct for the standard 70-kg male. Obviously considerable simplifications have been made, but the

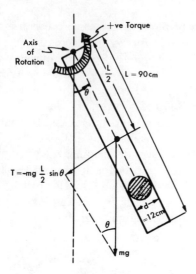

FIG. 8.2   Cylinder model of leg.

model may be refined to the analyst's pleasure and necessity.   However, the large inertial effect of the mass of the foot compensates at least partially for the tapering nature of actual legs.

(C) GRAVITY TORQUE

If we assume temporarily that there are neither viscous nor nonlinear retarding torques due to joint friction, tissue deformation, air friction, etc., then the net torque is only that resulting from gravity forces. This equals the product of the leg-weight component perpendicular to the leg ($mg \sin \theta$) and the moment arm $L/2$ (Fig. 8.2):

$$T = -mg \frac{L}{2} \sin \theta \qquad (8.4)$$

(D) INERTIA TORQUE

The right-hand term of Eq. (8.2) is often designated the *inertia torque*. To evaluate this we need the moment of inertia of the thin-cylindered leg swinging about its hip joint,

$$J = \frac{mL^2}{3} \qquad (8.5)$$

(E) DIFFERENTIAL EQUATION

By substituting Eqs. (8.4) and (8.5) in Eq. (8.2), there results the second-order differential equation of motion

$$\ddot{\theta}(t) + \frac{3}{2} \frac{g}{L} \sin \theta(t) = 0 \qquad (8.6)$$

Equation (8.6) is nonlinear because of the term sin θ, and it is usual practice to linearize for small deflections by using the approximation (Prob. 3.3)

$$\sin \theta \approx \theta \qquad \theta \text{ small and in radians} \tag{8.7}$$

In our particular case, if the half-pace is 38 cm (15 in.) with a 90-cm leg, θ equals about 23° or 0.4 rad whereas sin 23° equals 0.39 rad; thus the maximum error of linearization is about 2.5 percent and the average error much less. Consequently Eq. (8.6) becomes

$$\ddot{\theta} + \left(\frac{3}{2}\frac{g}{L}\right)\theta = 0 \tag{8.8}$$

(F) UNDAMPED SYSTEM RESPONSE

It is a classic result that this ideally conservative system responds to any initial conditions with an unending sinusoidal oscillation. As shown in Fig. 8.3, this takes the form

$$\theta(t) = \theta_0 \cos \omega_n t \qquad t > 0 \tag{8.9}$$

when the pendulum is released from rest with initial conditions

$$\theta(t) = \theta_0 \qquad \dot{\theta}(t) = 0 \qquad \text{at } t = 0 \tag{8.10}$$

Two initial conditions are now necessary because the system is second order.

The term ω_n is called the *undamped natural frequency* (in radians per second) of the second-order system and is one of three basic parameters. It may readily be evaluated by substituting Eq. (8.9) in Eq. (8.8). Therefore

$$\omega_n = \left(\frac{3}{2}\frac{g}{L}\right)^{\frac{1}{2}} \tag{8.11}$$

Consequently Eq. (8.8) may be rewritten in a standard form as

$$\ddot{\theta}(t) + \omega_n^2\theta(t) = 0 \tag{8.12}$$

From the known physical constants L and g ($= 981$ cm/sec²), we have $\omega_n \approx 4$ rad/sec, and thus the frequency f in double paces per second $f = \omega_n/2\pi = 0.64$ or 77 paces/min. Consequently, with

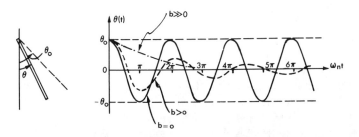

FIG. 8.3   *Initial-condition response of pendulum.*

a 3-ft pace the walking speed is 3.8 ft/sec or 2.6 mph, assuming that at each footfall a new half-cycle is started without delay. Since the army marches at around 120 paces/min at 4 mph, one can thus infer that extra muscle power must be expended to drive the system at a higher rate than its natural frequency. Indeed it has been found experimentally that the velocity for minimal energy expenditure per unit distance covered is either 2.25 or 3.5 mph depending on whether or not resting metabolism is deducted from the total energy-expenditure rate (see [8.1] and Sec. 15.2). Y. T. Li has demonstrated, furthermore, that an increased natural frequency and hence a higher speed can be obtained for the same oxygen consumption by using a torsion spring between the legs (Prob. 8.1).

(G) THE SYSTEM WITH DAMPING AND MUSCLE TORQUES

The results so far are plausible enough to suggest that energy-dissipation effects in relaxed walking are small. However, they must exist, and therefore a term

$$T_b = -b\dot{\theta} \qquad\qquad (8.13)$$

is now included, which is inevitably only a first approximation because it is linear. Again the negative sign indicates that the torque opposes the motion. This viscous friction causes the oscillatory response of Fig. 8.3 to decay, as shown dashed, with a terminal state of zero energy stored in the system. In steady walking the appropriate leg muscles must pump in enough energy to make up for this viscous dissipation as well as providing the various posturally required forces which the analysis has ignored. Therefore, a suitable time-varying torque function $T_m(t)$ is now added, so that Eq. (8.2) becomes

$$T_m - b\dot{\theta} - mg\,\frac{L}{2}\,\theta = J\ddot{\theta} \qquad\qquad (8.14)$$

or

$$J\ddot{\theta}(t) + b\dot{\theta}(t) + k\theta(t) = T_m(t) \qquad\qquad (8.15)$$

where the gravity coefficient has been defined as a generalized "spring" constant $k = mg(L/2)$. This expression represents a basic second-order linear system having input forcing and viscous damping in addition to the inertia and spring effects of the conservative unforced system of Eq. (8.12). In the energetic terms of Chap. 4 the spring stores PE, the inertia stores KE, and the viscous damping dissipates both. Figure 8.4a shows the block diagram for this basic form of system, which frequently occurs in dynamics. The block diagram is usually constructed by "solving" for the highest derivative, that is, by adding variables algebraically in the summing point to satisfy the form of Eq. (8.14).

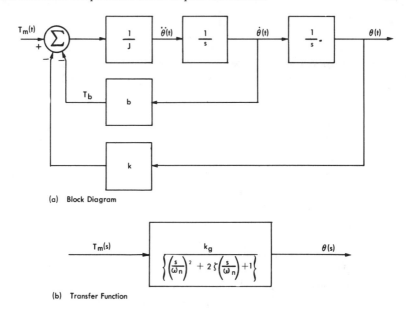

(a) Block Diagram

(b) Transfer Function

FIG. 8.4 Basic second-order system—from leg model.

(H) STANDARD FORM OF SECOND–ORDER SYSTEM; UNDAMPED NATURAL FREQUENCY AND DAMPING RATIO

If $b = 0$, it is clear that the previously defined parameter, the undamped natural frequency, equals

$$\omega_n{}^2 = \frac{k}{J} \qquad (8.16)$$

A second parameter is then conventionally defined to integrate the three original parameters J, b, and k for the general case when $b \neq 0$; this is called the *damping ratio* (or *factor*)

$$\zeta = \frac{b}{2\sqrt{Jk}} \qquad (8.17)$$

so that the standard form of Eq. (8.15) is

$$k_g T_m(t) = \frac{\ddot{\theta}(t)}{\omega_n{}^2} + 2\zeta \frac{\dot{\theta}(t)}{\omega_n} + \theta(t) \qquad (8.18)$$

where $k_g \triangleq 1/k$ and is defined as the steady-state gain of the system. It is the third parameter of this basic second-order system.

It is convenient to convert to transfer-function form before proceeding with the analysis of second-order response characteristics (Fig. 8.4b), and with the usual statement about zero initial conditions this can be written by inspection as

$$\frac{\theta}{T_m}(s) = \frac{k_g}{(s/\omega_n)^2 + 2\zeta s/\omega_n + 1} \qquad (8.19)$$

8.3 The Semicircular Canal

Higher animals have evolved sensitive transducers of translational and rotational skull motion in order to achieve both satisfactory stabilization of the eye and general postural control. Suitable dynamic analysis has been made in recent years of the semicircular canals, which are responsible for transducing rotational motion ([8.2] to [8.5], the present development following [8.4] by Jones and Milsum]).

The semicircular canals are small-bore circuits about 0.28 mm diameter in man, containing fluid which rotates slightly relative to the skull when the latter rotates in space (Fig. 8.5). This small relative motion deflects a watertight flap, the cupula, in a special compartment within the circuit, the ampulla. The cupula itself is a hair-cell transducer whose deflection produces appropriate neural signals. The small smooth bore of the canal ensures laminar flow of the contained fluid, the endolymph, so that the resulting viscous flow resistance is linearly dependent on velocity. Furthermore, the cupula acts as a weak spring tending to restore itself to zero deflection in the steady state. Therefore, the input-output relation between skull movement and cupula deflection results from the interaction of fluid inertia, viscous friction, and spring. Each inner ear has a set of three semicircular canals in approximately mutually perpendicular planes, so that all three rotational degrees of freedom can be sensed. Each set of canals has a common compartment called

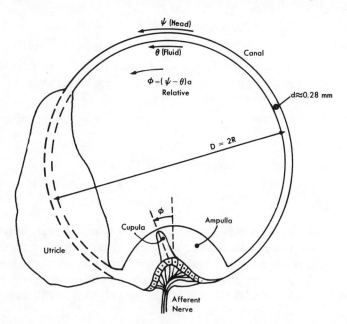

FIG. 8.5 The semicircular canal (diagrammatic). *(From Jones and Milsum.)*

the utricle, but it does not seem that the resulting hydrodynamic coupling is important. The following analysis is therefore based on angular motion in a single plane.

Consider the skull with absolute spatial position ψ, and the canal fluid with corresponding absolute spatial coordinate θ. When the skull accelerates, the fluid initially remains stationary so that a relative velocity builds up which produces a viscous frictional torque to accelerate the fluid in space. In turn, the spring acts to null any cupula deflection resulting and hence also provides an accelerating torque, and this effect is presumed for present purposes to be linear. Therefore Newton's second law states that

$$b(\dot{\psi} - \dot{\theta}) + k(\psi - \theta) = J\ddot{\theta} \tag{8.20}$$

where J = moment of inertia of the fluid in the thin tube ($= mR^2$)

b = torque per unit relative angular velocity

k = torque per unit relative angular displacement

However, although Eq. (8.20) correctly provides the absolute space relations, we are concerned with the relative angular displacement which produces the neural output. Therefore, we define ϕ as cupula deflection per unit relative angular displacement between skull and fluid, and $\phi = a(\psi - \theta)$, in radians, where a is merely a constant defining the area ratio between canal and ampulla. In consequence Eq. (8.20) becomes

$$J\ddot{\phi} + b\dot{\phi} + k\phi = (aJ)\ddot{\psi} \tag{8.21}$$

This is the linear second-order differential equation for the hydrodynamics and neural spring of the semicircular canal. The left-hand side is obviously identical with that for the free-swinging-leg model [Eq. (8.15)], and the right-hand sides are in fact also equivalent since both have the dimensions of torque. Furthermore, the systems are identical physically in combining PE- and KE-storage elements with viscous dissipation. Functionally, however, the likenesses end here because the characteristic parameters have quite different values. In particular, the canals have evolved with such small diameters that the viscous effect dissipates all energy from a step transient input before an oscillatory interchange can occur between the two energy-storage elements. The significance of this is explored in some detail in Sec. 8.5, and we merely note in passing now that the system is termed highly *overdamped*. Referring to Fig. 8.3, the corresponding initial-condition response is a steady decay back to zero, as shown chain-dotted.

The transfer function corresponding to Eq. (8.21) is

$$\frac{\phi}{\psi}(s) = \frac{k_c}{(J/k)s^2 + (b/k)s + 1} \tag{8.22}$$

where $k_c = aJ/k \triangleq$ steady-state gain. It should be noted that a rather mixed set of units is being used in that the dot notation for differentiation still appears in the transfer function to indicate an angular-acceleration variable.

Among other things this transfer function shows that a constant cupula *deflection* must eventually result from a sustained step input of skull angular *acceleration* of magnitude $\ddot{\psi}_1$:

$$\phi_\infty = k_c \ddot{\psi}_1 \qquad (8.23)$$

However, a steady angular acceleration of skull cannot in practice be maintained for long since the angular velocity would increase toward infinity, and certainly such inputs never occur naturally. Accordingly we redefine the transfer function in terms of angular velocity of the skull $\Omega \triangleq \dot{\psi}$ as input:

$$\frac{\phi}{\Omega}(s) = \frac{k_c s}{(J/k)s^2 + (b/k)s + 1} \qquad (8.24)$$

It is this transfer function whose frequency response is shown in Fig. 7.2, and which illustrates rather clearly the intended nature of the semicircular canal as an angular velocity transducer. For present analytical purposes, however, we work with the first version [Eq. (8.22)].

Although Eq. (8.22) is of the same form as Eq. (8.19), the parameter ω_n is not very meaningful because of the highly overdamped response characteristics of the semicircular canal, due to its hydrodynamic "design." In the overdamped case, however, a second-order system can be considered made up of two cascaded first-order lags, that is,

$$\frac{\phi}{\ddot{\psi}}(s) = \frac{k_c}{(T_1 s + 1)(T_2 s + 1)} \qquad (8.25)$$

as we now develop heuristically. Comparison of Eqs. (8.25) and (8.22) yields

$$T_1 T_2 = \frac{J}{k} \qquad T_1 + T_2 = \frac{b}{k} \qquad (8.26)$$

In this particularly highly overdamped case, it turns out that $T_1 \gg T_2$, so that

$$T_1 = \frac{b}{k} \qquad T_2 = \frac{J}{b} \qquad (8.27)$$

The values for man (part experimental, part calculated) as shown in Fig. 7.2 are approximately

$$T_1 = 10 \text{ sec} \qquad T_2 = \frac{1}{200} \text{ sec}$$

which confirms the validity of the approximation in Eqs. (8.27).

This approximation may also be justified by block-diagram manipulation, as follows. The inner viscoinertial loop (see, for example, Fig. 8.4a) may immediately be telescoped by using the rule of Eq. (5.28), to yield the block diagram of Fig. 8.6a, in which the

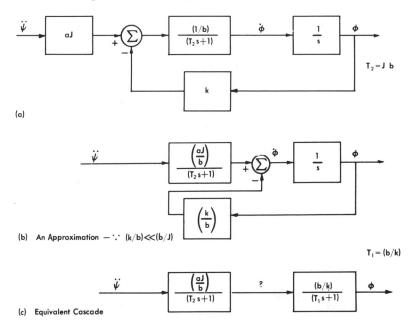

FIG. 8.6   *Manipulations of the semicircular canal's block diagram.*

first-order lag is that of the fast time constant $T_2 = J/b$. The feedback is determined by the cupula spring, and the static loop gain k/b ($= 1/T_1$) is very low. Hence, the resulting dynamics of this path are sluggish, and passing the signal $k\theta$ through the first-order lag $(1/b)/(T_2 s + 1)$ is essentially equivalent to multiplying it only by a pure gain $1/b$. Thus the outer loop may just as well be closed around the second integrator (Fig. 8.6b), including, however, this gain $1/b$. The resulting subsystem may now be collapsed to a second first-order lag in cascade with the first. In consequence, the overall system (Fig. 8.6c) is indeed the block diagram of Eq. (8.25), containing the overall gain k_c in two parts. Note, however, that the variable connecting the two blocks now has no direct physical significance.

    Although this technique is useful, it should be realized that the exact values of T_1 and T_2 can be computed easily enough, and that the same type of cascade results. Nonetheless, their expression as simple ratios of the basic parameters J, b, and k should simplify understanding. The two basic effects of this overdamped condition can be demonstrated usefully on a qualitative basis, although they will be established more rigorously later. Consider first the short-duration-type input in which the cupula has no time to deflect enough to produce significant spring forces. Equation (8.21) can then be simplified by omitting the $k\phi$ term and integrating, to yield

$$\frac{J}{b}\dot{\phi} + \phi = \frac{aJ}{b}\psi \tag{8.28}$$

which represents the familiar first-order system

$$\frac{\phi}{\psi} = \frac{k_2}{T_2 s + 1} \tag{8.29}$$

On Fig. 8.6c, "integrating once—for short-duration inputs" corresponds to the fact that for periods of less than about 2 sec, the second block acts as a pure integrator, as is shown by the step response for the first-order system (Fig. 6.3), for intervals short with respect to the time constant. Thus the canal acts as a fast transducer whose output *deflection* ϕ is proportional to the angular *velocity* of the skull $\dot{\psi}$, because the latter is the integral of $\ddot{\psi}$. This fast time constant T_2 in fact produces the upper break frequency in the semicircular-canal frequency response of Fig. 7.2.

The spring force is presumably necessary to ensure ultimate rezeroing of the system from whatever overloads it may suffer, in which case the flap may leak, and to limit long-duration deflections. As a result, however, there is another regime of operation when the angular acceleration of the head is zero, but when a steady angular velocity is present. In this circumstance the fluid's inertial torque becomes zero and angular-acceleration terms disappear in Eq. (8.21) to yield

$$T_1 \dot{\phi} + \phi = 0 \tag{8.30}$$

This is the unforced first-order system, and thus under sustained constant angular velocity the transducer fails because it exponentially "forgets" angular velocity in setting itself back to zero flap deflection. On Fig. 8.6c the effect is that if the input acceleration $\ddot{\psi}$ falls to and remains at zero, there rapidly ceases to be any output from the first, fast block; as a result the output of the second block decays on the slow time-constant exponential from whatever initial angular deflection it had when the acceleration ceased.

8.4 Transfer Functions and Impulse Responses of the Basic Second-order System

In Chap. 6 the important relation was established that the transfer function $H(s)$ has as its equivalent time-domain characteristic the impulse response $h(t)$. This section presents both characteristics for the second-order system:

$$H(s) = \frac{k}{(s/\omega_n)^2 + 2\zeta(s/\omega_n) + 1} \tag{8.31}$$

However, it was argued that for the semicircular-canal model with its large hydrodynamic damping a more appropriate, although equivalent, form of transfer function is

$$H(s) = \frac{k}{\left(\dfrac{T}{\alpha} s + 1\right)(\alpha T s + 1)} \tag{8.32}$$

where α^2 is the ratio of the two time constants. Equation (8.31) is the most widely used form, but the equivalence of Eq. (8.32) is easily established through the relations

$$\omega_n = \frac{1}{T} \qquad \zeta = \frac{1 + \alpha^2}{2\alpha} \tag{8.33}$$

There are four different domains of behavior, defined uniquely by ranges of the *damping-ratio* parameter ζ as presented in Table 8.1, the characteristics being explored more fully in Sec. 8.7. They are generally similar to the initial-condition responses already presented for different damping ratios of Fig. 8.3 except that they start differently (Figs. 8.8 and 8.9).

The classification of Table 8.1 enables one to proceed with frequency and transient response analyses, starting from such J, b, k equations as (8.15) and (8.22). However, it is probably better that most readers pause to learn some of the mathematical ideas behind this simplified tabulation. In particular, the ideas of *roots* in transfer functions and the plotting of such roots on the complex plane are very important conceptually. These are presented in the next two sections.

TABLE 8.1 BASIC SECOND-ORDER TRANSFER FUNCTIONS AND IMPULSE RESPONSES

Damping Ratio Range	*Transfer Function, $H(s)$*		*Impulse Response $h(t)$*
	The "Basic" $\omega_n - \zeta$ Form	*Equivalent Working Form*	
$\zeta = 0$		$\dfrac{1}{\left(\dfrac{s}{\omega_n}\right)^2 + 1}$	$\omega_n \sin(\omega_n t)$
$0 < \zeta < 1$	$\dfrac{1}{\left(\dfrac{s}{\omega_n}\right)^2 + 2\zeta\dfrac{s}{\omega_n} + 1}$	$\dfrac{1}{\left(\dfrac{s}{\omega_n}\right)^2 + 2\zeta\dfrac{s}{\omega_n} + 1}$	$\dfrac{\omega_n}{\sqrt{1 - \zeta^2}} e^{-\zeta\omega_n t}$ $\times \sin(\omega_n \sqrt{1 - \zeta^2}\, t)$
$\zeta = 1$		$\dfrac{1}{(Ts + 1)^2}$	$\left(\dfrac{t}{T^2}\right) e^{-t/T}$
$\zeta > 1$		$\dfrac{1}{(\alpha T s + 1)[(T/\alpha)s + 1]}$	$\dfrac{1}{T}\dfrac{\alpha}{\alpha^2 - 1}(e^{-t/\alpha T} - e^{-\alpha t/T})$

▼

8.5   Roots of the Numerator and Denominator Polynomials

(A) POLYNOMIAL TRANSFER FUNCTIONS

It has previously been mentioned that linear lumped models give rise to transfer functions having polynomials in the numerator and denominator, that is, of the form

$$H(s) = \frac{a_m s^m + \cdots + a_1 s + a_0}{b_n s^n + \cdots + b_1 s + b_0} \tag{8.34}$$

where $m < n$ for realizable physical systems.

Since s may be treated as an algebraic variable, both numerator and denominator are polynomials of s, and as such must have m and n roots just as do elementary algebraic equations. The roots of the polynomial are those values of s which make the polynomial (or certain of its derivatives) equal zero.

(B) POLES OF THE SYSTEM; CHARACTERISTIC FUNCTION

The roots of the denominator alone characterize the exponential and/or sinusoidal terms of the impulse response, as shown in the method of partial fractions (Sec. 6.8C). For this reason the denominator polynomial has special significance and is called the *characteristic function* of the system, and its roots are called the *poles* of the system.

(C) ZEROS

The roots of the numerator polynomial are concerned with the coefficients and phase of the response but not with its basic nature. They are called *zeros* of the system. Numerator polynomials arise in the system either inherently or because of imposed initial conditions (Sec. 6.7, for example), but must be of lower order than the denominator in realizable systems. Thus, for example, the idealized first-order lead-lag system is nonrealizable at sufficiently high frequencies (Sec. 7.9).

(D) FIRST—ORDER SYSTEM

Let us particularize the above ideas with those examples already treated. The basic first-order system

$$H(s) = \frac{k}{Ts + 1}$$

has only a first-order denominator, and this equals zero if

$$s = \frac{-1}{T} \tag{8.35}$$

which is then the value of the first-order pole. It is the negative inverse of the time constant that characterizes the impulse response, and in frequency response is the negative of the break frequency $\omega_b = 1/T$.

The lead-lag first-order system

$$H(s) = k \frac{\alpha Ts + 1}{Ts + 1}$$

adds a numerator root, or zero, at

$$s = \frac{-1}{\alpha T} \tag{8.36}$$

This zero does not change the actual time constant of the exponential time response but does affect the other constants (Fig. 7.25). Its effect on the frequency response is more explicit, however, because a new break frequency is added, at $\omega = 1/\alpha T$, which is a breakup rather than a breakdown.

(E) SECOND–ORDER SYSTEM

For the overdamped second-order case (Table 8.1), there are clearly two poles at

$$s = \frac{-1}{\alpha T} \quad \text{and} \quad \frac{-\alpha}{T} \tag{8.37}$$

Thus the impulse response has two corresponding exponential decay terms and the frequency response has two break(down) frequencies. When the underdamped second-order system is considered, however, one cannot find real s values which set the denominator equal to zero. This problem is, in fact, identically the classic algebraic one of finding the two roots x_1, x_2 of the quadratic function

$$f(x) = x^2 + \frac{b}{a} x + \frac{c}{a} \equiv (x - x_1)(x - x_2) \tag{8.38}$$

which is solved by setting $f(x) = 0$ and obtaining

$$x_{1,2} = \frac{-b \pm \sqrt{b^2 - 4ac}}{2a} \tag{8.39}$$

(F) DISCRIMINANT; COMPLEX ROOTS

The expression under the square root is called the *discriminant*. It has special significance because its square root is imaginary when $b^2 < 4ac$, although there is always a real part to the root $-b/2a$

as long as $b \neq 0$. Thus in general the roots of x are describable by complex numbers $x = u + jv$, as defined for the frequency response work [Eq. (7.12)]. By comparison between Eq. (8.38) and the characteristic function of the second-order system (Table 8.1),

$$f(s) = s^2 + 2\zeta\omega_n s + \omega_n^2 \qquad (8.40)$$

the two poles are clearly defined by

$$s_{1,2} = -\zeta\omega_n \pm \omega_n \sqrt{\zeta^2 - 1} \qquad (8.41)$$

However, the ζ and ω_n parameters are most significant when $\zeta < 1$ and the discriminant is negative, so that it is preferable to rewrite Eq. (8.41) in a way which directly illustrates the real and imaginary components:

$$s_{1,2} = -\zeta\omega_n \pm j\omega_n \sqrt{1 - \zeta^2} \triangleq \mathrm{Re} \pm j\,\mathrm{Im} \qquad (8.42)$$

This important relation demonstrates that only ζ determines whether the discriminant is positive or negative. In particular the roots are:

1. Negative real when $\zeta > 1$.
2. Repeated real when $\zeta = 1$.
3. Complex when $0 < \zeta < 1$.
4. Purely imaginary when $\zeta = 0$.

For J, b, k-parametered models [Eqs. (8.15) and (8.22)], the roots are

$$s_{1,2} = -\frac{b}{2J} \pm \sqrt{\left(\frac{b}{2J}\right)^2 - \frac{k}{J}} \qquad (8.43)$$

To summarize this section, the potentially oscillatory behavior of the second-order system is characterized mathematically by whether or not the poles are *complex*. First-order poles can only be real, so that the minimum order of denominator polynomial for complex poles to exist is two. However, the poles of high-order systems can all be real, depending only on the numerical values of the systems' parameters.

▼

8.6   Pole-Zero Plots on the Complex Plane

The previous section has established that lumped-parameter transfer functions can be equivalently expressed as a set of poles and zeros. Hence Eq. (8.34) can be written

$$H(s) = \frac{(s + z_1)(s + z_2) \cdots (s + z_m)}{(s + p_1)(s + p_2) \cdots (s + p_n)} \qquad (8.44)$$

where $-p_n$ represents the nth pole and $-z_m$ represents the mth zero. It is of conceptual value to plot the poles and zeros of a system upon the complex plane because the characteristic patterns of dynamic response in different regions are readily remembered. An additional consideration, of greater ultimate importance, is that this plotting provides the basis of the powerful *root-locus* technique by which the movement of system poles and zeros can be followed as parameters are modified within a closed loop (Sec. 11.7).

It should be noted that the Laplace-transform variable s is now exhibiting the properties of a complex variable. Originally its nature was not carefully defined, but in Chap. 7 it was shown that the frequency response is defined directly from the transfer function if s is replaced by the purely *imaginary* form $j\omega$. In transform theory, s is in fact a complex frequency variable defined by

$$s = \sigma + j\omega \qquad (8.45)$$

where σ and ω are respectively the real and imaginary parts.

The pole-zero configuration of a system may therefore be plotted on the complex s plane (Fig. 8.7). In Fig. 8.7a the pole and zero are shown for the basic and lead-lag first-order systems [Eqs. (8.35) and (8.36), respectively]. The impulse response of the basic system is sketched above the pole and is of course an exponential decay. The impulse response of the lead-lag system involves an impulse function and is not shown. This pole has been plotted in the left half-plane because its value is negative, but we now recall that the expanding population model [Eq. (2.8)] has a positive-valued pole at $+k$; its unstable impulse response therefore correlates with the pole being in the right half-plane. In fact, it is a general result that the whole right half-plane represents instability since as long as any one pole has any real part at all positive, the response grows exponentially.

The four pole configurations for the basic second-order system as functions of ζ are shown in Fig. 8.7b, together with sketched impulse response. The right-angle-triangle relation between the real and imaginary parts and the undamped natural frequency ω_n should be noted. In particular the *damped natural frequency* of the system ω_d is shown shortly to be specified by the imaginary part:

$$\omega_d \equiv \omega_n \sqrt{1 - \zeta^2} \qquad (8.46)$$

Similarly, the real part $-\zeta\omega_n$ corresponds to the inverse of the time constant of exponential decay for the impulse response's envelope. A radial line from the origin is called a *constant damping line* since the angle from the negative real axis is given by

$$\zeta = \cos\theta \qquad (8.47)$$

It should also be noted that a circle about the origin is at constant ω_n. For the overdamped case, the time constants $-\alpha T$ and $-T/\alpha$

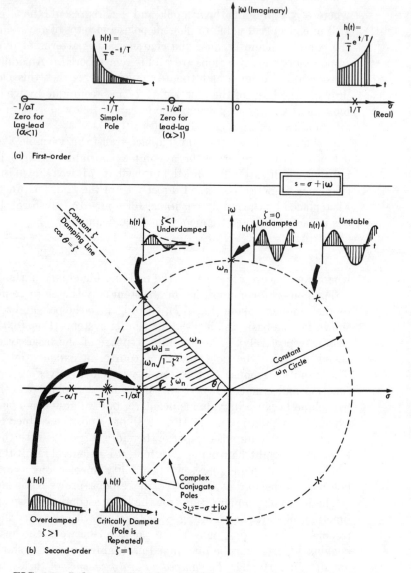

FIG. 8.7   Pole-zero plots—first and basic second-order systems.

are arranged either side of $-1/T$ (which also equals $-\zeta\omega_n$ for the $\zeta = 1$ condition).

The right half-plane is still a region of instability as sketched at one complex pole, and the imaginary axis is clearly the dividing line between stable and unstable systems; that is, it is a region of zero damping. If it helps him, the reader can validly look upon the sinusoidal inputs of the frequency response method as the impulse responses of a set of oscillators with their poles on the imaginary axis at various desired frequencies.

8.7   Impulse Response of Basic Second-order System

In Table 8.1 the Laplace-transform pairs have been given for the various versions of the basic second-order system, so that the impulse response formulas are already stated. In this section we examine these time-domain impulse responses in some detail, and in particular derive their analytical nature. The defining differential equation for the leg model is [Eq. (8.15)]

$$T_m(t) = J\ddot{\theta}(t) + b\dot{\theta}(t) + k\theta(t) \tag{8.48}$$

▼

(A) DERIVATION OF RESPONSE

Recall first that the impulse response is essentially equivalent to some initial-condition problem because the impulse input is completed before the system responds. Furthermore, the steady-state condition must be the same as the condition before the impulse is applied, at least for systems which do not store energy indefinitely. Finally, the solution must contain two exponential terms because the system is second order (Sec. 6.8). Therefore, we presume an impulse response of the form

$$\theta(t) = Be^{\beta t} + Ce^{\gamma t} \qquad t > 0 \tag{8.49}$$

where β, γ, A, and B are constants to be determined. We note immediately that since $\theta(\infty) = 0$, both β and γ must turn out to be negative.

The equivalent initial conditions for the unit impulse torque input can be determined by integrating Eq. (8.48) on both sides between $t = 0$ and $t = 0+$ to yield

$$1 = J\dot{\theta}(0+) + 0 + 0 \tag{8.50}$$

where $\theta(0)$, $\dot{\theta}(0)$, and $\ddot{\theta}(0)$ have all been assumed to be zero, although the method can cope equally well with nonzero values if necessary. Equation (8.50) states that immediately after the torque impulse the leg still has no angular deflection, but does start off with an "initial" velocity equal to $1/J$.

$$\theta(0+) = \theta(0) = 0 \qquad \dot{\theta}(0+) = \frac{1}{J} \tag{8.51}$$

The first value is placed in Eq. (8.49) and yields $B = -C$, while the second value is placed in the derivative of Eq. (8.49) and yields, after some manipulation, $B = 1/J(\beta - \gamma) = -C$. Finally, the $\theta(t)$ of Eq. (8.49), incorporating these values and its first two derivatives, are placed in Eq. (8.48),

$$JBe^{\beta t}\left(\beta^2 + \frac{b}{J}\beta + \frac{k}{J}\right) + JCe^{\gamma t}\left(\gamma^2 + \frac{b}{J}\gamma + \frac{k}{J}\right) = u_0(t) \qquad t > 0 \tag{8.52}$$

This last equation may look forbidding, but an important deduction can be made by direct physical reasoning. Provided both B and C are not zero, in which case the solution of Eq. (8.49) is trivial, and equally provided $\beta \neq \gamma$, the expressions in parentheses must both equal zero for $t > 0$ since the right-hand side, the impulse function $u_0(t)$ [$= T_m(t)$], is zero for all $t \neq 0$. Actually, β and γ are the two roots of a single quadratic expression specified by Eq. (8.43). Consequently, the four constants B, C, β, and γ of Eq. (8.49) have now all been evaluated in terms of the system parameters J, b, and k.

We have worked in the J, b, k domain to establish initial velocities, etc., but it is perhaps better to return now to the generalized parameters ζ and ω_n through Eqs. (8.42) and (8.43). Thus the impulse response [Eq. (8.49)] may be written down in the following very general form; the parameters of Table 8.1 are obtainable from it (except for $\zeta = 1$) after simple manipulations. The $1/k$ constant is particular to the leg system, however.

$$h(t) = \theta(t) = \frac{1}{k} \frac{\omega_n}{2j \sqrt{1 - \zeta^2}} [\exp{(-\zeta\omega_n + j\omega_n \sqrt{1 - \zeta^2})t}$$
$$- \exp{(-\zeta\omega_n - j\omega_n \sqrt{1 - \zeta^2})t}] \qquad t > 0 \qquad (8.53)$$

(B) IMPULSE RESPONSE PLOTS (SEE FORMULAS OF TABLE 8.1)

Underdamped. $\zeta = 0$ is obviously the undamped limit of the underdamped case, just before instability occurs. The undamped and underdamped cases are therefore plotted together on Fig. 8.8. Since the sine term (Table 8.1) can never exceed unity magnitude,

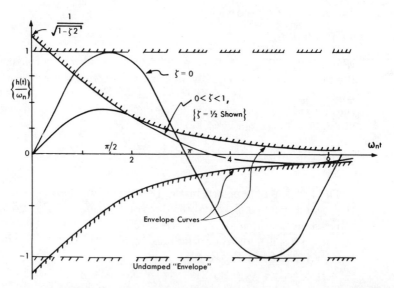

FIG. 8.8 *Underdamped impulse responses—basic second-order system.*

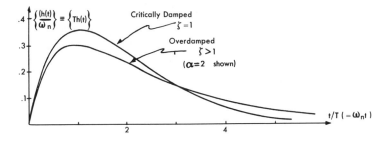

FIG. 8.9   Overdamped impulse responses—basic second-order system.

although its sign can alternate, the exponential decay term (with both positive and negative signs) provides the *envelope* within which the response decays.   Since ζ and ω_n are independent parameters, there are several meaningful ways to present the general trends, but here we choose a dimensionless time $\omega_n t$ which physically corresponds to maintaining constant the k/J ratio of the physical system, and furthermore means that the underdamped poles can only move around the constant $= \omega_n$ circle of Fig. 8.7b; similarly, a natural dimensionless ordinate is $h(t)/\omega_n$.

Critically Damped and Overdamped.   The $\zeta = 1$ case divides the overshooting impulse responses (underdamped) just discussed from those without overshoot (overdamped).   In the $\zeta = 1$ and $\zeta > 1$ cases sketched on Fig. 8.9, the same dimensionless time scale is possible because of the convenient way in which T was earlier defined. As the degree of overdamping is increased with constant k/J ratio, the maximum value of $h(t)$ decreases, but the "tail" lengthens in such a fashion that the area under each unit-gain impulse response is

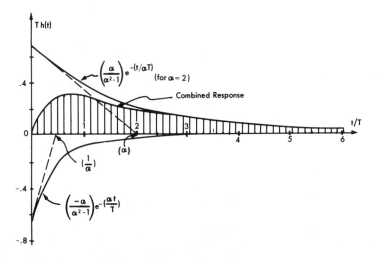

FIG. 8.10   Composition of overdamped impulse response.

constant (and equal to unity). For exemplary purposes, the over-damped case has been plotted with an α value of 2 so that the time constants are $T/2$ and $2T$. From Eqs. (8.33) the corresponding damping ratio is $\zeta = 1.25$, and in this case Table 8.1 yields

$$Th(t) = \tfrac{2}{3}[e^{-\frac{1}{2}(t/T)} - e^{-2(t/T)}] \qquad t > 0 \tag{8.54}$$

It is of conceptual importance that the resulting curve, shown in Fig. 8.9, is actually the difference between two decaying exponentials as shown in Fig. 8.10. This figure clearly shows that the tail of the impulse response consists essentially of the slow time constant's decay alone. This fact provides the basis for the graphical semi-logarithmic plotting method by which exponential terms are extracted from a given response (Sec. 10.4).

8.8 Step Response of the Basic Second-order System

Since it is not our purpose to present more mathematical manipulation than absolutely necessary, the unit step responses for this basic second-order system are tabulated directly in Table 8.2, thus providing some more useful transform pairs, and plotted on Fig. 8.11. The analytical forms can of course be found in the usual several

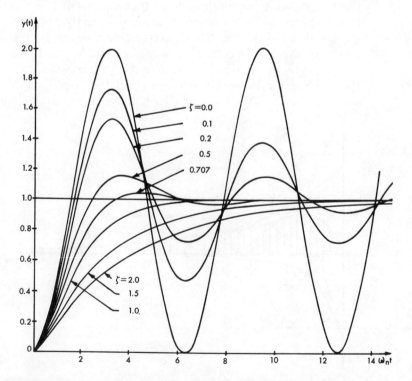

FIG. 8.11   Unit step responses of basic second-order system.

TABLE 8.2 STEP RESPONSES OF BASIC SECOND-ORDER SYSTEM

Damping Ratio Range	Working Form of Transfer Function $H(s)$	Transform Pairs	
		Transform of Step Response	Step Response
$\zeta = 0$	$\dfrac{1}{(s/\omega_n)^2 + 1}$	$\dfrac{1}{s[(s/\omega_n)^2 + 1]}$	$1 - \cos \omega_n t$
$0 < \zeta < 1$	$\dfrac{1}{(s/\omega_n)^2 + 2\zeta(s/\omega_n) + 1}$	$\dfrac{1}{s[(s/\omega_n)^2 + 2\zeta(s/\omega_n) + 1]}$	$1 - \left(\dfrac{e^{-\zeta\omega_n t}}{\sqrt{1 - \zeta^2}}\right)$ $\times \sin(\omega_n \sqrt{1 - \zeta^2}\, t + \theta)$ where $\theta = \arctan \dfrac{\sqrt{1 - \zeta^2}}{\zeta}$
$\zeta = 1$	$\dfrac{1}{(Ts + 1)^2}$	$\dfrac{1}{s(Ts + 1)^2}$	$1 - \left(1 + \dfrac{t}{T}\right)e^{-t/T}$
$\zeta > 1$ $\left(\zeta = \dfrac{1 + \alpha^2}{2\alpha}\right)$	$\dfrac{1}{(\alpha Ts + 1)[(T/\alpha)s + 1]}$	$\dfrac{1}{s(\alpha Ts + 1)[(T/\alpha)s + 1]}$	$1 + \dfrac{1}{\alpha^2 - 1}$ $(e^{-\alpha t/T} - \alpha^2 e^{-t/\alpha T})$

ways, and also by direct integration of the impulse formulas already given (Table 8.1).

▼

However, for the interested reader, the unit step response is now found by the partial-fraction technique. For the semicircular-canal example, the transfer function is given by Eq. (8.32):

$$H(s) = \frac{\phi}{\ddot{\psi}}(s) = \frac{k_c}{\left(\dfrac{T}{\alpha} s + 1\right)(\alpha Ts + 1)}$$

For unit step input in angular acceleration of skull, the cupula deflection is given by

$$\phi(s) = \frac{k_c}{s\left(\dfrac{T}{\alpha} s + 1\right)(\alpha Ts + 1)}$$

By the partial-fraction technique this equation is rewritten in a form displaying additive first-order terms in s which are easily inverse-transformable into the time domain. Therefore,

$$\phi(s) = \frac{k_c}{s\left(\dfrac{T}{\alpha} s + 1\right)(\alpha Ts + 1)} \equiv \frac{A}{s} + \frac{B}{\dfrac{T}{\alpha} s + 1} + \frac{C}{\alpha Ts + 1} \qquad (8.55)$$

A, B, and C are as yet undetermined constants, and several methods of determining them are available. Section 6.8 outlines a basic method for which the so-called *cover-up rule* provides a convenient shorthand technique. Consider, for example, the evaluation of B; mentally multiply both sides of Eq. (8.55) by its denominator term $(T/\alpha)s + 1$, and then replace all s terms by the corresponding pole value $s = -\alpha/T$. (The uninitiated reader should actually write out these steps.) As a result, the expressions containing A and C disappear and hence the remaining terms evaluate B:

$$B = \left[\frac{k_c}{s(\alpha T s + 1)} \right]_{s=-\alpha/T} = \frac{k_c T}{\alpha(\alpha^2 - 1)}$$

Similarly

$$A = k_c \quad \text{and} \quad C = - \frac{k_c T \alpha^3}{\alpha^2 - 1}$$

Placing these values in Eq. (8.55) yields

$$\phi(s) = k_c \left\{ \frac{1}{s} + \frac{T}{\alpha^2 - 1} \left[\frac{1/\alpha}{(T/\alpha)s + 1} - \frac{\alpha^3}{\alpha T s + 1} \right] \right\}$$

This equation can be inverse-transformed separately for each of the first-order expressions in s:

$$\phi(t) = k_c \left[1 + \frac{1}{\alpha^2 - 1} \left(e^{-\alpha t/T} - \alpha^2 e^{-t/\alpha T} \right) \right] \tag{8.56}$$

which checks the tabulations of Tables 8.2 and 6.1 except for the constant k_c particular to the canal system.

8.9 Frequency Response of the Basic Second-order System

All the basic results of Chap. 7 apply to the frequency response of the second-order system. On the rotating-vector diagram, however (Fig. 8.12), a new feature is the vector representing $\ddot{y}(t)$ at 180°

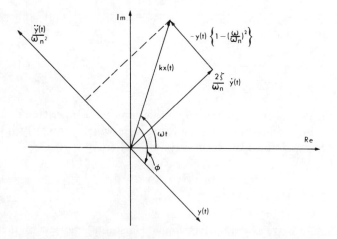

FIG. 8.12 Rotating vectors for basic second-order system.

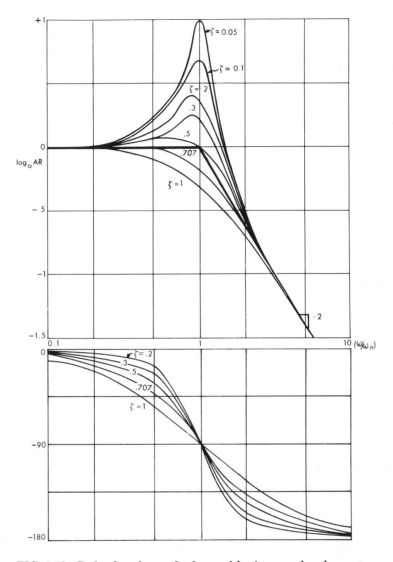

FIG. 8.13   Bode plots for underdamped basic second-order system.

phase advance to $y(t)$.   This same feature appears in the transfer function when s is replaced by $j\omega$.   Thus

$$H(j\omega) = \frac{k}{1 - (\omega/\omega_n)^2 + j(2\zeta\omega/\omega_n)} \qquad (8.57)$$

Notice that when the input frequency equals the undamped natural frequency $\omega/\omega_n = 1$, the real component disappears, hence the output lags the input by exactly $90°$, as shown in the Bode phase plots (Fig. 8.13).

(A) BODE PLOTS—$\zeta < 1$

High- and low-frequency asymptotes exist as before, but the slope of the high-frequency asymptote is now -2 logs/decade.

Low-frequency asymptote $\log |H(j\omega)| = \log k$

High-frequency asymptote $\log |H(j\omega)| = \log k - 2 \log \dfrac{\omega}{\omega_n}$

These asymptotes clearly intersect at $\omega/\omega_n = 1$. The new feature of the Bode gain plot is that the true curve does not necessarily fall below the low- and high-frequency asymptotes. In particular the AR $\to \infty$ as $\zeta \to 0$ at frequencies close to ω/ω_n of unity. This of course is the resonance phenomenon already mentioned. The exact value of the AR is calculated from Eq. (8.57) in the usual way [Eq. (7.32)]:

$$H(j\omega)| = \frac{k}{\{[1 - (\omega/\omega_n)^2]^2 + (2\zeta\omega/\omega_n)^2\}^{1/2}} \tag{8.58}$$

The phase varies between 0 at zero frequency and $-180°$ as $\omega \to \infty$, and is defined by [Eq. (7.33)]

$$\phi = - \arctan \frac{2\zeta\omega/\omega_n}{1 - (\omega/\omega_n)^2} \tag{8.59}$$

The following standard results for underdamped systems are worth placing together, but it is left to the reader to derive the expressions analytically:

1. For all values of damping ratio less than 0.707 the AR exceeds unity in some frequency range for $k = 1$. In particular the maximum AR, or *resonance*, is reached at the so-called *resonant frequency* ω_R. For $\zeta > 0.707$ there is no resonance.

2. The resonant frequency ω_R is less than the undamped natural frequency ω_n and also less than the damped natural frequency ω_d:

$$\omega_R = \omega_n \sqrt{1 - 2\zeta^2} \tag{8.60}$$
$$\omega_d = \omega_n \sqrt{1 - \zeta^2} \tag{8.61}$$

3. The maximum AR, at the resonant frequency ω_R, is as follows:

$$M_p = \text{AR}_{max} = \frac{1}{2\zeta \sqrt{1 - \zeta^2}} \tag{8.62}$$

4. The AR at the undamped natural frequency ω_n is

$$\text{AR}_{\omega_n} = \frac{1}{2\zeta} \tag{8.63}$$

where ϕ always equals $-90°$.

(B) BODE PLOTS—$\zeta \geq 1$

In the overdamped case the separated-time-constant representation is helpful. The Bode plots are found by superposition of the two first-order plots as already described in Sec. 7.7.

$$H(j\omega) = \frac{k}{\left(\dfrac{T}{\alpha} j\omega + 1\right)(\alpha T j\omega + 1)}$$

Hence

$$\log |H(j\omega)| = \log k - \tfrac{1}{2} \log\left[1 + \left(\frac{\omega T}{\alpha}\right)^2\right] - \tfrac{1}{2} \log [1 + (\alpha\omega T)^2] \tag{8.64}$$

and

$$\phi = -\arctan \frac{\omega T}{\alpha} - \arctan \alpha\omega T \tag{8.65}$$

In Fig. 8.14 exemplary plots are given for $\alpha = 1$ (repeated roots, $\zeta = 1$) and for $\alpha = 2$.

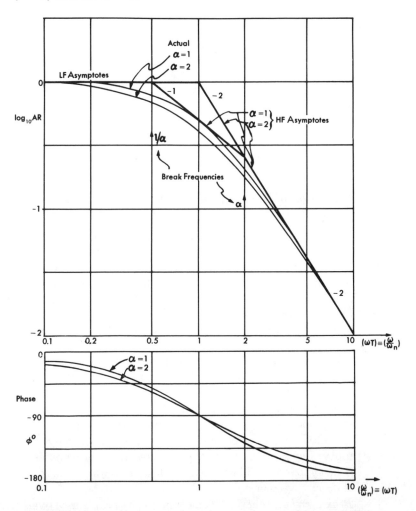

FIG. 8.14 Bode plots for overdamped basic second-order system.

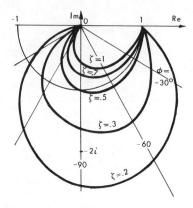

FIG. 8.15   Polar locus plot for basic second-order system.

The polar plots for various ζ are shown in Fig. 8.15.   Note that the $\zeta = 0.707$ condition provides the dividing line between those plots which lie within and outside the unit circle, which is consistent with the earlier statement on resonance.   Notice also that although the loci do enter the -90 to $-180°$ quadrant, there is no possibility of actual instability for $\zeta > 0$ since the -1 point cannot be "enclosed" by the trajectory (Sec. 7.7).

8.10   Correlation of Transient and Frequency Responses

Certain useful correlations are available between the transient and frequency responses of the basic second-order system.   These correlations are valuable because many complex systems also approximate to second order in their dynamic performance, at least within reasonable tolerance for some design purposes.   Figure 8.16 illustrates certain parameters of the log AR frequency and step transient response, in particular by using the following definitions:

Bandwidth ω_{bw}:   Already defined as the frequency at which the AR of the frequency response falls by $1/\sqrt{2}$ or -0.15 log compared with the low-frequency value.
Rise time T_R: The time to rise from 0.1 to 0.9 of the steady-state step response.   Rise time is also sometimes defined as the inverse of the maximum slope of the step response whose steady-state value is unity, but in many cases these values are approximately the same.
Settling time T_s: The time to settle within ± 5 percent of the steady-state step response.
Peak overshoot time T_p: The time to the peak (first) overshoot P in the step response.

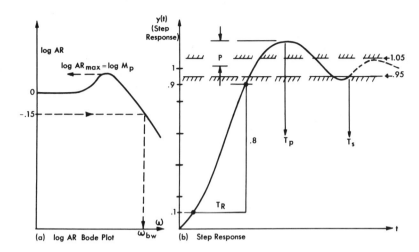

FIG. 8.16 *Correlation of transient and frequency response for basic second-order system.*

The following approximate relations then apply:

1. *Peak overshoot and AR*

For $P = 0.15 \qquad M_p \approx 1.3$ $\qquad\qquad\qquad\qquad$ (*8.66*)

2. *Overshoot—ζ and ω_n*

$$T_p = \frac{\pi}{\omega_n \sqrt{1 - \zeta^2}} \qquad P = \exp\left(-\frac{\pi\zeta}{\sqrt{1 - \zeta^2}}\right) \qquad (8.67)$$

3. *Rise time and bandwidth*

$$T_R \approx \frac{1}{\omega_{bw}} \qquad\qquad\qquad\qquad (8.68)$$

4. *Settling time and frequency*

$$\frac{3}{\omega_n} < T_s < \frac{5}{\omega_n} \qquad\qquad\qquad\qquad (8.69)$$

8.11 Other Second-order Systems

Previous sections have dealt with the basic second-order system

$$H(s) = \frac{1}{b_2 s^2 + b_1 s + b_0} \qquad\qquad\qquad (8.70)$$

to use yet another notation. Following the method of the lead-lag first-order system, one can define the canonical second-order system as

$$H(s) = \frac{a_2 s^2 + a_1 s + 1}{b_2 s^2 + b_1 s + b_0} \qquad\qquad\qquad (8.71)$$

Since there are five independent parameters in this transfer function, it is clear that the response characteristics cannot be simply and generally presented.   Furthermore, it has already been noted that such systems can be realized neither mechanically nor biologically, because the AR cannot in fact remain finite as ω tends to infinity.

One limited version of this system has already been presented in the sense that the meaningful input-output characteristics of the semicircular canal produce the transfer function

$$H(s) = \frac{a_1 s}{b_2 s^2 + b_1 s + 1} \qquad (8.72)$$

Although the resulting Bode plots are trivially altered in principle by this change from the form of Eq. (8.70) (addition of $+1$ slope to the log AR plot and of $+90°$ to the phase at all frequencies), the practical effect is clear and important in that the AR characteristic (Fig. 7.2) then points out that the canal must be intended as a velocity transducer.   Incidentally, this figure can now be completed with $+1$ and -1 low- and high-frequency asymptotes, respectively.

(A) TRANSFER–FUNCTION MODEL OF HUMAN IN MOTOR TRACKING

One rather more generalized second-order model has arisen in transfer-function analysis of the human operator.   The particular work of Ornstein [8.6] is directed toward automatic determination of the transfer function's coefficients under various test conditions; it is therefore important to the subject of adaptive control and is examined more fully in Sec. 15.4.   Consider a motor-coordination task in which an operator must try to track a spot on a visual-display screen by moving a second spot on the screen to match; the movement of the second spot is the result of initiating control through a lever which moves the spot, after passing through a dynamic black-box system. The human operator's performance can be studied as different types of dynamic systems are simulated, usually using an electronic computer.   One quasi-linear model which has been used fairly satisfactorily is

$$H(s) = \frac{a_1 s + a_0}{b_2 s^2 + b_1 s + 1}\, e^{-Ts} = \frac{y}{x}\,(s) \qquad (8.73)$$

The term e^{-Ts} represents the neural time delays within the motor-coordination loop, for which $T \approx 0.2$ sec; its dynamic characteristics are presented fully in Chap. 9.   The linear part of the transfer function is basically second order except for a derivative numerator term which represents the operator's need and ability to "anticipate" the input in order to achieve control which he judges to be "satisfactory"; it may be looked upon as providing a velocity transducer by visual tracking, just as the semicircular canal provides one hydro-dynamically.   For satisfactory tracking, proportional information a_0 is also necessary, however.   One interesting feature of the experiments referenced is the ability of the human operator to change the gain of his velocity channel a_1 in order to match changing process dynamics and methods of visual presentation.

9 High-order, Nonlinear, and Spatially Distributed Models

9.1 Introduction

Frequent emphasis has been placed upon the nonlinear and spatially distributed facts of life, and the modeling of a complex system, such as that of mammalian thermoregulation, illustrated some of these features (Chap. 3). On the other hand, the last four chapters have been concerned with obtaining two stylized dynamic response characteristics (transient and frequency) for first- and second-order linear lumped systems. It is hoped that these chapters have demonstrated that such abstract models are not trivial, and that understanding of their possible modes of behavior is an essential prerequisite to understanding more complicated systems. Unfortunately, the simple presentability of general response characteristics disappears as the order of linear lumped systems is increased, because each new parameter requires about a tenfold increase in the minimum number of curves which can characterize it graphically. Clearly, then, easily realizable systems have far too much complexity for useful presentation of characteristic solutions graphically, and we do not attempt to duplicate here the results of Chap. 8 for third-, fourth-, and higher-order systems. However, the analytical tools so far developed allow some pertinent remarks to be made on the qualitative nature of such responses (Secs. 9.2 and 9.3). Furthermore, the interesting special case of first-order nonoscillatory cascades is pursued (Sec. 9.5) in that they can be modeled increasingly accurately by a simple two-parameter model, as the length of the cascade and hence the order of the transfer function are increased.   Section 9.5 is not required reading, however.   In Sec. 9.4 the pure-time-delay element is considered in detail.

Turning to the consideration of *distributed linear models* instead of lumping, the basic solution forms are again quite reasonably simple, but there is considerably more manipulative and computational difficulty (Sec. 9.7).   Such modeling is proving necessary in some cases, and one recent successful example is presented.   Finally, turning to *nonlinear* modeling, one notes that a great simplification is lost because superposition of solutions is no longer valid.   Thus, for example, the actual amplitude of the input step or sinusoid becomes a parameter of the response pattern.   Furthermore, several parameters may be needed to describe any one nonlinearity, for example, "hysteresis" (Fig. 3.10).   In this chapter we apply some of the analytical tools of previous chapters to quantify the feature of biological sensing known as *unidirectional rate sensitivity* (Sec. 9.6). Even in this introductory work some surprising results appear.

9.2   The Pole-Zero Characterization of High-order Linear Lumped Systems

The first- and second-order systems merited extended consideration because of their basic dynamic and energetic natures.   The general system which the analyst faces will probably include several such subsystems, and the actual final order of the model is not normally inherently important.   Thus we never bothered to establish the total order of the thermoregulatory system in Chap. 3, although the lumped heat-conduction process itself was modeled as third order.   Computer simulations are therefore always indicated to obtain specific numerical solutions, or even to build up satisfactory models.   However, the pole-zero plot is one potentially powerful pencil-and-paper tool which allows us to predict something about the dynamic response of high-order systems.

(A) THIRD-ORDER SYSTEM

Initially, consider the example of a basic third-order system:

$$H(s) = \frac{1}{b_3 s^3 + b_2 s^2 + b_1 s + b_0} \tag{9.1}$$

Since complex roots always must occur in conjugate pairs, this system must be composed either of a simple third-order cascade with

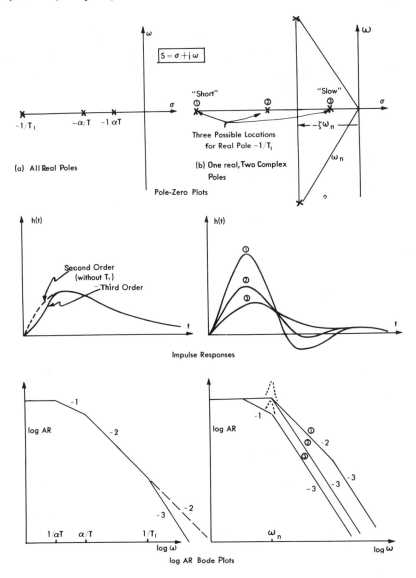

FIG. 9.1 *Third-order-system characteristics—overdamped and under-damped.* *(Not to scale.)*

three real roots or of one real and two complex roots (Fig. 9.1):

$$H(s) = \frac{1/b_0}{(T_1s + 1)\left(\dfrac{Ts}{\alpha} + 1\right)(\alpha Ts + 1)} \qquad (9.2)$$

or

$$H(s) = \frac{1/b_0}{(T_1s + 1)[(s/\omega_n)^2 + 2\zeta(s/\omega_n) + 1]} \qquad (9.3)$$

The Bode plots of Eq. (9.2) can quickly be found from the cascade frequency response technique (Sec. 7.7). The transient responses can be determined directly, although more tediously than by the Bode plots, by Laplace-transform techniques. They consist of three exponential decays (real and/or complex) describing the impulse response, plus whatever is the forced solution to the input function. The impulse and step responses exhibit the same general form as the overdamped second-order system, except that the added effect of time constant T_1 can only be to produce a slower system. In particular, as shown in Fig. 9.1, the impulse response of the basic third-order system has a zero first derivative at $t = 0+$, but a finite second derivative. The initial value of this second derivative is found by integrating twice the indicated differential equation:

$$\ddot{y}(0+) = \frac{1}{b_3} \tag{9.4}$$

The *initial-value theorem* (Sec. 6.8) provides the same result and also provides the general rule for the polynomial transfer function of the form

$$H(s) = \frac{a_m s^m + \cdots + a_1 s + a_0}{b_n s^n + \cdots + b_1 s + b_0} \tag{9.5}$$

This rule is that the lowest-order nonzero derivative of the impulse response is the $(n - m - 1)$st; for example, if $n = 2$ and $m = 0$, the first derivative is finite. Since the step function is the integral of the impulse function, it follows that only the $(n - m)$th and higher derivatives of the step response are nonzero. The *final-value theorem* is not usually needed. Instead, one merely equates derivatives to zero in the differential equation or equivalently places $s = 0$ in the transfer function to find the steady-state value of the unit step response. For the impulse response the steady-state value is always zero, by physical considerations, unless $b_0 = 0$ in Eq. (9.5), which is the special case of an integrator process (Sec. 11.5).

The underdamped third-order system of Eq. (9.3) can have various response characteristics dependent upon the relation between the real part of the complex roots $-\zeta\omega_n$ and the real (inverse) time constant $-1/T_1$. If the time constant is "short" compared with the underdamped system, the former's contribution is fleeting and small (Fig. 9.1); if it is of equal size, the response has the frequency of the underdamped case but is more damped; finally, if the time constant is "long," the underdamped response superimposes as a rapidly decaying oscillation upon it.

(B) FOURTH- AND HIGHER-ORDER SYSTEMS

One can qualitatively estimate the responses for four real roots and four complex roots, but no generalized picture is possible, especially for the other possible combinations. At this point one should obtain specific numerical solutions as necessary. The use of pole-zero plots can remain valuable, however, at least at the qualitative level.

▼

9.3   The Fast-pole Problem

It should be intuitively clear that as the time constant of a first-order component in a system becomes smaller, it should have less effect on the overall system response.   Typical fast poles which have been ignored in analysis thus far are the time constants of engine response in the car-acceleration model and the time constants of muscle torque buildup after receiving motoneuron impulses.

On the pole-zero plot a real negative pole moves out along the real axis toward minus infinity as the time constant decreases.   Consequently it would be convenient to establish a correlation between a pole's relative position and its dynamic effect.   In fact, by writing out the impulse response for the third-order system of Fig. 9.1, one can establish that the constant multiplier A of the fast exponential decay term $A e^{-t/T_1}$ decreases toward zero as $T_1 \to 0$.   We here obtain the same result with less manipulation by taking the overdamped second-order case (Table 8.1) and letting one pole remain constant at $-1/\alpha T \triangleq -1/T_2$ while the other, $-\alpha/T = -1/T_1 \triangleq -\alpha^2/T_2$, becomes progressively faster.   (T_2 merely provides a convenient "anchor" pole as the other moves away.)

$$h(t) = \frac{1}{T_2} \frac{1}{1 - (1/\alpha)^2} \left(e^{-t/T_2} - e^{-\alpha^2 t/T_2} \right)$$

It then follows directly that as $\alpha \to \infty$ the second exponential decays ever more rapidly with respect to the first, and the external constant tends to $1/T_2$.   Thus, in the limit,

$$\lim_{\alpha \to \infty} [h(t)] = \frac{1}{T_2} e^{-t/T_2}$$

which is correctly the impulse response of a first-order system.   In this case it is what has been left as the fast pole of a second-order system became infinitely fast.   As a general rule, if one pole is at least 10 times (often 5 times) faster than all other poles, its effect on the response is bound to be negligible, which means that this term can be left out of the analysis (see Fig. 8.10).   The Bode AR plot of Fig. 9.1 also illustrates that as the fast pole becomes faster it slides off the AR plot at the upper end of the frequency scale.   This has the same dynamic implication.

If there are fast complex poles, the above conclusions remain approximately valid provided the real part only is treated as the fast pole; however, the damping ratio must be large enough so that the fast oscillations decay quickly (for example, case 3 of Fig. 9.1).

9.4   The Pure-time-delay Element e^{-Ts}

(A) PHYSICAL NATURE

The pure time delay required for the action potential to be transmitted along the axon has already been mentioned. The *idealized pure time delay* is defined as an element which transmits without distortion any input signal presented to it, except for the stated pure time delay. No such idealized element can exist, of course, but the axon approximates it for the particular inputs of action potentials for which it has evolved. Axonal propagation may be regarded as a special case of the generalized wave propagation which occurs in electromagnetic and acoustic phenomena, neither of which involves mass translational motion of particles in the direction of the wave propagation.

On the other hand, the pure time delay also arises as the appropriate model for a phenomenon called *transportation delay* in which the whole "material" is involved in translational motion. For example, the blood transports nutrition and hormones which are exchanged at various sites after traveling some distance with the blood stream. These phenomena are only different in the sense that information and material mass are different, and from a functional analysis viewpoint there is often no difference at all.

(B) TRANSFER FUNCTION

The transfer function for the pure-time-delay element may be derived mathematically in several ways.   Consider first the impulse-function input of some material into the blood, the classical rapid intravenous injection:

$$x(t) = u_0(t)$$

If all input is carried unchanged to a distance L from the injection, it arrives with a time delay T equal to L/W, where W is the uniform mass-transportation velocity.   By definition the output $y(t)$ must be a unit impulse function; it does not, however, occur until time $t = T$, written

$$y(t) = u_0(t - T)$$

By definition

$$y(s) = \int_0^\infty u_0(t - T)e^{-st}\, dt \tag{9.6}$$

Since the impulse does not arrive until $t = T$, the limits of integration may be rewritten

$$y(s) = e^{-sT} \int_T^{T+} u_0(t - T)\, dt = e^{-sT}$$

The transfer function follows directly and the transform pair is (Fig. 9.2*a*)

$$
\begin{aligned}
H(s) &= e^{-sT} \\
h(t) &= u_0(t - T)
\end{aligned}
\tag{9.7}
$$

It can also be derived by using a sine-wave input and rotating-vector notation

$$x(j\omega) = e^{j\omega t}$$

The same sine wave is outputted, but lagging in time by an amount T:

$$y(j\omega) = e^{j\omega(t-T)}$$

so that

$$
H(j\omega) \triangleq \frac{y}{x}(j\omega) = e^{-j\omega T}
\tag{9.8}
$$

Equation (9.8) is identical with Eq. (9.7) except that it is in frequency response form. Since it represents a unit vector rotating as a function of frequency, its AR is unity for all frequencies, and its ϕ is always negative and proportional to frequency,

$$
\begin{aligned}
\phi &= -\omega T \\
|H(j\omega)| &= \text{AR} = 1 \qquad \log |H(j\omega)| = 0
\end{aligned}
\tag{9.9}
$$

As a general comment on the physical world, note that where passive transportation is concerned, diffusion or other dissipative phenomena rapidly smooth out such sharp peaks as the impulse function postulated above. It is because the axon is an active transmission line which continually regenerates its waveform that such degradation does not occur here also. Since the pure time delay is so intimately associated with wave phenomena which arise in distributed systems, it is perhaps not surprising that its transfer function turns out to correspond to an infinite number of poles, which in turn corresponds to an infinite number of lumps in the lumped model (see Sec. 9.4*E*).

(C) DYNAMIC RESPONSE CHARACTERISTICS

The transient response of this element is shown schematically in Fig. 9.2*b*. The frequency response exhibits clearly its most interesting characteristic—namely, linearly increasing phase lag combined with constant AR as frequency increases (Fig. 9.2*c*). The polar plot

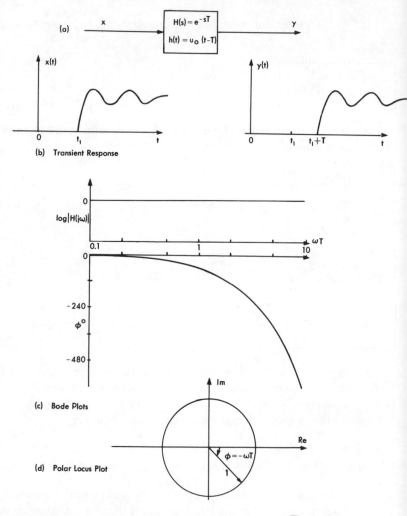

FIG. 9.2   The pure-time-delay element.

is obviously that of a unit rotating vector [Eq. (9.8)], as shown in Fig. 9.2d. A little consideration confirms that this element is destabilizing in a feedback system, since the open-loop polar-locus plot is always brought nearer to the -1 point by the increasing phase lag (which is not compensated for by decreasing AR at higher frequencies as are most other dynamic elements so far met).   The pure time delay is one of a class called *non-minimum-phase elements* because of this precipitous phase characteristic.   A minimum-phase element has the minimum-phase lag possible in a realizable system that has the given AR frequency characteristic.   Clearly, in this case the transfer function 1 has the same AR characteristic but zero phase, and is therefore the minimum-phase element corresponding to the pure time delay.

(D) THE PURE PREDICTIVE ELEMENT e^{sT}

There is satisfactory experimental evidence that biological systems can anticipate or predict certain inputs before they actually occur. For example, the ability to track visual input stimuli with considerably less phase lag than could be possible with a simple linear transfer function has been presented in Fig. 7.6. Obviously, the *pure predictive element* has the same transfer function as the pure time delay except that the sign of T is changed.

$$H(s) = e^{sT}$$
$$h(t) = u_0(t + T)$$

(9.10)

Such a transfer function is termed *unrealizable* in control and communications engineering because it cannot be performed perfectly unless the input waveform is known perfectly, in which case the problem is trivial. However, the engineer can design a system to perform prediction upon a random signal which is optimal in the sense of minimizing the average squared error of the output provided that certain statistical measures of the signal are either known or guessable. This is *optimal linear filtering*, established by Wiener and now considerably developed [see, for example, Truxal, 9.1]. Somewhat similarly, the CNS is apparently able to perform a certain amount of pattern recognition, especially on repetitive-waveform signals, so that necessary motor actions can be anticipated. Such an ability has obvious evolutionary advantages.

▼

(E) LUMPED APPROXIMATIONS FOR e^{-Ts}

The exponential nature of the time delay's transfer function makes it, for manipulative purposes, incompatible with the polynomial transfer functions that we have so far encountered; consequently it would be useful to obtain lumped parameter approximations. Furthermore, it is difficult to implement the pure time delay directly on the analog computer, where the digital computer's method of discrete storage is expensive for simple installations; on the other hand, RLC networks which are themselves lumped-parameter approximations can be cheaply constructed. Although a lumped network would need to be of infinite order to be exactly equivalent, there are some useful finite approximations which we now note.

The Padé polynomials provide a set of *all-pass* networks, that is, networks in which the AR remains unity at all frequencies although different phase characteristics may be obtained. The simplest Padé polynomial is a slight variation of the lead-lag network discussed in Chap. 5 et seq.:

$$H(s) = \frac{1 - \dfrac{T}{2}s}{1 + \dfrac{T}{2}s}$$

(9.11)

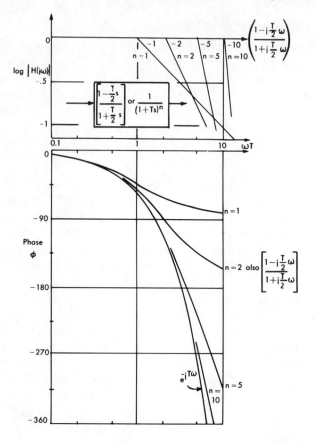

FIG. 9.3 Bode plots of time-delay lumped approximations.

The negative sign in the numerator produces a phase lag characteristic from the corresponding vector, instead of the lead characteristic of the lead-lag network (Fig. 9.3). As similar all-pass expressions of higher order than Eq. (9.11) are used, the phase characteristic approaches more closely to that of the pure time delay e^{-Ts}.

A second approximation, of particular interest for Sec. 9.5, is a simple cascade of equal-time-constant first-order systems. The approximation converges to a pure time delay if the individual time constant is decreased proportionately as the number of them increases:

$$H(s) = e^{-Ts} = \lim_{n \to \infty} \left[\frac{1}{\left(1 + \dfrac{Ts}{n}\right)^n} \right] \qquad (9.12)$$

The phase characteristic is shown in Fig. 9.3 for several n values. In this approximation the AR falls off from unity at high enough frequencies, but the break frequency and therefore the bandwidth increase with the number of cascaded units n. This bandwidth is a measure of the approximation's effectiveness. From the point of view of the pole-zero plot this approximation characterizes the pure time delay as a set of n negative real repeated poles at

$$s = -\frac{n}{T} \qquad (9.13)$$

which implies an infinite set at minus infinity for the exact case.

(F) SUMMARY

The pure-time-delay element occurs widely in biological systems, although diffusion often renders this model approximate. The exact transfer function is shown to be e^{-Ts} and to be destabilizing in feedback systems. Biological systems also appear to perform the opposite function of prediction in higher perceptual modes. The same reasons which make the pure time delay an approximate model also make such an ideal element expensive to construct in analog hardware. Various polynomial approximations simplify the construction task and often simplify the analysis problem in addition since the exponential form of transfer function is not particularly manipulable.

9.5 Simple Models for High-order Systems

It has become painfully obvious that there is an early limit to the complexity of systems which can be analytically manipulated and comprehended. In principle, of course, the electronic computer can largely overcome the problem by its numerical calculating ability. However, there always is advantage in simplifying high-order models if at all possible, and certainly high-order impulse responses typically look more or less similar to one of the several shown in Fig. 9.1. We now examine a few of the more useful approximations for complex transfer functions, but many readers may wish to omit the rather mathematical example of Secs. 9.5C and D for the present.

(A) SIMPLE MODELING OF HIGH–ORDER OSCILLATORY SYSTEMS

If the system has several resonant peaks in the frequency response due to pairs of underdamped poles, then no simpler model is possible which will demonstrate the resonances. However, the higher resonant frequencies are often of little interest in the system studies, because they lie outside the gain-frequency region where closed-loop instability may occur. In these cases a third-order model, such as that discussed in Sec. 9.2, or even a second-order model, suffices.

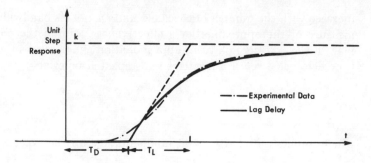

FIG. 9.4   The lag–delay model.

(B) THE LAG–DELAY MODEL

One of the major features of increasing-order systems is an increasing time lapse before significant response to an impulse appears. This is not a genuine time delay since not all derivatives of $h(t)$ are necessarily zero during the interval, but nonetheless it is often convenient to model the system as comprising one pure time *delay* $e^{-T_D s}$ and one other simple dynamic element.   The simplest second element is of course the basic first-order dynamic *lag* $1/(T_L s + 1)$— hence the term *lag-delay model* which is applicable for step responses without significant overshoot.   Figure 9.4 shows the step response of the model,

$$H(s) = \frac{k e^{-T_D s}}{T_L s + 1} \tag{9.14}$$

together with a suggested experimental step response for which the values shown for time delay T_D and time constant T_L would be suitable; best fitting is usually done by eye in this case.   If the step response exhibits significant overshoot, a second-order underdamped element may be substituted for the first-order lag, and so on.   This form of modeling based on the step response is largely empirical and dependent upon the analyst, so that only the above few guidelines are suggested at this time.

▼

(C) MODELING BY IMPULSE RESPONSE MOMENTS

Consider a heavy template cut out to represent the impulse response and supported in the horizontal plane on a knife edge normal to the t axis.   The center of gravity of this template may be found experimentally as the t value along which the template balances (Fig. 9.5). This center of gravity is a parameter which characterizes the impulse response to some extent, and it is defined as the *mean delay time* T_m. We shall now follow the development of Paynter [9.2].

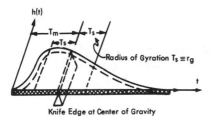

FIG. 9.5 *Moments of impulse response template.*

A second important measure characterizes the spatial distribution of template mass with respect to this center of gravity. This distribution is the moment of inertia about the knife edge as rotational axis, which for a given mass m is specified by its *radius of gyration r_g* (Fig. 9.5) from the defining formula

$$J = \int r^2 \, dm \triangleq mr_g{}^2$$

Our present interest arises because this radius of gyration of the impulse response's template characterizes its effective dispersion about the newly defined mean delay time. It also has the dimensions of time and therefore is now directly defined as the *mean dispersion time T_s.*

Mathematically the mean delay and dispersion times are related to the moments of the impulse response as follows:

$$M_1 \triangleq T_m A = \int_0^\infty t h(t) \, dt \quad \text{first moment} \qquad (9.15)$$

$$M_2 \triangleq T_s{}^2 A = \int_0^\infty (t - T_m)^2 h(t) \, dt \quad \text{second central moment} \qquad (9.16)$$

where

$$M_0 \triangleq A = \int_0^\infty h(t) \, dt \quad \text{zeroth moment} \qquad (9.17)$$

In general, higher-order moments exist, and must be specified if $h(t)$ is to be modeled *exactly*. Fortunately, however, T_m and T_s turn out to be sensitive parameters of many impulse responses, so that higher moments can often be ignored. Furthermore, all impulse responses can be normalized for unit area (or mass) $A = 1$, without loss of generality, by making the transfer function of the unit gain type, so that Eqs. (9.15) and (9.16) then provide direct definitions of T_m and T_s. It should be noted that since T_m and T_s can become large and ambiguous when the impulse response is oscillatory, the discussion is henceforth restricted to nonoscillatory processes.

The mean delay and mean dispersion times can be quickly evaluated numerically, given any arbitrary impulse response. They can also be computed analytically if the transfer function is known, and in fact can be written down by inspection for cascaded first-order systems such as discussed in Sec. 7.7, and with very little greater difficulty for polynomial-type denominators. Furthermore, they may be found by a slightly modified procedure when there are simple transfer functions in parallel rather than in cascade [9.2]. Even more remarkable is that, as the cascade tends toward infinite length so that the transfer function and pole-zero diagram become hopelessly complicated, the approximations become successively more accurate! The rule for finding T_m and T_s from an n-unit first-order cascade is: Given

$$H(s) = \prod_{l=1}^{n} \frac{1}{T_l s + 1} \qquad (9.18)$$

where $\displaystyle\prod_{l=1}^{n}$ means product of terms $1 \cdots l \cdots n$; then

$$T_m = \sum_{l=1}^{n} T_l \qquad (9.19)$$

$$T_s{}^2 = \sum_{l=1}^{n} T_l{}^2 \qquad (9.20)$$

Some of these results can be readily explored for the nth-order cascade of equal-time-constant systems. Here

$$H_n(s) = \frac{1}{(Ts + 1)^n}$$

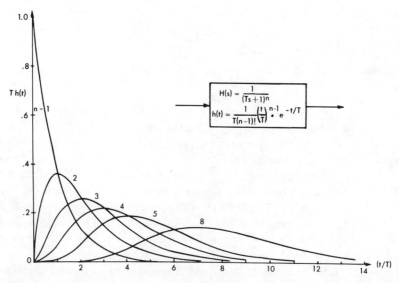

FIG. 9.6 *Impulse responses of* $\left(\dfrac{1}{Ts + 1}\right)^n$.

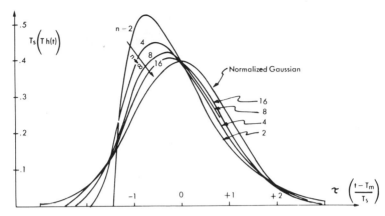

FIG. 9.7 *Normalized impulse responses of* $\left(\dfrac{1}{Ts+1}\right)^n$.

so that

$$T_m = nT \tag{9.21}$$
$$T_s = \sqrt{n}\,T \tag{9.22}$$

Now it may be shown that the impulse response of this cascade is

$$h_n(t) = \frac{1}{(n-1)!T}\left(\frac{t}{T}\right)^{n-1} e^{-t/T} \qquad t > 0 \tag{9.23}$$

and the successive shapes of these impulse responses and their correlations are very interesting, as shown in Fig. 9.6. Notice that the nth impulse response passes through its maximum value at $(n-1)T$ at the same time as the $(n-1)$st impulse response crosses it on the way down.

The important functional characteristic of these cascades is not revealed by the simple, direct time scale, since as n increases, $h_n(t)$ spreads more and more to the right and decreases in amplitude. If, however, they are all plotted to a suitable dimensionless scale (Fig. 9.7), decreasing asymmetry is revealed. Furthermore, as n tends to infinity it is clear that the impulse responses approach the other curve shown, which is the Gaussian probability density function of statistics. Interesting and important results of this asymptotic behavior are explored in Sec. 9.5D.

In summary, so far it has been shown how T_m and T_s may readily be found from experimental nonoscillating impulse response curves, and analytically from certain transfer functions. As the system's order increases, the normalized shape of the impulse response curve tends to become constant, and an analogy with the Gaussian curve of statistics can then be used to create a simple two-parameter model for complex systems, as explored below.

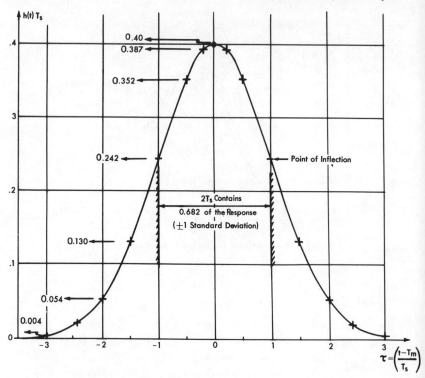

FIG. 9.8   The Gaussian impulse response.

▼

(D) THE GAUSSIAN IMPULSE RESPONSE

To obtain the symmetry of Fig. 9.7 the dimensionless time scale is carefully chosen as

$$\tau = \frac{t - T_m}{T_s} \qquad (9.24)$$

which includes both new measures and is also the dimensionless variable used to tabulate the normalized Gaussian curve.   This probability distribution is a widely used model for physical situations, and we shall return to it in Sec. 14.2.   For the present we only note that the time-domain form for a Gaussian impulse response is

$$h(t) = \frac{1}{\sqrt{2\pi}\ T_s}\ \exp\left[\ -\frac{1}{2}\left(\frac{t - T_m}{T_s}\right)^2\right] \qquad (9.25)$$

and that its corresponding transfer function is

$$H(s) = \exp\left[\ -sT_m + \frac{1}{2}\ (sT_s)^2\right] \qquad (9.26)$$

The advantage of the Gaussian response as a model for complicated systems is twofold:

Impulse Response. $h(t)$ is tabulated in statistics books and handbooks [e.g., 3.13], and so may easily be plotted once T_m and T_s are found [for example, from Eqs. (9.19) and (9.20)]. Furthermore, the few points (typically $\tau = 0$, ± 1, and ± 2) needed to construct it can easily be memorized (Fig. 9.8). Sixty-eight percent of the step response occurs between $\tau = \pm 1$, where it is worth recognizing that T_s is called the *standard deviation* of the Gaussian curve in statistics and symbolized σ. Similarly, T_m equals the *mean* μ. Although the idealized Gaussian model extends to $\pm \infty$ values of τ, the ordinate is negligible beyond ± 3, which simplifies construction greatly. Values of real time are obtained by multiplying the abscissa scale by T_s and adding T_m.

Frequency Response. The Gaussian transfer function has an extremely simple frequency response. Thus, replacing s by $j\omega$ in Eq. (9.26) yields, for the log AR and phase relations,

$$2.30 \times \log_{10} |H(j\omega)| = -\tfrac{1}{2}(\omega T_s)^2 \tag{9.27}$$
$$\phi = -\omega T_m \quad \text{rad} \tag{9.28}$$

The resulting Bode plots are presented in Fig. 9.9 (on different frequency scales). As far as phase is concerned, the Gaussian impulse response is identical with the pure time delay (Sec. 9.4). However,

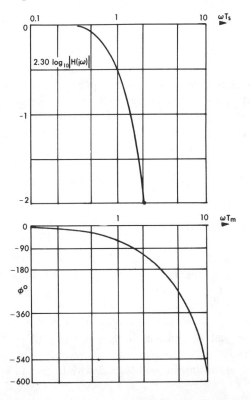

FIG. 9.9 *The Gaussian frequency response.*

it does not maintain the unity gain characteristic at all frequencies which made the pure time delay so destabilizing; instead the gain depends upon T_s only and is therefore independent of phase, except insofar as T_m and T_s are related physically for any given system. This independence of gain and phase with only two parameters makes the Gaussian function a powerful and flexible model for complex systems.

In summary, modeling by moments holds great promise but has not been greatly exploited yet. The particular form involving the Gaussian model of the impulse response is especially powerful for complex cascade and parallel systems of the nonovershooting type, because of the underlying dynamical-statistical analogy which makes the basic body of statistical theory available. The method perhaps provides a general hint that as successively more complex systems come to be treated, simpler forms of calculus may become available.

9.6   *Dynamic Nonlinear Systems*

Nonlinearity is a ubiquitous property of living systems. Some of these systems can be satisfactorily analyzed using the method of linearization (Sec. 3.5), but in general the nonlinearities must be included, as a result of which computer simulations are needed. Nevertheless, an understanding of linear analysis seems to be a prerequisite for the analyst because its results provide conceptual foundations for his thinking and act as a basic introductory vehicle for the study of nonlinear systems. Some recent examples of mathematical modeling that involve nonlinear aspects include the following biological systems: respiratory and cardiovascular control systems [9.3]; thermoregulatory control system (Chap. 3); and pupil-control system (Chap. 2). In this section we explore some dynamic performance aspects of *unidirectional rate sensitivity*, which seems to be a property exhibited by many biological systems or their receptors. The development essentially follows the work of Clynes [9.4, 9.5].

(A) A MATHEMATICAL MODEL FOR RATE SENSITIVITY

The steady-state response characteristics of receptors are usually nonlinear, for example, logarithmic (Sec. 2.6), but since present interest centers on the dynamic response, the steady-state response will henceforth be called *proportional* whether or not it is linear. As already discussed, the dynamic responses of receptors usually exhibit *adaptation;* that is, sudden changes of stimulus produce an immediate large response which subsequently decays. This response is generally similar to that of the lead-lag network ($\alpha > 1$ in Fig. 6.6),

$$H(s) = k\,\frac{\alpha Ts + 1}{Ts + 1} \qquad\qquad (9.29)$$

If the receptor "forgets" the input completely by adapting back to its original firing frequency (often zero), then it is *rate-sensitive* only, without a proportional part. A relevant linear transfer function is therefore that of the approximate differentiator,

$$H(s) = \frac{ks}{Ts + 1} \tag{9.30}$$

[It is readily shown that adding a proportional term k_1 to Eq. (9.30) recovers the form of Eq. (9.29).]

The linear form of Eq. (9.30) can be designated *bidirectional rate sensitivity* (BRS) in contrast to that of *unidirectional rate sensitivity* (URS). In the latter form, a response only ensues when there is a particular sign of input rate of change, usually positive. The transfer-function symbolism is now no longer strictly applicable, and in particular, algebraic manipulatability is reduced. Nevertheless, the symbolism remains useful; thus Eq. (9.30) becomes

$$H(s) = \frac{y}{x}(s) = \Omega \frac{ks}{Ts + 1} \tag{9.31}$$

where Ω is 1 when the input derivative $\dot{x} > 0$ and 0 when the input derivative $\dot{x} < 0$. The equivalent describing differential equation is

$$\begin{aligned} T\ddot{y} + y &= k\dot{x} \qquad \dot{x} > 0 \\ T\dot{y} + y &= 0 \qquad \dot{x} < 0 \end{aligned} \tag{9.32}$$

A reasonably general abstraction may now be made to describe a receptor which exhibits bidirectional *proportional* response and unidirectional *rate* response:

$$H(s) = \frac{y}{x}(s) = k_1 + \Omega \frac{k_2 s}{T_2 s + 1} \tag{9.33}$$

Since this equation cannot be manipulated into more concise form, the total response must be found as the sum of the two component responses.

Clynes suggests that URS can arise from the basic nature of biological information channels. Thus, when a single channel releases chemical agents to transmit information at various points (synapses, motor end plates) the rate of release may be presumed proportional to the stimulus strength. Then, since direct *negative* release of the chemical is not possible, concentrations can only decay relatively slowly through normal ongoing chemical reactions (unless a second channel is brought into action which releases a second chemical to nullify the first more quickly). Therefore, in order for the channel to exhibit BRS, a high concentration is necessary (and hence presumably a high firing frequency is also necessary). This high d-c level is obviously expensive metabolically, and therefore is not evolutionarily advantageous unless BRS has functional importance for the system concerned. It is therefore interesting that cutaneous cold receptors, for example, do indeed exhibit this latter characteristic (Sec. 3.8). On the other hand, Clynes has studied complete reflex and autonomous systems which exhibit URS, rather than the receptors themselves (Sec. 9.6C). In such cases both the proportional and the URS terms may need to be modeled by transfer functions which include further dynamic lag expressions in the denominators, presumably due to such phenomena as muscle-force dynamics, etc.

(B) DYNAMIC RESPONSE CHARACTERISTICS OF THE URS ELEMENT

The dynamic response characteristics of the URS element alone [Eqs. (9.31) and (9.32)] are sufficiently interesting when revealing dynamic inputs are chosen that they are presented here in some detail. In

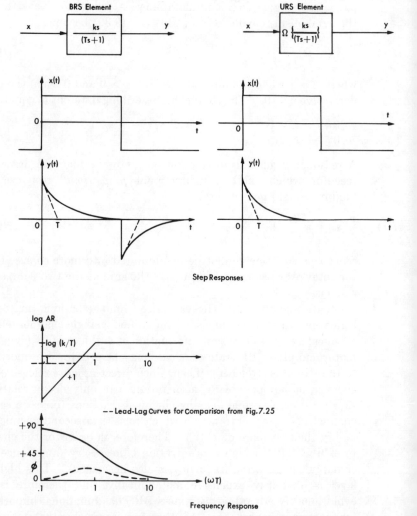

FIG. 9.10 Dynamic responses of basic BRS and URS elements.

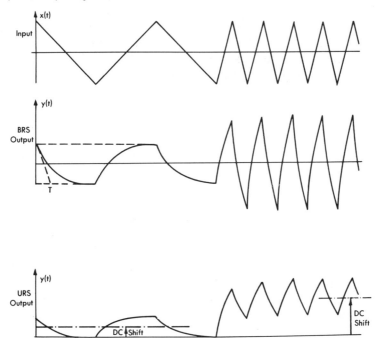

FIG. 9.11 *Triangular waveform responses of BRS and URS elements.* (*From Clynes.*)

order to demonstrate that the responses are indeed rather special, however, the comparative results of the BRS element [Eq. (9.30)] are presented at the same time. One point which emerges from this sort of analysis is the importance of finding *revealing* dynamic forms of input, and of course relating these to "real life" if possible. With complicated dynamic systems, one can seldom casually predict the response to some untried input, even though the input may turn out to represent an important operating mode of the system. In particular, one cannot reveal the dynamics of even linear systems with steady-state tests, and this point is doubly important in nonlinear systems.

Since the basic positive and negative step response of the BRS element is slightly different from the lead-lag elements, it is shown in Fig. 9.10. That of the URS is the same except that it does not respond to the negative step. The frequency response of the BRS is as shown, and differs from the basic lead-lag element (shown dashed in Fig. 9.10, from Fig. 7.25) in having no d-c gain (as $\omega \to 0$). The frequency response of the URS element cannot be obtained analytically from Eq. (9.31) by the direct means of Chap. 7. If the rate-sensitive element is intended to be a differentiator, it should respond to a ramp input with a step output. However, in the approximate differentiator this steady d-c response is only achieved after an interval (the classic first-order step response). If the ramp is alternated as a slow triangular wave (Fig. 9.11), the BRS element responds with successive exponentially decaying "rises"; the URS element, on the other hand, is only forced during the positive ramp, and exponentially decays toward zero during the negative ramp, as an unforced initial-condition response. This slow triangular input reveals the first important difference of the URS, namely, that a symmetric input has produced a d-c level in the output, symbolically due, one may say, to the *rectifying* quality of unidirectional sensitivity. At such slow frequency, where the d-c level is reached each time before the new ramp of opposite sign commences, the d-c shift is proportional to the slope of the ramp.

When the triangular-wave frequency is increased so that the transient response is not completed before the next ramp appears, the BRS output begins to look very similar to the input waveform itself, and is approximately symmetric (actually antisymmetric). On the other hand, the initial URS output is no longer symmetric even in the previous limited sense; in particular, the forced rising response is generally steeper than the unforced decaying response, so that the overall response moves away from the base line, on a curve defined by the input and system parameters, to a new d-c level, which is again proportional to the slope of the ramp. In this equilibrium condition antisymmetry is reestablished (Fig. 9.12). The half-period of the wave in the example shown has been made equal to the time constant of decay. As the time of this half-period is further reduced, the response waveforms of both the BRS and URS increasingly approach the triangular, but the d-c shift of the URS element continues to increase, at least until it reaches some physiological saturation value at which it remains constant. At sufficiently high triangular-wave frequency both waveforms therefore supply proportional response. This is not surprising in view of the relevant Bode plots of Fig. 9.10, which show constant gain and zero phase at high frequencies, that is, the characteristic of a simple

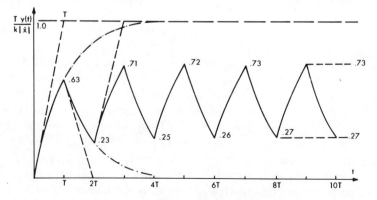

FIG. 9.12 Development of the d-c shift of the URS.

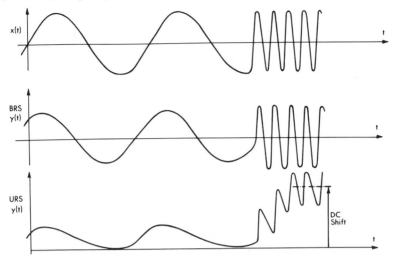

FIG. 9.13   Sinusoidal waveform response of BRS and URS elements.
(*After Clynes.*)

gain $H(j\omega) = k$.   Since the input waveform is repetitive and uniform the URS does not lose any information by failing to process half the wave, except a loss of 2 in gain; on the other hand, it provides a large d-c shift proportional to slope, that is, proportional to frequency for constant amplitude, in the very regions of frequency where the BRS element fails to act in a rate-sensitive manner at all. Thus this rather imperfect-looking transfer function turns out to have highly desirable dynamic characteristics.

When the input waveform is sinusoidal, somewhat similar characteristics appear except that the output waveform of the linear BRS element is also sinusoidal, whereas that of the URS element is asymmetric and therefore has harmonic content (Fig. 9.13).   Again the d-c shift phenomenon extends the frequency range for which the URS element is sensitive.   For the triangular wave it was noted that the d-c shift is proportional to maximum slope, viz., velocity of the input.   This is true of other inputs also, so that the d-c shift is proportional to the product of amplitude and frequency of the input waveform.

Finally, the BRS and URS elements have distinctly different forms of pulse response (Fig. 9.14).   The BRS response is small and rapidly over.   This can be explained as the result of superimposing positive and negative steps in rapid sequence, and because of superposition the two responses cancel out almost perfectly.   The URS, on the other hand, cannot respond to the negative-going step, and hence the pulse response becomes equivalent to the positive step response.   Furthermore, as already noted, a negative pulse produces an identical but slightly delayed response of the same polarity as the positive pulse.

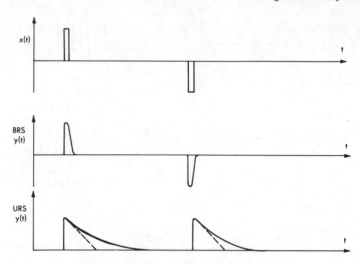

FIG. 9.14   Pulse responses of BRS and URS elements.   *(From Clynes.)*

(C) APPLICATION OF URS MODELING IN PUPIL–CONTROL SYSTEM

Clearly this mathematical modeling is of no use to biologists unless it helps to explain observable phenomena, and therefore we briefly reconsider the pupil system of Chap. 2 to illustrate one successful application.   Indeed, it is worth repeating that the motivation in URS work has been to find mathematical, dynamic, nonlinear models which would explain observed responses.   Then, as always happens with a fruitful model, its manipulation is useful in comprehending the system and providing predictions which can be checked experimentally.

In the intermediate-frequency range around 1 cps when neither long-term retinal adaptation nor short-term flicker fusion is effective, Clynes was able to demonstrate that the following transfer function provided good correlation of predicted and observed dynamic responses:

$$H(s) = \frac{be^{-T_5 s}}{1 + T_3 s} + \Omega \frac{ase^{-T_4 s}}{(1 + T_1 s)(1 + T_2 s)} \tag{9.34}$$

where $a = 1.2$ $T_3 = 1.0$ sec
        $b = 0.24$ $T_4 = 0.3$ sec
        $T_1 = 0.4$ sec $T_5 = 0.6$ sec
        $T_2 = 0.46$ sec

In view of the surprising results which emerged from even the simplest URS element in the previous section, we can hardly expect to understand this system in detail without considerable further effort.   The first term represents a "proportional" response of the

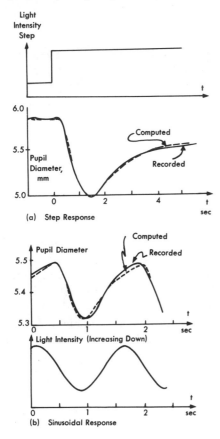

FIG. 9.15 Dynamic responses of pupillary-light reflex and its URS model. *(From Clynes.)*

pupil to light-intensity changes, and as already noted in Chap. 2, the gain b of the compensation is small.   The denominator first-order lag and the pure delay term constitute an example of the lag-delay model mentioned in Sec. 9.5, and result from the fact that "proportionality" does not result instantaneously.   The second term represents URS but requires a second dynamic lag before the responses correlate.   Since this transfer function is between light input and pupil reaction, however, there is no implication that such a lag occurs in the retinal receptors.   The proving of this model is demonstrated by the close agreement between predicted and actual dynamic responses as exemplified by the step and sinusoidal responses of Fig. 9.15.   Clynes presents other interesting responses, such as the large rate-sensitive reflex which occurs when square-waveform light steps of opposite sign are separately inputted to each eye; in a linear system the response in this case would be zero since the total light illumination is constant.

(D) SUMMARY

Nonlinear models of systems are generally rather specific, and the superposition advantages of linearity disappear. This section attempts to give some feel for nonlinear system modeling by discussing URS, which does seem to contain inherent generality as far as biological systems are concerned.

▼
9.7   Spatially Distributed Dynamic Systems

(A) INTRODUCTION

Only lumped models have been considered so far, but it has been emphasized that physical systems are always distributed spatially as well as temporally, so that partial differential equations should ideally be used. The process of lumping, by which ordinary differential equations are derived from partial differential equations, has been explained for the example of heat conduction in Sec. 3.4. Another example of partial-differential-equation phenomena involves wave propagation, for example, in the neuron, in heart-muscle contraction, in peristalsis, and in blood flow in the arterial tree. In the next section we consider a further class.

(B) THE TRANSPORTATION–EXCHANGER PROBLEM

The transportation-exchanger problem arises in the situation in which a material in bulk (usually fluid) is transported along some conduit, and exchange of material or energy occurs along the conduit. As a first example, recall that much of the heat transfer achieved during thermoregulation is done by the blood which bathes the various regions. This is clearly a spatially distributed process, but a lumped modeling was justified primarily on the basis of its relatively short systemic circulation time. However, this does not always apply, and another example on which we shortly concentrate concerns the exchange of materials from the bloodstream with the various organs and tissues of the body. The criterion of whether lumping is permissible or not depends largely upon the exchange-dynamics speed as compared with the bulk-transportation velocity. In this respect, the point-to-point passage along the conduit of the exchange material involves the pure-time-delay transfer function e^{-Ts} discussed in Sec. 9.4, and the usual simple lumping procedures do not incorporate this delay inherently in their model.

Consider, then, the rather general case of flow of fluid (say the blood) in a conduit (say an artery or organ), during which some property (say concentration or temperature) varies as a result of exchange across the conduit boundary. Analysis can be initiated

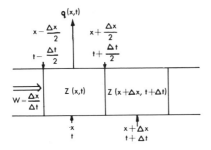

FIG. 9.16   The distributed transportation-exchanger.

from the viewpoint of an observer either sitting on the pipe and watching the fluid go by or moving with the fluid; the latter method is used here. The simplest technique is to imagine the fluid packaged into a set of adjoining segments of length Δx (Fig. 9.16). We now make the assumption that there is no exchange axially between these packages, although exchange is permitted across the conduit wall. This assumption is not necessary but does simplify matters because it effectively precludes axial diffusion, which often really is small. One biological exemplification is bolus flow, but generally laminar flow also approaches it.

We replace the presumably smooth velocity of bulk transmission W by a set of discrete jumps Δx, each lasting Δt, with obviously the implied restriction that

$$W = \frac{\Delta x}{\Delta t} \qquad (9.35)$$

During the stationary periods Δt the appropriate peripheral exchanges occur. Whimsically, one can imagine a train of subway cars stopping every car's length to exchange passengers. Consider, then, a particular segment occupying the mean position x between $x - \Delta x/2$ and $x + \Delta x/2$, whose property of interest is expressed as $Z(x,t)$. This segment must become that occupying the mean position $x + \Delta x$ at the mean time $t + \Delta t$, and it is therefore described as $Z(x + \Delta x, \, t + \Delta t)$. A physical constraint that may now be applied is that the property Z, at $x + \Delta x$ and $t + \Delta t$, must be less than that at x and t by the amount which flows out through Δx during Δt under the appropriate rate-of-flow law $q(Z, \text{etc.})$ (Fig. 9.16). Therefore

$$\frac{Z(x + \Delta x, \, t + \Delta t) - Z(x,t)}{\Delta t} = -q(x,t) \qquad (9.36)$$

Since we expect that a partial differential equation will result, the left-hand side of Eq. (9.36) must be carefully examined. It is, in fact, the *total time derivative* of the function $Z(x,t)$ in the limit as $\Delta t \to 0$ and $\Delta x \to 0$ because both independent variables have been changed incrementally.   Analytically, one now utilizes the Taylor's series result for a function of two variables [Eqs. (3.26) and (3.27)]:

$$dZ = \frac{\partial Z}{\partial x} dx + \frac{\partial Z}{\partial t} dt$$

and hence, dividing by dt, the total time derivative is

$$\frac{dZ}{dt} = \frac{\partial Z}{\partial x}\frac{dx}{dt} + \frac{\partial Z}{\partial t} = W\frac{\partial Z}{\partial x} + \frac{\partial Z}{\partial t} \tag{9.37}$$

This equation asserts that the total time derivative of Z, dZ/dt, depends not only upon the partial time derivative $\partial Z/\partial t$ but also on the partial spatial derivative $\partial Z/\partial x$ multiplied by the space-time derivative dx/dt, which from Eq. (9.35) is W.   Equation (9.37) may also be derived by obtaining the partial derivatives separately, as follows: In Eq. (9.36) add and subtract the term $Z(x + \Delta x, t)$,

$$\frac{Z(x + \Delta x, t + \Delta t) - Z(x + \Delta x, t)}{\Delta t} + \frac{Z(x + \Delta x, t) - Z(x,t)}{\Delta t} = -q(x,t) \tag{9.38}$$

The first expression in parentheses obviously represents the partial time derivative $\partial Z/\partial t$, in the limit as $\Delta t \to 0$, whereas a second manipulation, premultiplying by $\Delta x/\Delta x$, is required to reveal the second parenthesized expression as the partial spatial derivative $\partial Z/\partial x$ multiplied by dx/dt as $\Delta x \to 0$, $\Delta t \to 0$.   Hence Eq. (9.37) results as before, and Eq. (9.36) becomes

$$\frac{\partial}{\partial t}[Z(x,t)] + W\frac{\partial}{\partial x}[Z(x,t)] = -q(x,t) \tag{9.39}$$

(C) BLOOD–HEAT TRANSFER

So far the development has been rather general.   Consider therefore the blood-heat exchanger problem in which the property Z represents thermal stored energy, defined here as

$$Z = c\rho\, dv\, \theta = c\rho a\, dx\, \theta \tag{9.40}$$

where c = specific heat of blood
ρ = density of blood
a = blood-vessel cross-sectional area
dv = segmental volume = $a\, dx$
θ = temperature of blood

and where q represents convective heat-flow rate through the vessel walls.

$$q = hp\, dx\, (\theta - \phi) \tag{9.41}$$

where p = blood-vessel perimeter
h = heat-transfer coefficient
ϕ = tissue temperature

Hence Eq. (9.39) becomes

$$\frac{\partial \theta}{\partial t} + W \frac{\partial \theta}{\partial x} = \frac{-h}{c\rho} \frac{p}{a} (\theta - \phi) \qquad (9.42)$$

Note that for a lumped bloodstream heat-exchange model, $\partial\theta/\partial x = 0$ and one immediately reverts to a single-energy-storage first-order equation such as was written for the skin in thermoregulation. Also note that if $q = 0$ (the well-insulated leakless pipe), any time changes in θ at some initial x are propagated with the appropriate velocity as corresponding distance changes. The solution of Eq. (9.42) is not pursued here, but we should note that it is only valid when mass does not also transfer itself from the bloodstream. It is with this latter problem of extravascular mass exchange that we now concern ourselves a little more deeply.

(D) INTRAVASCULAR AND EXTRAVASCULAR MASS EXCHANGE

This model follows Goresky [9.6] concerning the determination of extravascular storage volumes from the liver sinusoids as an indirect method for determining liver composition. His experiments consisted of dilution studies using several radioactively labeled substances, especially including the red cells, water, urea, albumin, and inulin. The red cells are confined within the sinusoids' capillaries, whereas all the other materials distribute themselves temporarily into varying amounts of extravascular volume. Eventually all labeled materials reappear in the circulation, and typical response curves are shown in Fig. 9.17 for recovery rate at the output (hepatic vein) after an impulse input (rapid intravenous injection in the hepatic portal vein). Notice the interesting likeness to the cascaded impulse response curves of Fig. 9.6; in detail, however, the

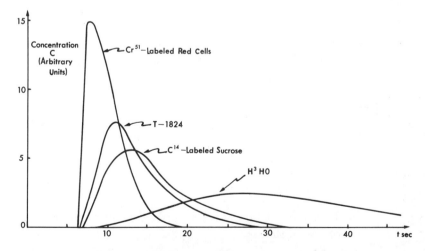

FIG. 9.17 Hepatic venous blood concentration—time patterns. (*From Goresky.*)

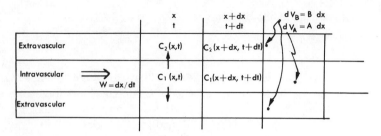

FIG. 9.18   The liver's extravascular storage model—one sinusoid.

analogy does not prove particularly useful. The initial period before any response occurs is due of course to the velocity distance, or pure time delay, across the liver.

Initially, analysis was attempted on a lumped two-compartment basis, that is, assuming all extravascular exchange occurred at one site; for some labeled materials, however, this method produced self-contradictory results in terms of the computed extravascular storage volumes, concentrations, and flow rates. Consequently a distributed exchange model was developed, including one important restriction which seemed experimentally justified, that of *flow-limited diffusion*. In flow-limited diffusion, instantaneous equilibration occurs between local extravascular and intravascular concentrations, on the assumption that the presentation of material due to axial bulk flow is slow compared with radial "diffusion" exchange with the small extravascular compartment (small, that is, for each liver sinusoid). The intravascular and extravascular compartment volumes per incremental length dx are assumed to remain constant at $A\,dx$ and $B\,dx$, respectively (Fig. 9.18). Hence, the variables for Eq. (9.39) become

$$Z(x,t) = \text{amount of label in package} = A\,dx\,C_1(x,t)$$

where $C_1(x,t)$ is the intravascular concentration of a given label. Also, the net extravascular flow is stored in the segmented compartment B; therefore

$$q = B\,dx\left[\frac{\partial C_2}{\partial t}(x,t)\right]$$

where $C_2(x,t)$ is the extravascular concentration of a given label. Consequently Eq. (9.39) becomes

$$\frac{\partial C_1}{\partial t} + W\,\frac{\partial C_1}{\partial x} + \frac{B}{A}\,\frac{\partial C_2}{\partial t} = 0 \qquad\qquad (9.43)$$

Finally the restriction of instantaneous equilibration remains to be incorporated, that is,

$$C_1(x,t) = C_2(x,t) \qquad \frac{\partial C_1}{\partial t} = \frac{\partial C_2}{\partial t}$$

so that Eq. (9.43) becomes

$$\frac{W}{1 + \gamma} \frac{\partial C_1}{\partial x} + \frac{\partial C_1}{\partial t} = 0 \qquad (9.44)$$

where $\gamma = B/A$.

It is shown in [9.6] that the solution of this equation is

$$C_1(x,t) = \psi\left(x - \frac{Wt}{1 + \gamma}\right) \qquad (9.45)$$

where $\psi(x,0)$ is some initial-concentration profile, which in this case is zero everywhere except for an impulse around $x = 0$. Equation (9.45) represents a wave which travels down the sinusoid with velocity W if there is no extravascular volume (if $\gamma = 0$) and with a reduced velocity $W/(1 + \gamma)$ otherwise, involving consequently the pure-time-delay element. However, no time spread can result in the outflow, especially when $\gamma = 0$, which is not compatible with the results of Fig. 9.17. Goresky therefore combines a large number of sinusoids of varying length to produce the observed diffusion effect in the output, and is able to show that compatible concentration-volume relations result from this model for all labels. There seems no need to present this further and rather complicated development here, however, since the main point has been made about spatially distributed modeling.

(E) SUMMARY

This section has been concerned with the development of a partial-differential-equation model for a rather general process in which the spatial distribution of an exchange phenomenon becomes important. Such a model cannot be reduced to simple transfer-function notation because of the presence of both spatial and temporal derivatives, although Laplace transformation can still be applied with respect to *either* temporal *or* spatial derivatives. When the output at only one point is important, the solution is often simplified. If a closed-loop study is to be performed containing such a distributed system as this process, rather special techniques have to be used, and Paynter [4.2] has outlined some general solution methods using pure-time-delay elements.

10 Computer Simulations and Numerical Solutions

10.1 Introduction

Conceptual modeling provides the abstract background with which the analyst pursues the basic method of scientific discovery, i.e., preliminary data, hypothesizing (modeling), predicting, gathering more experimental data, rehypothesizing, and iterating as necessary. The phases of prediction and correlation with experimental data have as an essential requirement the obtaining of *numerical solutions*. These numerical solutions may appear in different functional ways depending upon the postulated problem, and furthermore the mathematical forms of the problem may differ. In this chapter an attempt is made to introduce the main methods of numerical analysis and problem simulation, with most of the discussion centering around the use of general- and special-purpose electronic computers. However, this chapter cannot be expanded into a working manual for computer programming, and an extensive bibliography is given at the end of the book.

In Sec. 10.2 the various functional forms in which dynamic systems problems arise are briefly considered. Some analytical and graphical techniques are then introduced in Secs. 10.3 and 10.4. Computation and simulation on the analog computer are considered in Sec. 10.5, including a brief introduction to the working principles of electronic analog computers. In Sec. 10.6 sufficient introduction is given to the methods of numerical analysis so that solution of differential equations on the digital computer can be understood. Finally, in Sec. 10.7 the distinction between instruments and computers is explored, together with some comments on special-purpose averaging computers.

10.2 The Functional and Mathematical Nature of Dynamic Systems Problems

(A) GENERAL

Figure 10.1 shows the basic configuration for the general linear dynamic system. This general system may itself be either simple or very complex, and may have both multivariable inputs and multivariable outputs, but for manipulative convenience we restrict attention to one input $x(t)$ and one output variable $y(t)$. The relation between input and output is expressed by either the transfer function $H(s)$ in the complex frequency domain:

$$y(s) = H(s)x(s) \tag{10.1}$$

or the convolution integral in the time domain:

$$y(t) = \int_{-\infty}^{\infty} x(\sigma)h(t - \sigma)\,d\sigma \tag{10.2}$$

where $h(t)$ is the impulse response of the system. A meaningful problem in analysis arises when two of these functions, x, y, and H (or h), are specified, and the third one is to be found; in principle this can always be done.

(B) FUNCTIONAL NATURE OF PROBLEMS

Functionally three different problems arise depending on which two of the three functions are given.

The Output, or Forward-analysis, Problem. In this most direct version of the problem the input x and the dynamics of the system H or h are given, and the output y is to be calculated. This output problem is the third or prediction step of the scientific method

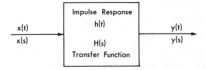

FIG. 10.1 **The linear dynamic system.**

outlined above, and it is the one with which this chapter is mostly concerned.

The Reception, or Input-discovery, Problem. In the inverse version of the above problem, the output function and the system dynamics are known, so that the input function is to be determined. This is essentially the problem of translating the output of all instruments and communications channels. Wherever possible the well-designed instrument and communication channel has sufficiently fast dynamics that the correction to be applied may merely be made as a steady-state calibration. However, in many experimental circumstances the dynamics of the input function and system response are of the same time and frequency scales, and the required analysis is not then trivial. The catheter-correction problem provides a frequently encountered example, but another rather general model is examined for a particular case in Sec. 10.3.

The Identification, or Transfer-function Discovery, Problem. In this third problem both input and output records are available and it is the intervening system dynamics which must be determined. This is the central problem of scientific discovery, and in general is only as tractable as the investigator's ability to control the input function and to conceive models for the system dynamics, that is, to carry out planned experiments. *Identification* is the defining word usually used in control-engineering literature, but perhaps *transfer-function discovery* or dynamic discovery conveys the essential idea more strongly.

Transfer-function discovery may be carried out with any of the three basic classes of inputs: transient, periodic, and statistical. The work may proceed in two somewhat different ways:

1. *Computer simulation:* Computer simulation is a mathematical model-building exercise in which various functional forms and parameter values are tried until the model's input-output characteristics match the experimental data. An important recent development is determining parameter values automatically by an adaptive control technique when the structure of the system can be presumed a priori (Sec. 15.4).

2. *Analytical-graphical techniques:* Analytical-graphical techniques usually involve determining either the frequency response or the time response to suitable transient inputs. From the Bode plots of the frequency response an analytical transfer function is derived by drawing in best-fit asymptotes. From the transient responses, under certain restrictions, exponentially decaying and sinusoidal functions may be extracted which reveal the essential features of the impulse response (Sec. 10.4).

In some practical cases it is not permissible to inject suitable input signals for identification purposes since these would upset normal operation. Fortunately, it is still possible to infer something of the system dynamics by analysis of ongoing input-output records. Since the operating signals in such cases are usually statistical in nature, special averaging techniques must be adopted (Secs. 10.7 and 14.4).

The output-prediction, input-discovery, and identification problems have all been discussed under the restraint of linear lumped systems. In general, neither Eq. (10.1) nor Eq. (10.2) is valid when nonlinearities are included, and a differential-equation description must then be used. However, sometimes it is possible to separate linear and nonlinear components for analytical purposes, in which case transfer functions are still useful, as shown in Sec. 9.6, for example.

(C) MATHEMATICAL NATURE OF PROBLEMS

The need for careful distinction between mathematical classes of a system model has partly disappeared with the advent of electronic computers able to obtain numerical solutions to most definable dynamic problems of interest. Furthermore, even when the analytical solution to a particular system with given input is known, a computer will often be programmed to solve the problem numerically from the describing differential equation and the input function, rather than from the analytic solution. However, it is worthwhile to distinguish certain broad classes of mathematical representation, some of which have already been noted.

Sets of Algebraic Equations. In principle, sets of algebraic equations arise in connection with lumped systems having no energy-storage elements; therefore they are nonexistent in practice. However, stable systems do settle down under constant-forcing input to definite energy levels, so that algebraic equations describe the steady-state result. There is no requirement on linearity (see, for example, the graphical solution for the pupil-control system in Fig. 2.21).

Sets of Ordinary Differential Equations. Lumped energy-storing models produce sets of ordinary differential equations, and if the models are linear then so are the mathematics which describe them. In the unusual case when nonlinearities all occur in cascade fashion with the linear elements, analysis is not unduly complicated. However, the nonlinearities are frequently present inside closed loops so that iterative solution methods become necessary in practice.

The mathematical nature of the problem is considerably affected by the forcing or boundary conditions. Thus, throwing a ball into the air at an arbitrary angle and velocity provides an *initial-value* or *trajectory* calculation which may be more or less simple according to the system concerned, but in any case is straightforward since only the initial forcing conditions and the system dynamics are relevant.

However, by the apparently trivial difference of asking that the ball pass through a certain distant point in space with certain dynamic conditions, as in ball games, one converts from an *initial-condition* problem into a *boundary-value* problem which is much more difficult to solve mathematically. Perhaps this should not surprise us, however, because the baseball player has to practice extensively to solve it reliably also. Both on the computer and on the baseball diamond the dynamic system (muscles, ball, winds, etc.) is subject to many trial input conditions (throwing angle and velocity) until the correct result ensues; in both situations this result may then be stored for future use, and biologically we say that "practice makes perfect."

Boundary-value problems have so far been ignored because of this terminal-value restraint. However, there are many other biological situations where preprogrammed and well-practiced actions probably take place without much feedback, representative examples being the eye saccades and the finger movements in rapid piano playing. Actually, it is not yet clearly established whether proprioceptive feedback is important in such terminal-value problems, but the evidence suggests not, especially in the case of rapid motor actions. Practice has presumably therefore made one able to predict the end result of a given forcing action and hence to be able to grade this forcing before release. We should notice that in the learning stage of a skilled action the CNS has been successively testing the input-output behavior of the system, the *forward-analysis problem*, in order ultimately to solve its *input-discovery problem*.

Sets of Partial Differential Equations. In spatially distributed systems, energy storage is a function of the independent variables of space and time, so that partial differential equations result. In the general case such systems are also nonlinear, and furthermore both the initial-condition and the boundary-value types of configuration occur.

10.3 Analytical Techniques

In this section a few principles regarding analytical solution techniques are discussed, as distinct from the graphical and computer techniques of succeeding sections.

(A) THE OUTPUT, OR FORWARD–ANALYSIS, PROBLEM

The basic analytical techniques are differential-equation theory and Laplace-transform techniques, as already introduced.

(B) THE INPUT–DISCOVERY PROBLEM

A meaningful transform relationship for this case is obtained by inverting Eq. (10.1):

$$x(s) = \frac{y(s)}{H(s)} \triangleq H^{-1}(s)y(s) \tag{10.3}$$

That is, the ratio of $x(s)$ and $y(s)$ is described by the inverse transfer function $1/H(s) \triangleq H^{-1}(s)$. If $y(s)$ is known analytically, $x(s)$ follows, and hence $x(t)$ can be determined. If $y(t)$, rather than $y(s)$, is the known output variable, then some extra manipulation is necessary. This basic idea is now applied to a biological example, where it has proved useful.

The subject of intestinal absorption rates is of considerable interest, and various tracer studies have been carried out. Silverman and Burgen [10.1] used radioactive magnesium (Mg^{28}) in the study which we now present. This dynamic analysis problem is complicated because any Mg^{28} absorbed across the gut into the plasma is then subject to the plasma's own basic "disappearance" dynamics, due to excretion, storage, etc. Therefore, measuring the plasma concentration after an oral impulse ingestion of Mg^{28} produces the impulse response of the complete system, rather than that of intestinal absorption alone. However, an impulse input of Mg^{28} can also be administered intravenously to obtain the impulse response of the plasma disappearance, called the *plasma curve* (Fig. 10.2a). If the two effects of intestinal absorption and plasma disappearance can be considered uncoupled, then we can treat the system as com-

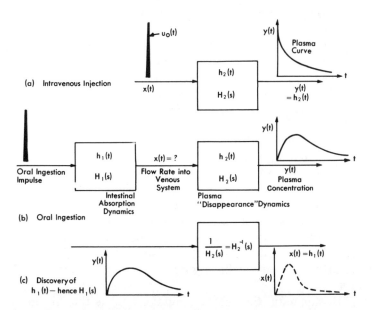

FIG. 10.2 The input-discovery problem—for gastrointestinal absorption.

prised of two cascaded dynamic elements (Fig. 10.2b); fortunately, this does appear to be the case.

We define the cascaded transfer functions as $H_1(s)$ for intestinal absorption and $H_2(s)$ for plasma disappearance; $h_2(t)$ is therefore the plasma curve. After the plasma curve has been recorded it can be decomposed into a sum of exponentials (Sec. 10.4), and the typical result in this case is

$$h_2(t) = Ae^{-t/T_1} + Be^{-t/T_2} + C \tag{10.4}$$

After Laplace transformation

$$H_2(s) = \frac{AT_1}{T_1 s + 1} + \frac{BT_2}{T_2 s + 1} + \frac{C}{s} \tag{10.5}$$

which is the partial-fraction expansion of the following form of transfer function:

$$H_2(s) = \frac{a_2 s^2 + a_1 s + a_0}{(b_2 s^2 + b_1 s + b_0)s} \tag{10.6}$$

At present it is not worthwhile to write out the equivalences between the constants of Eqs. (10.5) and (10.6), but we should note that in some cases a_2 and a_1 would both equal zero. It can be observed from Fig. 10.2b that our problem is to obtain the intestinal absorption response to an impulse function of oral ingestion rate, in other words to find $x(t)$, which identically equals the impulse response $h_1(t)$. From Eq. (10.3) we have

$$x(s) = H_2^{-1}(s)y(s) = \frac{(b_2 s^2 + b_1 s + b_0)s}{a_2 s^2 + a_1 s + a_0} y(s) \tag{10.7}$$

Therefore, $x(s)$ and hence $x(t)$ can be obtained in the usual manner of the forward-analysis problem, since the new input $y(t)$, the oral response, is an experimentally recorded function.

The tedious numerical computations implied can fortunately be avoided very simply by utilizing the analog computer to solve Eq. (10.7). Thus this equation says that $x(t)$ is the output when the dynamic system defined by transfer function $H_2^{-1}(s)$ is subjected to the input $y(t)$. Figure 10.3 shows the block diagram which solves Eq. (10.7), and the analog-computer implementation is almost identical (Sec. 10.5). If one does not include the final differentiation, or s box, the output is the integral of the impulse response, that is, the step response of the intestinal absorption dynamics $H_1(s)$. The input function $y(t)$ should preferably be recorded on a magnetic tape recorder when obtained as the output from the experiment, since it can then be inputted directly into the analog-computer simulation of Fig. 10.3.

In summary, the analog computer renders the "reversed" technique for the input-discovery problem relatively trivial, although some manual work is necessary to extract the plasma-curve exponen-

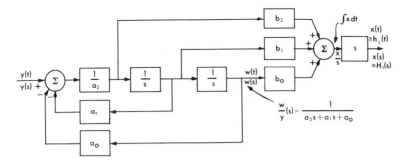

FIG. 10.3 Block diagram for H_2^{-1} (s) (see Fig. 10.2).

tials and thence to convert to transfer-function form. This manual task could be eliminated by such automatic hill-climbing techniques for transfer-function discovery as described in Sec. 15.4.

10.4 Graphical Techniques

(A) THE OUTPUT, OR FORWARD–ANALYSIS, PROBLEM

Graphical techniques for construction of basic transient and frequency response of systems in terms of system parameters have been presented in Chaps. 6 to 9. In the frequency domain the advantage of Bode plots is that the log AR and phase values for systems in cascade can be directly added graphically. In the time domain the convolution integral permits graphical techniques to be applied in principle, but few simple methods seem available. These results are for linear systems, and although there are a number of techniques for incorporating nonlinear elements in linear analysis, they are almost all rather limited and certainly too complex for presentation here.

However, one simple graphical technique merits demonstration now, that for obtaining transient responses for arbitrary nonlinear input functions and systems. It is sometimes called the *slope-line technique*, and is actually a graphical implementation of the numerical analysis used in digital computation (Sec. 10.6). We demonstrate it only for a first-order system, although it can be extended to higher-order systems [10.2, 10.3].

For the car example of Fig. 2.4a, consider the nonlinear propulsive relation, the dotted curve $F_e(x_1,V)$, and the nonlinear resistance relation $F_r(V)$. Newton's second law states that

$$F(t) \triangleq F_e - F_r = m\frac{dV}{dt} \approx m\frac{\Delta V}{\Delta t}$$

and if Δt is kept small,

$$\Delta V = \frac{\Delta t}{m}F \quad \text{or} \quad \frac{\Delta V}{F} = \frac{\Delta t}{m} \tag{10.8}$$

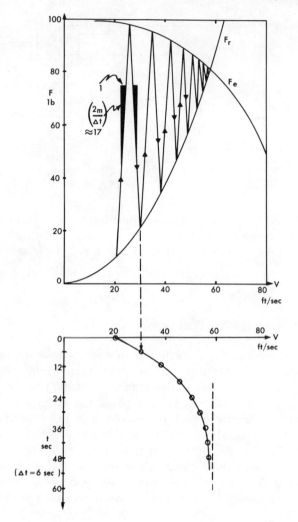

FIG. 10.4   Slope-line technique for nonlinear first-order system.

The $F_e - F_r$ value at any time is the distance between their curves
(Fig. 10.4), and Eq. (10.8) therefore defines the ΔV accomplished in
each Δt.   A repetitive graphical construction for ΔV increments in
Δt follows if the slope $\Delta t/m$ is changed to $\Delta t/2m$, so that the slope line
starting at the given $V(t = 0)$ and $F_r(t = 0)$ values is reflected back
off the F_e curve at $0 + \Delta t/2$ to reach the F_r curve again at Δt.   The
value of ΔV is the change in V measurable along the abscissa, and is
directly replottable on a $V - t$ graph.   The advantage of the
method is the ease with which Δt can be modified if the V changes
occur too slowly or too rapidly for the accuracy which suits the
analyst.   For exemplary purposes, we start at $V(0) = 20$ ft/sec, so
that $F(0) \approx 90$ lb.   Assuming $m = 50$ slugs (that is, weight
$W = 50$ g $= 1,610$ lb), a suitable value of Δt is 6 sec, giving a slope
of $2m/\Delta t \approx 17$, and therefore an initial jump of $\Delta V \approx 10$ ft/sec in

the first interval. V builds up as

$$V(n\,\Delta t) = V[(n-1)\,\Delta t] + \frac{\Delta t}{m}\left[F\left(\frac{n-1}{2}\,\Delta t\right)\right] \tag{10.9}$$

and this is a basic form of difference expression for numerical analysis which we shall reencounter in Sec. 10.6.

The slope-line technique provides a low-accuracy quick look at the time responses of relatively simple nonlinear systems. The forcing function may vary in time also in any arbitrary manner as long as it can be shown on the characteristics of Fig. 10.4; extension of the method is direct for a second-order system, but is probably not justified for higher-order systems.

(B) THE TRANSFER–FUNCTION DISCOVERY PROBLEM

The usual end of a dynamic experiment is that the analyst has some transient, some frequency, or some statistical response curves. In the dynamic discovery problem he wishes to infer the dynamic characteristics of the system, and usually it is desirable to express these analytically. Since the biologist is most familiar with obtaining transient response curves, this subject is considered first.

Extraction of Multicomponent Exponential Decay Curves.
It has been shown that a lumped model with n storage elements gives rise to an nth-order differential equation, and hence that the step and impulse responses are made up of n exponentials. In this section the problem is faced in reverse; that is, a given transient response is recorded, and the problem is to best-fit it with a sum of exponentials and hence to obtain the transfer function. In the typical *indicator-dilution-curve* experiments, a reasonably small number of exponential decays $2 < n < 4$ fits the curve as well as the accuracy of data permits; in consequence the order of the corresponding multicompartment model is n or $n+1$, dependent upon one's viewpoint (see Fig. 10.6).

The direct graphical method extracts the exponential decays successively from the slowest up to the fastest, provided they are adequately separated, and is most easily explained by an actual example. We use here the results of Matthews [10.4], who injected I^{131}-labeled plasma proteins into the central compartment of the open-mammillary system (see also [10.5] to [10.8]). Consider, then, the plasma-activity curve of Fig. 10.5, recorded after rapid injection of I^{131}-labeled albumin into a human, that is, the impulse response of the system. A logarithmic ordinate is chosen so that any single exponential decay should be revealed as a straight line (Fig. 2.9). Now although the response is curved initially, it does eventually tend to a straight line, suggesting that one can approximate this part of the response by the slowest exponential decay term, because all

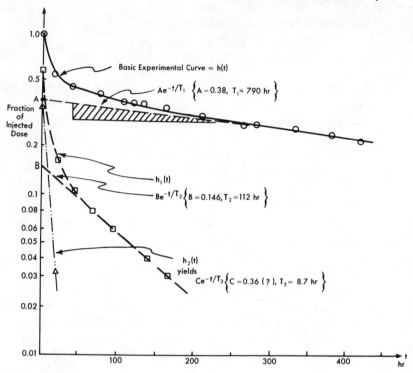

FIG. 10.5 *Extraction of exponentials from impulse response.* (*From Matthews.*)

the "faster" exponentials have decayed to zero by this time (see, for example, Fig. 8.10). Analytically we write the plasma-activity impulse response as

$$h(t) = Ae^{-t/T_1} + Be^{-t/T_2} + Ce^{-t/T_3} \qquad (10.10)$$

where only three exponentials are used in anticipation of the actual results which follow, and where $T_1 > T_2 > T_3$. In the large t region (in base 10 logarithms)

$$\log_{10} h(t) = \log_{10} A - 0.434 \frac{t}{T_1} \qquad (10.11)$$

holds. Hence T_1 equals 0.434 of the time required for $h(t)$ to decrease by one decade and can easily be measured from the graph. In the example, $A = 0.38$ and $T_1 = 790$ hr. Some numerical work is now needed to prepare new data points for

$$h_1(t) = h(t) - Ae^{-t/T_1}$$

which is then superposed on Fig. 10.5. Again, as t grows large, $h_1(t)$ behaves as if only a single exponential were present, and hence the form of Eq. (10.11) also specifies B and T_2; in this case, $B = 0.146$ and $T_2 = 112$ hr. After a further repetition a straight line is

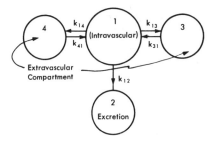

FIG. 10.6 *A four-compartment (three-exponential) model for system of Fig. 10.5 (From Matthews.)*

obtained, so that the process ends in this case with $C = 0.36$ and $T_3 = 8.7$ hr. The total experimental curve of Eq. (10.10) therefore becomes

$$h(t) = 0.38e^{-t/790} + 0.146e^{-t/112} + 0.36e^{-t/8.7} \tag{10.12}$$

and a corresponding transfer function may be obtained whenever needed, as exemplified by Eqs. (10.4) to (10.6); see Fig. 10.6.

The following general points should be made concerning the method:

1. The method provides a check on the assumed separation of the time constants; for example, the time constants in this case are satisfactorily separated by factors of 7 and 13. However, the method also produces separated time constants in cases where the original data are obtained from a continuum of exponential processes [10.9], so that its results are not guaranteed to be uniquely descriptive of a single system.

2. If there are actually n repeated roots, the logarithm of the corresponding impulse response component [see Eq. (9.23)] becomes

$$\log h(t) = -\log[(n-1)!T] + \left[(n-1)\log\frac{t}{T}\right] - \frac{t}{T}$$

If there are damped oscillatory roots in some other physical models than dilution curves, these should be visually rather obvious.

3. The accuracy deteriorates badly as the faster exponentials must be successively considered. This is usually due to inadequate data in the fast time region; for example, in Fig. 10.5 the first data point is at 20 hr or after nearly 2.5 fast time constants T_3.

4. A further cause of inaccuracy for the fast exponentials is the accumulation of previous errors. For example, Eq. (10.12) fails to satisfy the basic consistency check that at $t = 0$, $h(t = 0)$ must equal 1.

5. Some of the inaccuracy problems discussed above can be alleviated by an analytical technique which converts the transient response into the frequency domain before extracting the system parameter values [10.10]. Furthermore, increasingly sophisticated computer

programs are available [10.11, for example] which reduce the need for performing such analysis graphically.

Transfer Function from the Experimental Frequency Response. An experimentally obtained frequency response for a given system contains essentially the same information as the impulse response, but such a form of input-output analysis has not been widely used by biologists. If the system is not essentially nonlinear, there may often be advantages to a frequency response approach; for example, the relatively smooth disturbances of the sinusoidal form may be more acceptable. Furthermore, with the advent of on-line averaging computers (Sec. 10.7), signal-noise ratios can be improved so greatly that very small amplitude sinusoids may be used.

Once the data have been recorded and presented on Bode plots, a transfer function must be invented which best-fits the data, and this is done by constructing asymptotes to fit straight-line sections of the log AR plot. Of course, inaccuracies arise when the poles and zeros are not well separated, just as in the method of extracting exponentials. Resonances, however, provide good evidence for establishing values of natural frequency and damping ratio. Since the phase curve is not independent of the log AR curve in ordinary linear systems, a consistency check on the postulated transfer function is possible, but when they do not match, it is possibly due to the presence of a time-delay element.

10.5   Analog Computation and Simulation

(A) INTRODUCTION

The electronic analog computer provides a simple tool for use in solving computational and conceptual problems. It is considered here before the electronic digital computer because an understanding of analog techniques often aids comprehension of the physical basis behind the numerical techniques used in digital-computer solutions. This is partly because the analog-computer method actually solves the differential equations of a problem by setting up an analogous electronic, but thoroughly physical, circuit. In contrast, the digital-computer method is abstract in that it is restricted to simple arithmetic operations utilizing successive sample values of the physical variables. The word *simulation* is often used for an analog-computer solution method, since the given physical system is simulated electronically through the translating medium of a mathematical model. Thus, if the swinging leg is simulated on the analog computer and if it and a real leg are set swinging together, the voltage of the appropriate amplifier will indeed oscillate synchronously with the leg displacement and with proportional magnitude.

Analog computers are ideally suited for systems of intermediate order of complexity, such as have been discussed in this book, and usually their accuracy matches well that of the biological data available for the simulation. They are useful in all three classes of dynamic problems discussed in Sec. 10.2, that is, output prediction, input discovery, and identification. We consider only the conventional prediction or forward-analysis mode in this introduction, but its role in the input-discovery problem has already been implied in Sec. 10.3, and its important role in dynamic discovery is treated more fully in Sec. 15.4. Although this section provides a short introduction to the analog computer, the subject is too large to be dealt with satisfactorily in this book. The reader therefore should consult the Bibliography for basic books on the subject.

(B) ANALOG SIMULATION OF DIFFERENTIAL EQUATIONS

The analog computer "solves" a differential equation by setting up an electronic model whose several voltage variables are proportional to the physical variables described by the differential equation. The differential equation has been consistently illustrated by a block diagram, which turns out to bear an essentially one-to-one relation to the computer connection diagram. We illustrate the required procedure with the example of the semicircular canal (block diagram of Fig. 8.4a, also Fig. 8.6). Each of the noncommon arrowed lines represents a physical variable, although not every line can necessarily be physically measured with ease. On the analog computer each of these variables is indeed represented physically by a wire on the patchboard (see Fig. 10.7). The value of the voltage on each wire at any time is proportional to the value of the corresponding physical variable, where analog computers work typically in permissible voltage ranges of ± 100 volts.

Fig. 10.7   Typical analog computer.   (Courtesy Electronic Associates, Inc.)

We now consider the three basic linear operations performed by the blocks themselves:

1. *Summation:* In the one circular box, several input variables must be added algebraically to provide the output variable.
2. *Integration:* In the blocks marked here $1/s$, or in earlier diagrams $\int (\ ) \, dt$, integration must be performed on the incoming variable to produce the output variable.
3. *Multiplication by a constant:* The constant coefficients of the defining ordinary differential equation (8.21),

$$J\ddot{\phi} + b\dot{\phi} + k\phi = aJ\ddot{\psi} \qquad (10.13)$$

are represented by blocks on the block diagram (Fig. 8.6). In each case a time-varying voltage must therefore be multiplied by a constant quantity.

These three operations are in fact the only ones necessary to solve linear, constant-coefficient ordinary differential equations, namely, the standard linear lumped systems for which we have been using transfer functions. However, many nonlinear operations can also be performed on the analog computer, especially multiplication and division of variables, function generation, and logic and memory functions.

Integration versus Differentiation. When Eq. (10.13) is arranged for solution as in Fig. 10.10a, the configuration is called *solving for the highest derivative*, and two auxiliary equations of integration are implicitly involved:

$$\ddot{\phi}(t) = \frac{1}{J} \left[aJ\ddot{\psi}(t) - b\dot{\phi}(t) - k\phi(t) \right]$$
$$\dot{\phi}(t) = \int \ddot{\phi}(t) \, dt \qquad (10.14)$$
$$\phi(t) = \int \dot{\phi}(t) \, dt$$

Mathematically the process can be inverted by solving Eq. (10.13) for $\phi(t)$ instead and then performing two successive differentiations to obtain and feed back $\dot{\phi}(t)$ and $\ddot{\phi}(t)$. However, it is painfully obvious to the electronics engineer that his integration circuits are inherently more accurate than differentiation circuits, since the latter tend to amplify any noise or other inaccuracies present whereas the former tend to average them out.

(C) ANALOG–COMPUTER LINEAR COMPONENTS

The Operational Amplifier. The basic active unit of the electronic analog computer is the operational amplifier, which can perform two of the three operations just discussed, namely, summing and summing-integration. Its functional operation depends upon feedback and is of sufficiently general interest to present now in skeleton form.

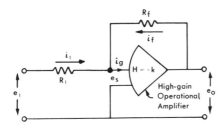

FIG. 10.8   The operational amplifier.

A pseudo circuit diagram is shown in Fig. 10.8.   The two essential passive components are the resistor R_1 in the input path and the resistor R_f in the feedback path.   The triangular segment symbolizes the active amplifying element, whose circuit details are not presently of vital concern since it can be assumed that the amplifier acts simply as a static inverting element of high gain:

$$10^6 < k < 10^8 \tag{10.15}$$

where $e_0 = -ke_s$.

This element is therefore rather naturally termed a *high-gain d-c amplifier*.   The node where input and feedback paths join is called the *summing junction*.   The amplifier's input impedance is sufficiently high that the input current i_g is negligible for calculation purposes, but it is not identically zero because otherwise the amplifier would have no input.   Hence

$$\frac{e_1 - e_s}{R_1} + \frac{e_0 - e_s}{R_f} = 0 \tag{10.16}$$

Now since e_0 has a finite operating range, such as ± 100 volts, Eq. (10.15) shows that the maximum value of e_s is of the order $\pm 10^{-4}$ volt.   Thus, e_s may be ignored as far as calculations are concerned, because the design accuracy is only within perhaps 10^{-2} volt.   Therefore

$$e_0 = -\left(\frac{R_f}{R_1}\right) e_1 \tag{10.17}$$

If $R_f = R_1$ a simple inverter results, that is, a voltage follower. However, this achievement is not trivial since the output power level can be very much amplified over the input level without serious loss of dynamic response, although admittedly this is not explicitly stated in the equation.   This type of "generalized-displacement" follower with power amplification exemplifies the classical tracking servomechanism, although the operational amplifier is a very simple example.

In practice, operational amplifiers utilize either a capacitor or a resistor in the feedback path, and also have multiple input resistors (Fig. 10.9a).   Two main input-output roles result, which are most

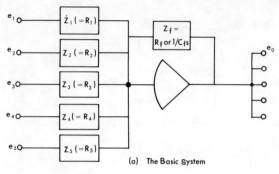

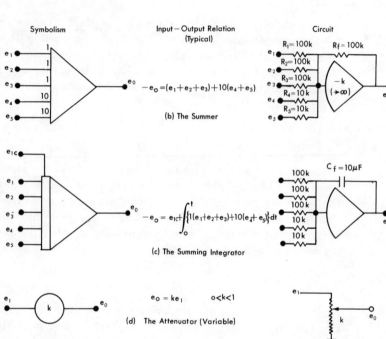

FIG. 10.9   *The operational amplifier—and the three linear operational elements.*

easily analyzed by using the operational notation for a generalized impedance:

$$Z(s) = \frac{e}{i}(s)$$

The generalized input-output relation is then

$$-e_0(s) = Z_f(s)\left[\frac{e_1}{Z_1}(s) + \frac{e_2}{Z_2}(s) + \cdots + \frac{e_N}{Z_N}(s)\right]$$

$$= \sum_{i=1}^{N} \frac{Z_f}{Z_i} e_i(s) \qquad\qquad (10.18)$$

and it should be noted that sign inversion always occurs.

Summer. The summer is the multi-input generalization of the inverter [Eq. (10.17)] and follows from Eq. (10.18), where both impedances are those of resistances (Fig. 10.9*b*).

$$-e_0(s) = R_f \left[\frac{e_1(s)}{R_1} + \frac{e_2(s)}{R_2} + \cdots + \frac{e_N(s)}{R_N} \right] \tag{10.19}$$

$$-e_0(t) = R_f \sum_{i=1}^{N} \frac{e_i(t)}{R_i} \tag{10.20}$$

The R_f/R_i ratios are called the *input gains* and typically have values of 1 and 10. The single output variable e_0 is made available on several terminals.

Summing-Integrator. When a capacitor instead of a resistor is used in the feedback path, the summer is converted to a summing-integrator (Fig. 10.9*c*). In consequence the input gains of the summer are converted into integrating rate constants $1/T_i$ of the summing-integrator, although the numerical value remains the same. Thus

$$-e_0(s) = \frac{1}{C_f s} \left[\frac{e_1(s)}{R_1} + \cdots + \frac{e_N(s)}{R_N} \right] = \sum_{i=1}^{N} \frac{e_i(s)}{T_i s} \tag{10.21}$$

where $T_i = R_i C_f$, or

$$-e_0(t) = \sum_{i=1}^{N} \frac{1}{T_i} \int_0^t e_i(t) \, dt + e_{ic}(t = 0) \tag{10.22}$$

In a typical case we have

$$R_i = 100 \text{ kilohms} \qquad C_f = 10 \text{ } \mu\text{f} \qquad \text{so that} \qquad T_i = 1 \text{ sec}$$

That is, if 1 volt is the input, the output increases by 1 volt/sec. Initial conditions $e_{ic}(t = 0)$ may be placed onto integrating amplifiers by precharging the feedback capacitor appropriately before the solution is run.

Attenuator. Multiplication by a constant is performed basically by grounded potentiometers, and the symbolism is shown in Fig. 10.9*d*. Since the element is passive, the constant k must lie between 0 and 1.

$$e_0 = ke_1 \qquad 0 < k < 1 \tag{10.23}$$

Multiplication by a larger-than-unity constant is achieved by using a larger-than-unity input gain of an amplifier.

(D) SIMULATION OF THE SEMICIRCULAR CANAL

Reconsider, now, the differential equation for the semicircular-canal transfer characteristics between skull's angular acceleration $\ddot{\psi}$ as input and cupula deflection ϕ as output Eqs. [(10.14)]. In Sec. 8.3

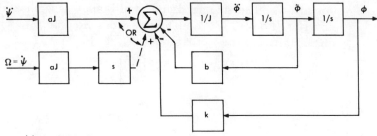

(a) Acceleration Input

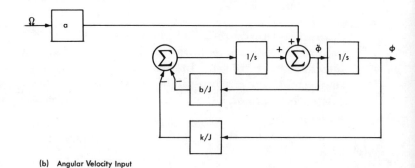

(b) Angular Velocity Input

(c) & (d) Analog Computer Circuits for (a) and (b) Respectively

FIG. 10.10 From block diagram to analog-computer circuit—semi-circular canal.

the intended use of the canal was shown to be as a transducer of the skull's angular velocity Ω $(= \dot{\psi})$, so that the equation is alternatively written

$$J\ddot{\phi} + b\dot{\phi} + k\phi = aJ\dot{\Omega} \tag{10.24}$$

Then with Ω as input the basic block diagram requires differentiation to be performed in order to feed $\dot{\Omega}$ into the system (Fig. 10.10a). This is incorporated easily on block diagrams, but on the analog computer only approximate differentiators are practical, so that the operation is usually avoided if possible.   In fact, this differentiation is not essential since the input may be fed in after the first integration of $\ddot{\phi}$ (Fig. 10.10*b*).   Thus

$$\dot{\phi} - a\Omega = - \int \left(\frac{b}{J} \dot{\phi} + \frac{k}{J} \phi \right) dt$$

J, b, and k need not be shown individually, but only their two independent ratios b/J and k/J.   Their values are readily found from
the ratios $T_1 = b/k = 10$ sec and $T_2 = J/b = \frac{1}{200}$ sec, as $b/J = 200$
and $k/J = 20$.   Since the attenuators shown as producing b/J and
k/J can in fact only provide values less than unity, the gains of
magnitude 10 and 100 must be made up in the available amplifier
inputs surrounding the respective potentiometers.   The configuration of the analog-computer connection diagram is not normally
affected by these details of numerical scaling, however.

Several different circuits would solve the given equation equally
well, but some of them would usually be wasteful of amplifiers.   A
basic rule is that only one integrator is essential for each order of the
ordinary differential equation or transfer function, even when the
latter's numerator is as high-order as the denominator (Fig. 10.3).
For a basic second-order system a total of three amplifiers (two
integrators and one summer) is the minimum number possible.   The
corresponding analog-computer circuits are shown for angular
acceleration and angular velocity inputs in Fig. 10.10*c* and *d*,
respectively.   They differ from the block diagram only in symbolism
and because of the sign inversion properties of operational amplifiers.

(E) MAGNITUDE AND TIME SCALING

Unfortunately, transference to the analog computer has not yet
been completed because the numerical values of the physical variables have not been matched to the permissible range of analog-
computer voltages.   Furthermore, the speed of solution has not
necessarily been matched to either the computer's speed or that of
the desired recording and observing equipment.   These are the
problems of magnitude and time scaling which are not particularly
difficult but do require some ability to estimate the general dynamics
of the system.   A satisfactory development requires more space
than justified in this introductory section, and the reader is referred
to the Bibliography.

(F) OPERATING THE COMPUTER

In typical simulations such as the thermoregulatory and unidirectional rate-sensitivity investigations already described, the analog
computer is used to build up a model whose responses fit actual
experimental data increasingly well.   Since modern analog computers have high-speed repetitive-operation facilities which allow
transient or even periodic responses to be repeated frequently (20 to

50 times/sec), the response appears as a stationary picture on an oscilloscope (Fig. 10.7). If relevant parameters are manually varied the modified responses appear essentially instantaneously, so that a broad range of parameter values can be quickly explored. When a solution of permanent interest is obtained, it may be recorded on a magnetic tape unit or photographed on the oscilloscope; alternatively, the computer may be switched to a so-called *real-time mode* and the solution recorded by pen recorders of the x-y or strip-chart type, typically in 1 to 10 sec.

Although this conceptual use of the analog computer represents perhaps its brightest potential, it can also be used for straightforward calculation of the response of a system which is already definitively modeled; for example, given a satisfactory thermoregulatory model, one might wish to know how a subject would react to various imposed environmental conditions. Another important use is in the area of man-machine studies, and here the emphasis is necessarily on real-time simulations; for example, there is much interest in the instrumentation and dynamic handling characteristics necessary for a man to control various unconventional vehicles, such as vertical-takeoff and -land aircraft, submarines, and even bicycles.

(G) SUMMARY

Analog computers can conveniently handle problems ranging from single dynamic elements to complete autonomous control systems. Their most valuable role is as a conceptual tool for mathematical model building, in which case they are used side by side with physiological experiment. They also have calculating and man-machine simulator roles. Their accuracy is in the 0.01 to 1.0 percent error range. It is worth emphasizing that analog-computer installations are sufficiently flexible regarding cost and complexity that they can be fruitfully employed as the private tool of individual researchers.

10.6   Digital Computation and Numerical Analysis

(A) INTRODUCTION

The electronic digital computer performs simple arithmetic operations and logical functions on numerical data which are stored in its memory, under the control of a stored program of instructions. The high speed with which such operations can be performed (typically 10^6 faster than by a human), combined with a low error rate (at least 10^6 better than human), permits calculations of great numerical tedium and complexity to be routinely and reliably handled. The simple logical abilities, particularly that of comparing two numbers and choosing the next step in the calculation on the basis of which is larger, enable programs to be automatic and rudimentarily intelligent.

Because of its limitation to arithmetic functions, the digital computer certainly cannot solve ordinary-differential-equation models by the direct-analogy method of the analog computer, but instead must resort to an approximate step-by-step technique in which each change of each variable over a small increment of time is separately calculated and then stored for further use. Paradoxically, however, this *approximate* method of the digital computer can be made as accurate as one wishes by using small increments of time and by carrying sufficient significant figures (larger word length) through the arithmetic calculation, whereas the exact mathematical method of the analog computer is unable to achieve lower error than about 10^{-2} percent due to inherent physical limitations of the equipment. Of course, additional word length and consequent accuracy involve extra processing time in digital computers, but this may represent a penalty which is gladly paid when the accuracy is important.

Current digital computers employ a *serial* method of calculation because only one calculation proceeds instantaneously, whereas the analog computer uses a *parallel* method in that all variables are simultaneously operated upon, just as in any natural system. This apparently inherent weakness of digital computation is at the same time a source of strength since rapid advances in technology have made this simple operation increasingly fast and reliable, so much so that the digital computer can sometimes exceed the analog computer's speed in system simulations. Furthermore, paralleled operations of addition and multiplication are now being developed. However, each method retains certain advantages, and hybrid computers are proving advantageous also in special cases.

(B) NUMERICAL ANALYSIS

In this book we are largely concerned with describing dynamic models by systems of differential equations, whose solution on the digital computer involves the *finite-differencing* or the *numerical-analysis* technique. Therefore, the following development is limited to this problem, and to solution of the semicircular-canal example by the method developed; however, it must be realized that there are many other commonly arising computational problems which the digital computer can also solve.

The numerical-analysis method produces discrete sets of points as the output data, rather than the continuous curves of the analog-computer method, for example, the various input-output functions of the thermoregulatory system. Of course, smooth curves can be obtained by continuous interpolation between discrete points. At this stage a comment is relevant about the digital method of evaluating the various functions which arise during computation. With relatively simple functions such as the polynomial

$$f(x) = a_0 + a_1 x + a_2 x^2 + \cdots$$

the evaluation procedure is quite straightforward. However, with transcendental functions the procedure is no longer direct, and it is for these cases that such methods as the Taylor's series expansion into polynomial form are necessary (Sec. 3.5). Thus, for example, the expansion of sin x is given by

$$\sin x = x - \frac{x^3}{3!} + \frac{x^5}{5!} - \cdots$$

The accuracy of evaluating sin x can be improved by including a larger number of terms of the series, and thus successive evaluations are typically made with extra terms until the latest result does not differ from the previous one by more than a specified amount.

(C) FINITE–DIFFERENCE TECHNIQUES FOR DIFFERENTIAL EQUATIONS

Digital solution fundamentally involves converting the differential equations to a set of algebraic difference equations—hence the general term *finite-difference methods*.

A Finite-difference Definition of the Derivative. The crude definition of a derivative given in Eq. (2.7) results in

$$x_{n+1} \approx x_n + \Delta t \frac{dx_n}{dt} \tag{10.25}$$

where $x_n \triangleq x(n \Delta t)$, etc.

This particular finite-difference relation provides the simplest possible basis for "marching out" a solution of a first-order ordinary differential equation from a known initial condition $x(0)$. It is called Euler's method [see, for example, 10.12].

Standard Form of Set of Differential Equations. The differential equations constituting a model of a particular dynamic system should be put into a standard form before the finite-difference form [Eq. (10.25)] is applied. Here we deal initially only with a single variable by taking the semicircular canal's relation [Eq. (10.24)]

$$a\ddot{\psi} = \frac{d^2\phi}{dt^2} + \frac{b}{J}\frac{d\phi}{dt} + \frac{k\phi}{J} \tag{10.26}$$

In this case the angular acceleration of the skull $\ddot{\psi}$, rather than the velocity or position, will be treated as the direct input variable for analysis purposes. On the other hand, the relevant output variable is the cupula position ϕ rather than its velocity or acceleration. The equation is second order in ϕ, but Eq. (10.25) only provides a finite-difference relation for a first-order derivative. Therefore, Eq.

(10.26) is converted into a *standard form* of two first-order equations by defining two new variables:

$$x_1 \triangleq \phi$$
$$x_2 \triangleq \frac{dx_1}{dt} = \frac{d\phi}{dt} \tag{10.27}$$

We now arrange Eq. (10.26) to solve for the second-derivative term in ϕ, as in analog-computer solutions, and hence by using Eq. (10.27)

$$\frac{dx_2}{dt} = a\ddot{\psi} - \frac{b}{J} x_2 - \frac{k}{J} x_1$$

and   (10.28)

$$\frac{dx_1}{dt} = x_2$$

Equations (10.28) constitute the standard form for solution by the finite-difference technique. Solutions can be obtained by manual computation if desired, just as directly as by digital computer.

In this standard form each first-order derivative appears only once, constituting the left-hand side of one of the equations. In general, the standard form may be written

$$\frac{dx_i}{dt} = f_i(x_1, \ldots, x_i, \ldots, x_m, u_1, \ldots, u_j, \ldots, u_n, t) \qquad t > 0 \tag{10.29}$$

with initial conditions $x_i = x_i(0)$; $t = 0$; $i = 1, 2, \ldots, m$; $j = 1, 2, \ldots, n$ for the usual open-ended transient problem. The x_i are the variables defining the state of the system, of which some (or some combinations thereof) are the outputs of interest, and the u_j are the input-forcing functions. Since the general model may contain k equations of lth order, the corresponding standard form will contain kl first-order equations.

The Recurrence Equation for Numerical Solution. The standard form of the system equation produces a set of recurrence equations by incorporating the finite-difference relation [Eq. (10.25)] for the derivatives. Thus, Eq. (10.29) becomes

$$x_{1,n+1} = x_{1,n} + \Delta t f_{1,n} \qquad n > 0$$
$$x_{2,n+1} = x_{2,n} + \Delta t f_{2,n} \qquad n > 0 \tag{10.30}$$

with

$$x_1 = x_{1,0} \quad \text{and} \quad x_2 = x_{2,0} \qquad n = 0$$

where in this case

$$f_{1,n} = x_{2,n}$$
$$f_{2,n} = a\ddot{\psi}_n - \frac{b}{J} x_{2,n} - \frac{k}{J} x_{1,n} \tag{10.31}$$

The solutions for $x_{1,n+1}$, $x_{2,n+1}$ are now methodically obtained by starting from the initial conditions $x_{1,0}$ and $x_{2,0}$ and incorporating the skull-acceleration program $\ddot{\psi}_n$ of current interest.

The Incremental Step Width Δt. One problem of immediate practical importance which must be decided is the choice of step size Δt. Intuitively one expects that the accuracy will increase as Δt decreases, but taking a smaller Δt than necessary for the desired accuracy involves an increase in the number of computations. Furthermore, since the number of steps necessary for a given analysis period increases inversely as the step size, the small round off error at each step due to finite word length may propagate sufficiently to produce significant overall solution errors. However, the digital computer operates at high speed and can be arranged to carry a large number of decimal places in its calculations. Consequently the analyst who has a computer available can choose a conservatively low Δt much more happily than if he had to do the numerical work himself. Nevertheless, it is a good principle not to use much more computation time than necessary, so that some preliminary knowledge about the dynamic characteristics is obviously desirable. In the present example, the overdamped nature of the semicircular canal's response produces interesting results which justify some further study.

Calculating the Step Response of the Semicircular Canal. The transfer-function form of Eq. (10.26) is

$$H(s) = \frac{\phi}{\ddot{\psi}}(s) = \frac{aJ/k}{(J/k)s^2 + (b/k)s + 1}$$

By using the parameter values of Sec. 10.5 with $a = 1$ assumed for convenience, and with the overdamped approximation of Fig. 8.6, the transfer function becomes

$$H(s) = \frac{\frac{1}{20}}{\left(\frac{s}{200} + 1\right)(10s + 1)} \tag{10.32}$$

We consider a unit step input of skull angular acceleration, for example, $1°/\sec^2$, and thus from Table 8.2 for $\zeta > 1$,

$$\phi(t) = \frac{1}{20}\left[1 + \frac{1}{2{,}000 - 1}\left(e^{-200t} - 2{,}000\ e^{-t/10}\right)\right] \tag{10.33}$$

where

$$T_2 = \frac{T}{\alpha} = \frac{1}{200} \qquad T_1 = \alpha T = 10$$

A significant dynamic feature of this step response is that the fast exponential decays 2,000 times faster than the slow one. Furthermore, its coefficient is 1/1,999 and therefore negligible in practice, except for its effect in the first 20 msec, a period of about four fast time constants. For most practical purposes, one would therefore be happy to obtain a step response which ignores the first few milliseconds hence Eq. (10.33) would be written as

$$\phi(t) \approx \tfrac{1}{2}0(1 - e^{-t/10}) \tag{10.34}$$

which is consistent with the approximate fast transfer function given in Sec. 8.3,

$$H(s) = \frac{\tfrac{1}{2}0}{10s + 1} = \frac{aJ/k}{(b/k)s + 1} \tag{10.35}$$

These features present the numerical analyst with a minor dilemma, because it is fairly clear that the time increment Δt must be less than any time constant whose response is to be included. For example, as one arbitrary rule we could require

$$T \geq 10 \, \Delta t \tag{10.36}$$

in which case we presently need

$$t \approx 0.0005 \text{ sec} \approx 10^{-3} \text{ sec}$$

say, if the initial response is to be revealed. Since the overall response occupies about 50 sec, use of this small increment would require 50,000 steps. If the computations are done by hand, the analyst soon appreciates that he can profitably switch to a larger Δt after the first 50 msec or so, for this constant input at least. Similarly, a digital-computer program can be so programmed provided that one knows enough about the system's dynamics to predict this need beforehand.

The Δt Increment and Stability. We now pursue the problem of increasing the Δt in order to learn more of the finite-difference process. On the basis of a 50-sec solution time and a 10-sec time constant of interest, it would be hoped that a Δt of perhaps 1 sec would prove satisfactory. We consider, then, the evaluation of Eqs. (10.30) and (10.31) with a unit step input of $\ddot{\psi}$ at $n = 0$, for the first few time increments; if

$$x_{1,0} = 0 \qquad x_{2,0} = 0$$

and using $a = 1$, then

$$f_{1,0} = x_{2,0} = 0 \quad \text{and} \quad f_{2,0} = a\ddot{\psi}_{\text{step}} - \frac{b}{J} x_{2,0} - \frac{k}{J} x_{1,0} = 1 - 0 - 0 = 1$$

$$x_1(\Delta t) \triangleq x_{1,1} = x_{1,0} + \Delta t \, f_{1,0} = 0 + 0 = 0$$
$$x_2(\Delta t) \triangleq x_{2,1} = x_{2,0} + \Delta t \, f_{2,0} = 0 + \Delta t = \Delta t$$

$$f_{1,1} = \Delta t \quad \text{and} \quad f_{2,1} = 1 - \frac{\Delta t}{T_2}$$

where $T_2 = J/b$

$$x_{1,2} = (\Delta t)^2 \tag{10.37}$$

$$x_{2,2} = \Delta t \left(2 - \frac{\Delta t}{T_2} \right) \tag{10.38}$$

etc. Since $x_2(t) \triangleq d\phi/dt$, the correct solution of Eq. (10.38) should yield nonnegative values for the step response; this follows because the slope of the step response equals the impulse response, which is always nonnegative in overdamped second-order systems. The term of Eq. (10.38) in parentheses yields some sort of *stability criterion* for the solution, in fact, because if Δt is large enough for the term to become negative, incorrectly oscillatory solutions result. An independent demonstration has been made that since $T_2 (= J/b)$ is the fast time constant, a necessary condition for solution stability is $\Delta t/2 < $ fast T. This of course is not a rigorous proof, nor in particular is it a statement of the sufficient conditions for stability [see, for example, 10.12]. Furthermore, the exact conditions depend upon the finite-difference method used as well as upon the system's differential equations in any given case. The presently important point is that any instability resulting from choice of a large Δt has been introduced by the finite-difference method itself, and has no necessary connection with stability of the physical system. The unfortunate conclusion is that, for numerical solution of a given set of differential equations, one cannot simply ignore the fast dynamics in selecting the desirable Δt, since solution instability may result.

However, a stable solution can be achieved with a larger Δt if the differential equation itself is suitably modified to omit the fast dynamics. In the present example, for instance, we noted that it is generally reasonable to ignore the fast-time-constant expression of Eq. (10.33), and the corresponding differential equation has the standard form

$$\frac{d\phi}{dt} = \frac{1}{T_1} \left(\frac{\ddot{\psi}}{20} - \phi \right) = f(\ddot{\psi}, \phi) \tag{10.39}$$

so that the recurrence formula (10.25) becomes

$$\phi_{n+1} = \phi_n + \Delta t\, f_n = \phi_n + \frac{\Delta t}{T_1} \left(\frac{\ddot{\psi}_n}{20} - \phi_n \right) \tag{10.40}$$

$$\phi_{n+1} = \phi_n \left(1 - \frac{\Delta t}{T_1} \right) + \frac{\Delta t}{T_1} \frac{\ddot{\psi}}{20} \tag{10.41}$$

In this case Eq. (10.41) predicts that ϕ_{n+1} oscillates if $\Delta t \geq T_1$. However, T_1 is now the slow time constant, because the fast dynamics have been removed from the differential equation. Since $T_1 = 10$ sec, Δt can perhaps be made about 1 sec, in which case only 50 steps are needed instead of 50,000.

Numerical Solution for First-order System. The recurrence relation [Eq. (10.41)] is common to all basic first-order systems.

TABLE 10.1 SOLUTION OF FIRST-ORDER STEP RESPONSE

n	t_n, sec	ϕ_n	$\dfrac{\dot{\psi}_n}{20}$	$T_1 f_n = \dfrac{\dot{\psi}_n}{20} - \phi_n$	$\dfrac{\Delta t f_n}{} = \dfrac{\Delta t}{T_1}\left(\dfrac{\dot{\psi}_n}{20} - \phi_n\right)$	ϕ exact [Eq. (10.34)]
0	0	0	1	0.05	0.005	
1	1	0.005	1	0.045	0.0045	
2	2	0.0095	1	0.0405	0.00405	
3	3	0.01355	1	0.03645	0.003645	
4	4	0.017195	1	0.032805	0.003280	
5	5	0.020475	1	0.029525	0.002952	0.01967
6	6	0.023427	1	0.026573	0.002657	
7	7	0.026084	1	0.023916	0.002392	
8	8	0.028476	1	0.021524	0.002152	
9	9	0.030628	1	0.019372	0.001937	
10	10	0.032565	1			0.03166

Note that if T_1 has a negative value, as in population growth, ϕ_{n+1} is always larger than ϕ_n, so that the solution tends to infinity.

Any desired input program of $\ddot{\psi}(t)$ can of course be investigated, and for illustrative purposes we now compare exact and finite-difference solutions for a unit step input of $\ddot{\psi}(t)$. Further, we take $T_1 = 10$ sec and $\Delta t = 1$ sec, so that Eq. (10.41) becomes

$$\phi_{n+1} = 0.9\phi_n + \frac{1}{200} \qquad (10.42)$$

The steady-state result emerges immediately, since $\phi_{n+1} \triangleq \phi_n \triangleq \phi_\infty$, and therefore $\phi = \frac{1}{20}°$, for step magnitude in $\ddot{\psi}$ of $1°/\text{sec}^2$. Table 10.1 follows the standard-form method for Eqs. (10.39) and (10.40) in order to illustrate the general method, rather than the obviously shorter method available from Eq. (10.42) in this circumstance. The reader can use this Eq. (10.42), however, to check the results given. Note that the word length in this case has been chosen as six digits, with roundoff. The errors in this solution, as compared with the exact solution, amount to about 4 percent at $t = 5$ sec and 3 percent at $t = 10$ sec. This inaccuracy is not due to the six-digit word length, but to the finite-difference method and the size of Δt, as study of the table readily shows. On a digital computer the problem could be rerun with decreasing size Δt until the results converged.

More Accurate Finite-difference Approximations. The Taylor's series expansion provides a relevant formula for comparing various finite-difference approximations:

$$x_{n+1} = x_n + \Delta t \frac{dx_n}{dt} + \cdots + \frac{(\Delta t)^m}{m!} \frac{d^m x_n}{dt^m} + \cdots \qquad (10.43)$$

Thus, Euler's method [(Eq. 10.25)] is seen to ignore the second derivative and higher terms of the expansion. If all terms were considered, the computation of x_{n+1} would become precise but difficult to use. More powerful finite-difference methods than Euler's are universally used, however, by retaining some higher-order derivative terms than the first. The series is then said to be truncated at a certain order, and the resultant error in evaluating x_{n+1} is called the *truncation error*. The order of the method is defined as being 1 less than the power of the first term dropped in Eq. (10.43), and a typical value for digital computation is fourth order. The subject is beyond our present scope, and moreover, the finite-difference recurrence formula for a particular computer is usually pre-programmed for the standard form of differential equations. It is therefore available to the investigator as a "package," and all he need do is select the order of the method and the incremental time. Furthermore, in some programs the Δt is selected automatically to satisfy a specified value of truncation error.

Several points should be noted:

1. Of the various methods available, not all are "self-starting"; that is, extra initial conditions must be supplied, or there must be evaluation by power series, or there must be evaluation by making several small Δt steps with Euler's method.
2. The particular class of methods called *Runge-Kutta* does not generally introduce extraneous solutions [e.g., 10.12].
3. Higher-order truncation error involves more computation per step, but the step increment can be larger for the same accuracy. There is no overall formula for optimal trade-off among these three parameters (Δt, error percentage, computation), but as noted already, fourth-order formulas are often used.

Partial Differential Equations. In deriving the finite-difference form of a derivative Eq. [(10.43)], the independent variable can as well be spatial as temporal. Hence, all partial differential equations can be broken down into difference equations in the manner described here except that the details become more complicated (see Problems).

(D) PROGRAMMING THE PROBLEM FOR THE DIGITAL COMPUTER

The previous section was necessary to demonstrate the basic approach to solving differential equations utilizing step-by-step methods. The technique can of course be carried out manually, and indeed this was the only general method available until the advent of electronic computers. In principle, the actual programming of the given problem for a digital computer is rather trivial because the finite-difference technique is the heart of the method. However, as inevitably happens with fast, expensive machines, there are many detailed "tricks of the trade" which increase efficiency; in addition, and perhaps rather more importantly, there are certain logical techniques to learn with regard to such features as internal checking of solution accuracy, exploring various parameter values, etc., which enable the digital computer to run through its calculations automatically. Since this book cannot be a manual for digital-computer programming, however, the present treatment is limited to a presentation of the flow diagram, which is to the digital computer what the wiring circuit is to the analog computer.

Consider, then, the flow diagram for the semicircular-canal example, reverting to the second-order model for generality. Let us assume that:

1. Some skull angular acceleration program $\ddot{\psi}(t)$ is available either in analytical form or as a set of data points.
2. The cupula response $\phi(t)$ is to be found for a range of a, b, J, and k parameters.
3. The results are to be tabulated in some such format as Table 10.1.
4. The response can be assumed complete at $T = n\,\Delta t$, or alternatively, one can allow the response to run until the computer finds that it has reached a steady state defined by $|\phi_{n+1} - \phi_n| \leq \epsilon$, where ϵ is some small value specified by the analyst.

The resulting flow diagram is shown in Fig. 10.11 and is rather obvious to read, although usually less obvious to generate. It should particularly be noted that the feedback loops now refer to the logical sequence of performing the incremental steps or the parameter changes, rather than to the feedback variables of the physical problem as shown in Fig. 10.10. The heart of the computational program resides in the box marked "Calculate $x_{1,n+1}x_{2,n+1}$" because here the finite-difference method is programmed to work on the appropriate data.

This flow diagram is general enough for essentially all computers. However, the computer cannot in fact be instructed in quite such a straightforward fashion, and here we should mention the two basic ways in which the necessary communication is achieved:

Machine-language Programming. Machine-language programming is largely reserved for the professional programmer who works with the detailed internal command structure of the machine and thereby hopefully minimizes the operational time needed for a given program. However, such niceties are not normally employed by the scientific or biological analyst who has a problem to solve.

Compilers. Compilers are programs which translate from some simple problem-oriented "source" language to machine language. This machine-language, or "object"-language, program is then run to produce the desired numerical solutions. Sentences like $t_n = t_n + \Delta t$ are allowed (Fig. 10.11). Compiler programs of a wide variety have been developed, of which the most used today are Algol and Fortran. Many subroutines are available for any given computer which can be incorporated into the program as complete packages, for example, complete differential-equation-

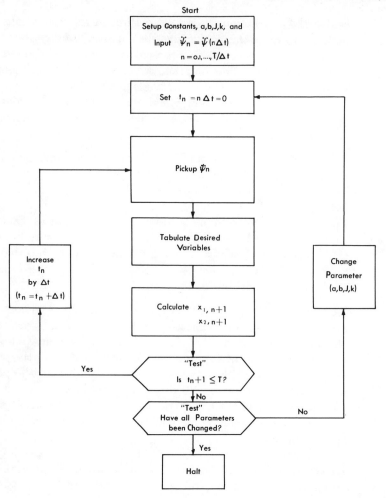

FIG. 10.11   Flow diagram for digital-computer solution—semicircular canal.

solving routines.   Compilers also exist which allow dynamic systems to be programmed directly from the analog-computer-oriented block diagrams (for example, from Fig. 10.10).

Finally, the following assorted points regarding digital-computer programming should be made:

1. *Input and output:*   Punched cards, punched paper tape, and typewriters are the basic forms of input and output; however, in large installations the data may be transferred off-line to or from magnetic-tape units as intermediate storage.   For output, high-speed printers and plotters are also available.

2. *Floating-point arithmetic:*   Although the digital computer operates with typically a 10-decimal digit word length, it is still quite possible for inaccurate solutions to build up if the decimal point is not suita-

bly located, or for other reasons. The consequent nuisance of scaling can be avoided, however, by an automatic use of floating-point arithmetic. In this pseudologarithmic operation a number is designated by, first, the significant digits and, second, a number which defines the location of the decimal point. Normally, however, the programmer need not know anything about this technique apart from specifying that his program shall operate in floating-point mode. In contrast, there are no floating-point techniques commercially available on the analog computer.

3. *Computer "anatomy":* The label *electronic brain* has been bestowed so often on the digital computer that one inevitably must say a few words in this regard. Insofar as both the brain and the computer have incoming data which are processed in accord with some stored program to produce relevant outputs, they are indeed similar. However, much more than this is normally implied. Current digital computers have one central arithmetic processor, one or more storage units with which the processor communicates, and appropriate synchronizing and control hardware. In contrast, the CNS undoubtedly has multiple processing and memory units, as well as circuits with considerable plasticity or adaptation. The analog computer performs parallel processing rather as the CNS does, but on the other hand, it has no general-purpose memory sites. Obviously all electronic-brain analogies should be treated with care, as is done for example in the excellent introductory treatment of Wooldridge [10.13].

The above comments may suggest to the reader that electronic computers should turn to more hybrid techniques. In fact, there are already many examples of this in the biological and scientific field, some of which are mentioned in the next section. One interesting modification of digital programming is the *incremental method.* In incremental computers, often called *digital differential analyzers,* only the increment of each variable during the Δt period is carried forward and processed, storage registers being provided to accumulate the total current value of each variable. The digital differential analyzer is programmed somewhat similarly to the analog computer since parallel operation can be provided, but the accuracy of digital computation can be maintained. In the present context this computing method is interesting because the carrying forward of only increments corresponds to providing information about rates of change [Eq. (10.25)]. This can be compared to the experimentally observed facts that many receptor channels seem to be largely rate-sensitive, and that data processing in the retina, for example, seems largely concerned with extracting rate-of-change information regarding the visual scene for onward transmission to the CNS. The obvious advantage of incremental or rate-of-change information channels is that their required capacity, in the information-theory measure of bits per second, can be greatly reduced.

10.7   Special-purpose Computers and Instrumentation

In practice there is no clear dividing line between an instrument and the computer which processes the information yielded by the instrument; similarly, the various biological receptor organs apparently do much "computing" in order to minimize both communication-channel capacity and the work required of the CNS.   The problem of definition is difficult because all measuring instruments inherently tend to filter the measured signal in some fashion.   Often this filtering arises from the inevitably sluggish characteristics of the instrument, resulting in some loss of information content in the signal.   When this is undesirable, a computer may be needed to correct such distortion, and one common biological example is the need to correct catheter curves for the volumetric capacity of the instrument itself.   If a small correction computer is incorporated as part of the same basic catheter package, obviously one should regard the package merely as an instrument, even though a computer is actually included.

A second aspect arises when the filtering effect of some physical process provides the information with which a desired parameter value must be calculated.   For example, in the *dilution-curve technique* some measurable variable is introduced into the system at one point, and the dilution curve of the variable at some other point typically allows one to calculate the volume of the intervening organ [10.8, 10.14].   A spatially distributed model for this case is presented in Sec. 9.7, but current techniques are usually based on a first-order lumped model, which therefore assumes ideal mixing.   Special computers are now available for the automatic analysis of cardiac output based on this technique, but the combined computer-instrument system is probably best regarded as a single instrument.

A third aspect arises when the biological output variable measured comprises both the desired signal itself and a "noise" component.   An obvious example of this is the whole experimental field of neurophysiology and neuropsychiatry, and in particular the stimulus-response technique in both single-cell and electroencephalographic work.   The ideal output of a computer-instrument system, then, is that of the signal alone, but unfortunately this can only be done on an averaged basis; that is, successive responses over the period between successive stimuli are added with the result that the noise component tends to average out.   In this way the signal-noise ratio may be improved by several orders of magnitude, provided the two components are originally uncorrelated.   This averaging can be done by manual or automatic processing of the data after the experiment is completed (i.e., off-line), but a new experimental field opens up when the averaging can be done simultaneously with the experiment (i.e., on-line), since then the experimenter can modify

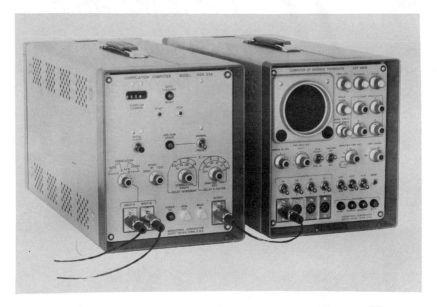

FIG. 10.12 *An averaging and correlating computer.* (*Courtesy Mnemotron Corporation.*)

his tests as he goes along, in the manner shown to be desirable by the emerging results. Initially such averaging computers were developed by biological experimenters, but solid-state machines are now commercially available. They are typically of oscilloscope size, and they output either directly to the oscilloscope or via paper tape, etc. (Fig. 10.12). Since analysis of averaging computation requires a knowledge of statistical signals, quantitative examination is deferred until Sec. 14.4.

11 Principles and Techniques of Feedback Control

11.1 Introduction

The qualitative nature of biological control systems, and particularly of the pupillary-light reflex, has already been presented to provide the reader with some orientation and motivation. However, in order to approach the task of quantitative analysis logically, it has been necessary to build up first a number of basic techniques for dynamic analysis. These techniques do not depend upon the system being closed-loop, and indeed the reader should have acquired an understanding of how such tools can be applied to his open-loop components or systems.

We now can turn to the quantitative exploration of closed-loop control systems using analytical techniques and electronic computers. For didactic purposes it is again desirable to progress logically from simple to complex linear systems in developing the dynamic characteristics of control systems. Unfortunately, nature does not seem to have evolved closed-loop systems for which simple linear models are very plausible, although in previous chapters we have been able to justify linear models for several of the different process and transducer components. In any case, it seems that an analyst needs to understand certain basic linear results before he can tackle complex systems satisfactorily. Therefore, in successive sections the dynamic and stability characteristics of first-, second-, and higher-order closed-loop systems are considered; frequency domain techniques and the root-locus method are applied to the design of compensation or control action in such systems; and finally maximum-effort nonlinear control and its biological significance are touched upon. After this preparatory work the reader should be able to tackle the analysis of realistic nonlinear biological control systems with the aid of electronic computers, as discussed in Chap. 12.

Before continuing, the reader should preferably review Sec. 2.7 dealing with the generic structure of control systems, but the following provides a brief summary. The purpose of a control system (Fig. 11.1) is to cause some physical system either to follow a reference input signal slavishly (servomechanism) or to maintain itself constant at some desired reference level (regulator or homeostat), despite the action of either uncontrollable disturbance inputs or system parameter changes. Servomechanisms are exemplified biologically by neuromuscular systems, and regulators by the homeostatic control of variables in the *milieu interne* of the body. Essential functional elements within the control loop are process, feedback transducer, "comparator," controller, and actuator (effector), although not all elements need necessarily exist separately. Thus, biological "comparators" do not always occur explicitly on the block diagram, but their equivalent action is produced through various nonlinear characteristics (Secs. 2.6, 2.7). The feedback signal is often at a low power level and may need to be transmitted over a considerable distance. The effector usually provides most of the power amplification necessary within the loop, although other components can also be amplifiers. Finally, whereas the controller usually provides compensation for process sluggishness in engineering systems, this function often seems to be performed by the feedback transducer in biological systems.

11.2 Basic Closed-loop Relations

Some important closed-loop relations can be written to combine the loop's various dynamic components, especially for linear systems where transfer functions provide a convenient shorthand symbolism. Since superposition applies in linear systems, the effects of the reference input R and disturbance input D can be considered independent of each other (Fig. 11.1). In Sec. 5.7 the basic method of block-diagram manipulation was demonstrated, and hence we now write down without proof the relevant closed-loop relations. The error E of the system is by definition between R and controlled output C:

$$E = R - C \tag{11.1}$$

but only appears directly in the loop of Fig. 11.1 if the feedback transducer $H_4(s)$ has a unity transfer function [or unless R is suitably

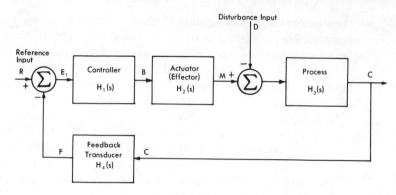

FIG. 11.1   Basic configuration of feedback-control system.

prefiltered by $H_4(s)$ also]. It should be noted that although capitalized variables are being used, this is not for any very strong reason, except that most of the manipulative work is with transfer functions $H_i(s)$. In general, moreover, we shall not define separate symbols for time- and s-domain variables.

After some manipulations the following closed-loop transfer-function relations are obtained for the basic single-variable control system of Fig. 11.1:

Controlled output C—as function of reference and disturbance inputs

$$C(s) = \left[\frac{H_1 H_2 H_3}{1 + H_L}(s)\right] R(s) - \left[\frac{H_3}{1 + H_L}(s)\right] D(s) \quad (11.2)$$

Pseudo-error E_1 (equals error E when $H_4 = 1$)

$$E_1(s) = \frac{1}{1 + H_L} R(s) + \frac{H_3 H_4}{1 + H_L} D(s) \quad (11.3)$$

Manipulated variable M $\quad M(s) = \dfrac{H_1 H_2}{1 + H_L} R(s) + \dfrac{H_L}{1 + H_L} D(s) \quad (11.4)$

where $H_L(s) \triangleq H_1 H_2 H_3 H_4(s)$
$\qquad\quad \triangleq$ open-loop transfer function
$1 + H_L(s) \triangleq$ characteristic function of closed loop

The denominator of each transfer function is the *characteristic function*, which determines the poles of the closed-loop system and hence the dynamics of the system's response. The numerator equals the product of transfer functions and arithmetic signs in the direct forward path from the input of interest to the output of interest.

One aspect of these dynamics which is very important in closed loops is that of *stability*, to be discussed later. The block diagrams of many control systems contain multiple inner loops, but in general they can be converted to the canonical form of Fig. 11.1, so that all the above relations are still valid. This is not the case when there are multivariable inputs and outputs, but we omit consideration of them at present. Although these equations look rather meaningless in their generality, one nontrivial result can be immediately derived;

namely, assuming that each $H_i(s)$ has a steady-state value k_i, these equations provide directly the steady-state values of C, E_1, and M. It is seen that the common denominator now becomes $1 + k_l$, that is, 1 plus the loop gain k_l. Assuming, for example, that $k_4 = 1$, the steady-state error equals $1/(1 + k_l)$ of the reference input, and $k_3/(1 + k_l)$ of the disturbance input.

11.3   Basic Closed-loop Dynamics—First-order System

(A) GENERAL

We consider first the dynamics which result from closing the loop around a first-order process, because these are easily understood and because they illustrate several important characteristic effects of feedback. A first-order (at least) process arises when forces are used to control the velocity of an inertial system, because of the integration between acceleration (proportional to force) and velocity. Hence, it is convenient to return to the man-machine problem of car driving as the formal explanatory vehicle for first-order control-system dynamics.

(B) TRANSFER FUNCTION OF SYSTEM COMPONENTS

Consider the task of a driver on a highway with varying speed limit, traffic, and gradients. Posted speed limits and his relation to the car ahead constitute the main reference inputs to his speed-control system, whereas the variable gradient and wind forces represent typical disturbance inputs. This is a man-machine system, but the analysis we shall apply would be equally valid with an automatic car pilot. An idealized configuration is used, as follows (Fig. 11.2):

Plant. The process or plant is the vehicle, and the transfer function relates the output velocity V to the relevant inputs, the

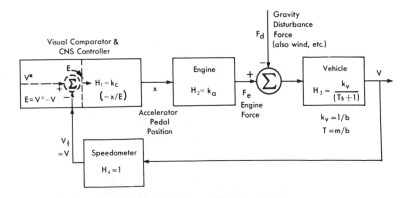

FIG. 11.2   *Manual speed control of vehicle.*

manipulating force of the engine $F_e(x)$, and the disturbance input of resistive force F_d due to gravity, where $F_e(x)$ is again assumed to depend only on the accelerator-pedal position x (Fig. 2.4a). The resistive forces due to still air and road contact depend on the vehicle's output V and are therefore considered part of the plant dynamics:

$$F_b = bV$$

where linearity is also assumed. Hence Newton's second law yields

$$F_e(x) - F_d(t) = m\frac{dV}{dt} + bV$$

and the corresponding transfer relation is

$$V(s) = \frac{k_v}{Ts + 1} [F_e(s) - F_d(s)] \tag{11.5}$$

where $T = m/b$ and $k_v = 1/b$. Thus the vehicle's transfer function is

$$H_3(s) = \frac{k_v}{Ts + 1} \tag{11.6}$$

with the usual understanding that initial conditions can be incorporated whenever they are nonzero.

Comparator, Feedback Transducer, and Controller. The driver provides the error comparator by visual inspection of the speedometer and mental comparison of the result with the currently desired reference speed V^*. The speedometer and driver-perception combination is assumed to have unity gain and negligible dynamics:

$$H_4(s) \equiv \frac{V_f}{V} = 1 \tag{11.7}$$

That is, speed changes are displayed by the speedometer and perceived instantaneously by the CNS of the driver, where they are compatibly compared with the internally stored desired speed. The point regarding compatibility becomes more obvious if we consider an automated version, where the desired reference speed (picked up by a telemetering system from the highway) would appear in the system as, perhaps, a voltage. The compatibility requirement on the feedback signal is, then, that it must also be a voltage and of the same scale—volts per mile per hour, for example. In consequence, the comparator outputs the true error E:

$$E = V^* - V \tag{11.8}$$

into the CNS (or automatic pilot), which is then translated into a suitable command signal for the actuator.

As already mentioned, the controller is the unit whose transfer function the control engineer can normally manipulate in order to synthesize a satisfactory overall system. However, no such single component necessarily appears in biological control systems, and in particular, dynamic compensation is often provided elsewhere. Furthermore, there is some evidence from man-machine experiments that man is most effective as a machine controller when only called upon to operate as a "pure gain" device, that is, when he is not asked to perform differentiations or integrations in his CNS ([11.1] and Sec. 15.4). However, most machine-control tasks are set up so that man does provide both anticipation and memory, so that in detail this statement is a simplification. Furthermore, man operates in man-machine systems as an intermittent type of controller, but development of this point is conveniently deferred until Secs. 11.8 and 12.3. For present purposes, therefore, the controller is taken as a pure gain element,

$$H_1(s) \equiv \frac{x}{E} = k_c \qquad (11.9)$$

Actuator. The actuator closes the loop back onto the plant. The actuator input is the accelerator-pedal position x as determined by the driver's foot. The actuator itself is the engine, producing propulsive force $F_e(x)$ as output, and it clearly provides a large power amplification. We assume that the engine force responds essentially instantaneously to accelerator position as input command, and also that the response is linear:

$$H_2(s) \equiv \frac{F_e}{x} = k_a \qquad (11.10)$$

(C) CLOSED–LOOP TRANSFER FUNCTION

The resulting closed-loop transfer function can be obtained after straightforward manipulation:

$$V(s) = \frac{k_l}{1 + k_l} \frac{1}{T_l s + 1} \left[V^*(s) - \frac{F_d(s)}{k_a k_c} \right] = \frac{\dfrac{k_l}{1 + k_l} V^*(s)}{T_l s + 1} - \frac{\dfrac{1}{1 + k_l} \dfrac{F_d(s)}{b}}{T_l s + 1} \qquad (11.11)$$

where k_l = loop gain = $(k_c)(k_a)(k_v)(1) = k_a k_c / b$
T_l = closed-loop time constant = $T/(1 + k_l)$

The steady-state loop gain k_l is obviously an important parameter of the closed-loop system, to be discussed later. It should be reemphasized that all inner loops must be *collapsed* to evaluate this parameter, as has already been done here for the inner feedback loop containing the $F_b = bV$ term (Fig. 2.4c).

(D) OPEN–LOOP TRANSFER FUNCTION

The characteristics of the closed loop can be best understood by comparison with corresponding open-loop characteristics. Open-loop speed control of cars has already been mentioned in Sec. 2.3, and in practice it is exemplified by early designs in which the throttle control was placed on the steering column, so that open-loop control would exist between the intermittent hand settings. The open loop is conveniently obtained by cutting the feedback path, in this case the speedometer output (Fig. 11.2). The forward path to the actuator and process is not now driven by the error E, but instead directly by the reference input V^*; alternatively, one can say that E numerically equals V^*. In consequence the open-loop transfer function is

$$V = \frac{k_a k_c/b}{Ts + 1} V^* - \frac{1}{Ts + 1} \frac{F_d}{b} \qquad (11.12)$$

In this rather artificial case of open-loop control, the gain $k_a k_c/b$ should clearly equal 1 since the control system has not been correctly calibrated unless V comes to V^* in the steady state. Therefore

$$V = \frac{1}{Ts + 1} \left(V^* - \frac{F_d}{b} \right) \qquad (11.13)$$

The open-loop and closed-loop transfer functions [Eqs. (11.13), (11.11), respectively] have the same functional form, but important dynamic differences arise because their gains and time constants are different.

(E) STEP RESPONSE—OPEN AND CLOSED LOOPS

The step responses of the open and closed loops are found from Eqs. (11.13) and (11.11) as follows:

Open-loop $\quad V(t) = \left(V_1^* - \frac{F_{d1}}{b} \right) (1 - e^{-t/T}) \qquad t > 0 \qquad (11.14)$

Closed-loop $\quad V(t) = \left(\frac{k_l}{1 + k_l} V_1^* - \frac{1}{1 + k_l} \frac{F_{d1}}{b} \right) (1 - e^{-t/T_i}) \qquad t > 0$

$$(11.15)$$

where V_1^* is the magnitude of step input in V^* and F_{d1} is the magnitude of step input in F_d. These responses are plotted separately in Fig. 11.3a and b for the reference and disturbance inputs, respectively, using an arbitrary loop-gain value of 4. The closed-loop response to the reference step starts off much faster than that of the open loop, because the closed-loop effect has reduced the time constant approximately in proportion to the loop gain. On the other hand, the closed-loop response crosses back under the open-loop curve because it is asymptotic to a proportion $k_l/(1 + k_l)$ of the desired response. The increased speed of response is one of the important and desirable dynamic effects of negative feedback. Of

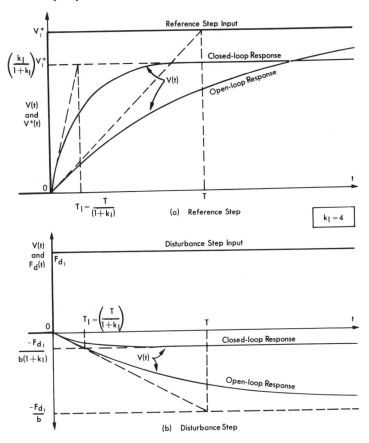

(a) Reference Step

$k_I = 4$

(b) Disturbance Step

FIG. 11.3 *Step responses of first-order control system.*

course, the same initial response can be obtained from the open-loop configuration by initially increasing the gain, but then this gain must later be reduced so that the desired steady-state output is achieved; hence negative feedback may be regarded as a way of automatically changing this gain. However, negative feedback apparently has an undesirable steady-state *droop* effect of failing to bring the controlled output to the desired reference-input level. We shall later see that this is readily overcome by incorporating integral control action as well as proportional control k_c, and in fact this action is used by normal drivers who certainly have no difficulty bringing the achieved speed up to that desired. In some cases, when the loop gain is known, the desired steady-state response can also be obtained by premultiplying the reference input in the ratio $(1 + k_l)/k_l$ so that the closed loop receives a larger value of reference input.

The disturbance response curves for both open and closed loop start off with the same slope, but since the closed loop has the faster time constant T_l, already noted, the response rapidly decays to a steady-state value which is smaller by the factor $1 + k_l$ than the open loop. This represents the desirable disturbance-compensation effect of the closed loop, which cannot be achieved at all by the open loop except to a small extent by the vehicle's inherent passive dynamics.

Effect of Loop Gain. If the loop gain is decreased toward zero the feedback effect is progressively eliminated and the closed-loop responses tend toward the open-loop response, as illustrated in Fig. 11.4 for loop gains of 4 and 1. [The fact that the steady-state levels of these reference responses do not actually tend toward the open loop's steady state (Fig. 11.3a) is due to a detailed problem of manipulation, namely, that the open-loop gain has been set at 1, as representing a basic standard, whereas in the closed-loop case we are letting it tend toward zero.]

Measurable Disturbances. The above results demonstrate again that control of unmeasurable disturbances cannot be achieved without some form of feedback. However, if the disturbance is measurable, the designer is in an easier position because this information can be fed into the comparator-controller to initiate compensatory action before performance has already deteriorated, as has been illustrated already for automatic aperture control of cameras (Sec. 2.10, Fig. 2.28a). In similar fashion, alert drivers do anticipate gradient changes and hence initiate compensatory accelerator-position changes before any change of car speed has occurred which would bring the basic feedback loop into play.

Summary. Finite loop-gain feedback systems can achieve neither perfect steady-state reference "tracking" nor perfect disturbance compensation. The dynamic response of the closed loop is quicker

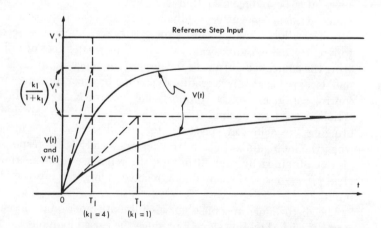

FIG. 11.4 ***Effect of loop gain on closed-loop step responses.***

than that of the open loop, and increasing the loop gain improves both these performance aspects. However, improvement is not achieved without penalty, as examined below.

(F) THE MANIPULATED VARIABLE

Any dynamic improvements achieved by the closed loop must ultimately be effected by the actuator output, in this case the engine force. One should therefore immediately suspect that large values of the manipulated variable will result from increasing the loop gain, and that this will be an inevitable penalty for the improved dynamic performance of the linear system.

Open loop: Working directly from Fig. 11.2, with the V_f path disconnected, and recalling that we set $k_a k_c / b = 1$, there results

$$F_e(s) = k_a k_c V^*(s) = b V^*(s) \qquad (11.16)$$

Closed loop: By noting that $F_e = k_a k_c (V^* - V)$ and using Eq. (11.11),

$$F_e(s) = \frac{k_l}{1 + k_l} \left[\frac{Ts + 1}{T_l s + 1} b V^*(s) + \frac{1}{T_l s + 1} F_d(s) \right] \qquad (11.17)$$

Equation (11.16) shows the expected result that no engine force is called upon to compensate for disturbance effects in the open loop. The closed loop is different, however [Eq. (11.17)], and in particular the transfer function to reference inputs is that of a lead-lag network with $\alpha = 1 + k_l$.

The corresponding step responses are

$$\text{Open loop} \qquad F_e(t) = b V_1^* \qquad t > 0 \qquad (11.18)$$

$$\text{Closed loop} \qquad F_e(t) = \frac{k_l}{1 + k_l} [b V_1^* (1 + k_l e^{-t/T_l})$$
$$+ F_{d_1}(1 - e^{-t/T_l})] \qquad t > 0 \qquad (11.19)$$

They are illustrated in Fig. 11.5, and can be explained qualitatively by following signals around the loop. When a step of V_1^* is imposed, there is an immediate large jump in error and hence in manipulated variable, which subsequently decays to a steady level as the output rises and the error falls. On the other hand, when a step disturbance force is applied, no error is produced until after the step has passed around the loop, through the vehicle's dynamic lag.

Economics and Saturation. The initial value of manipulated variable required for a reference step V_1^* is $k_l b V_1^*$, which is a factor k_l larger than the maximum requirement for the open loop, and is one price that must be paid for the smaller time constant T_l achieved by closed-loop control. Thus, although the required F_e rapidly falls to a relatively low steady-state value as t increases, the design requirement for the *linear* actuator is that it must supply the maximum value at whatever cost, even if it occurs infrequently and for

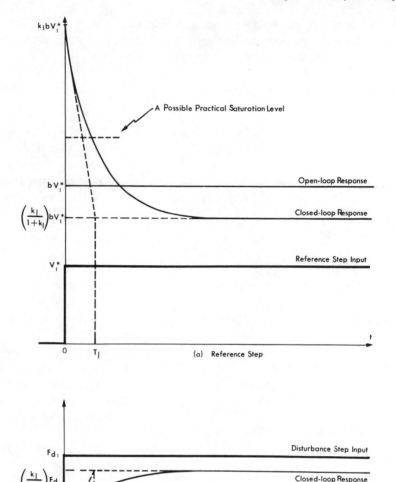

FIG. 11.5   Manipulated-variable step responses of first-order control system.

brief periods.   Obviously, a practical designer looks very carefully at the trade-off possibilities between the desirability of a low maximum value of F_e and the desirability of good dynamic performance in the closed loop.   It should be noted further that the high-peak-rated actuator not only costs more but tends to be larger, so that its own dynamics may become sluggish rather than that of a pure gain k_a, thus partially defeating the dynamic improvement originally sought.   One might therefore suspect that a more economical operation would result from using a saturating or relay-type system, that is, an actuator with a maximum output level such as sketched in Fig. 11.5. The maximum effort would need to be used for a longer period, but the actual dynamic penalty might be small, while the cost and bulk would be reduced.

(G) THE ERROR VARIABLE

For completeness the error transfer functions and step responses are also presented.

$$\text{Open loop} \qquad E = \frac{Ts}{Ts + 1} V^* + \frac{1}{Ts + 1} \frac{F_d}{b} \qquad (11.20)$$

$$\text{Closed loop} \qquad E = \frac{1}{1 + k_l} \left(\frac{Ts + 1}{T_l s + 1} V^* + \frac{1}{T_l s + 1} \frac{F_d}{b} \right) \qquad (11.21)$$

The transfer function $Ts/(Ts + 1)$ between V^* and E in Eq. (11.20) has already been met as an approximate differentiator [Eq. (9.30)], but it is also a special case of the lead-lag family:

$$H(s) = k \frac{\alpha Ts + 1}{Ts + 1} \equiv k\alpha \frac{Ts + 1/\alpha}{Ts + 1}$$

and in the limit as $\alpha \rightarrow \infty$ and $k \rightarrow 0$, so that $k\alpha = 1$ throughout,

$$H(s) \equiv \frac{Ts}{Ts + 1} \qquad (11.22)$$

The step responses are

$$\text{Open loop} \qquad E = V_1^* e^{-t/T} + \frac{F_{d1}}{b} (1 - e^{-t/T}) \qquad t > 0 \qquad (11.23)$$

$$\text{Closed loop} \qquad E = \frac{1}{1 + k_l} \left[V_1^*(1 + k_l e^{-t/T_l}) + \frac{F_{d1}}{b} (1 - e^{-t/T_l}) \right] \qquad t > 0 \qquad (11.24)$$

The plots of Fig. 11.6 are complementary to those of Fig. 11.3a, both showing that zero steady-state error is not achieved with finite loop gain.

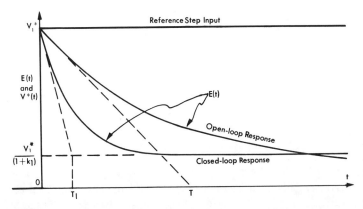

FIG. 11.6   Error-variable step responses of first-order control system.

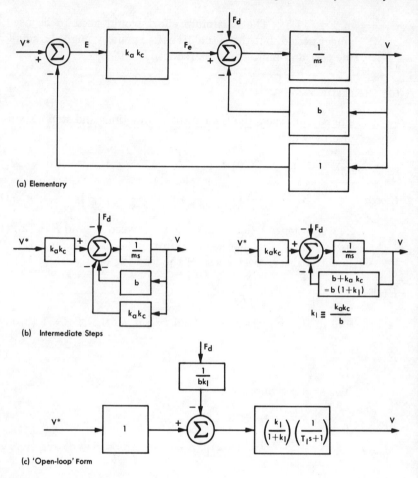

FIG. 11.7   Block-diagram manipulations for first-order control system.

(H) BLOCK DIAGRAMS AS AN ANALYTICAL AID

Block diagrams and their manipulations are useful for understanding the dynamic behavior of feedback systems.   Thus Fig. 11.7 illustrates the elementary block diagram, and successive manipulations clearly show that the viscous effect b and the feedback-controller-actuator effect $k_a k_c$ are dynamically complementary; that is, only their sum determines the net effect.   In physical detail, however, the effects comprise quite different phenomena; for example, b is passive and $k_a k_c$ active.

(I) EFFECT OF LOOP GAIN ON THE OPEN–LOOP POLE

One effect of closing the loop around a first-order model of car and driver has been to reduce the original time constant T of the vehicle to $T/(1 + k_l)$.   Thus, for all the transfer functions which have been derived relating loop variables, there is the same single pole represent-

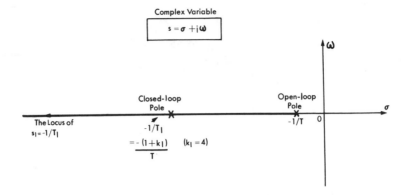

FIG. 11.8 *Locus of closed-loop pole for first-order control system.*

ing the closed-loop behavior, namely, $s_1 = -(1 + k_l)/T$. Clearly this pole tends toward the open-loop pole at $k_l = 0$, and when plotted on the complex plane (Fig. 11.8) it must travel toward minus infinity along the real negative axis as k_l is increased. The resulting system is always stable unless k_l increases negatively, in which case the pole moves to the right from $-1/T$. In particular, when the magnitude of negative k_l exceeds 1 the pole moves into the right half-plane and therefore produces an unstable system—which is plausible because a negative loop gain combined with a negative comparator action constitutes positive feedback. Figure 11.8 sketches the trajectory or locus of the closed-loop pole in relation to the open-loop pole as k_l is varied, which constitutes a simple example of the powerful root-locus method (Sec. 11.7).

This simple linear model predicts a very fast pole indeed as k_l is sufficiently increased. In the car example previously considered, $T \approx 25$ sec (for example from Fig. 10.4b), so that if k_l is made 99, this time constant would nominally be reduced to $T_l = 0.25$ sec. Obviously a first-order model would no longer suffice since this time constant is of the same magnitude as the driver's reaction time and as the engine's time constant (time constant for engine force response after pedal is depressed). Since we concluded in Sec. 9.3 that a pole could be ignored only as long as it was about 10 times faster than all slower poles, it is now clear that increasing the gain in a closed loop is likely to reduce the time constants of basic poles sufficiently that previously fast poles must also be included in the analysis. It should also be noted that the k_l value of 99 just quoted would place impractical transient requirements on the manipulated variable of engine force, as well as on the relevant gears, tires, etc.

(J) FREQUENCY RESPONSE

The Bode plots illustrating the closed-loop effects on controlled output [Eqs. (11.11) and (11.13)] are readily constructed, as shown in Fig. 11.9. The AR plots for V/V^* and V/F_d have their high-frequency

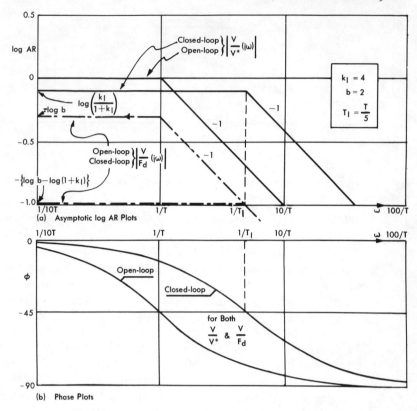

FIG. 11.9   Bode plots for first-order control system.

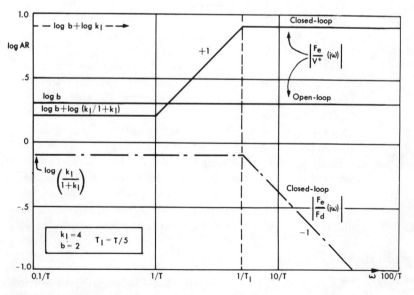

FIG. 11.10   Manipulated-variable Bode plots for first-order control system.

asymptotes shifted to the right by increased loop gain and, therefore, decreased T_l. However, for V/V^* the magnitude increases with k_l, whereas for V/F_d it is reduced, both effects being desirable. The Bode plots for the manipulated variable F_e are shown on Fig. 11.10 and clearly illustrate the high bandwidth requirement resulting from increasing loop gain.

(K) SUMMARY OF CLOSED–LOOP EFFECTS—FIRST–ORDER SYSTEM AND PURE GAIN CONTROLLER

1. Disturbance effects are attenuated in approximate proportion to the loop gain k_l.
2. The open-loop pole is speeded up in approximate proportion to the loop gain.
3. The closed loop, with pure gain control only, cannot reduce steady-state errors to zero for either reference or disturbance inputs, but reduces them in approximately inverse proportion to the loop gain.
4. The maximum transient values required of the manipulated variable increase in approximate proportion to loop gain.
5. The idealized first-order control loop never becomes unstable however high the loop gain, provided the feedback is negative; however, if k_l is sufficiently increased other previously ignored dynamic effects must be considered in the analysis. In practice all loops do indeed become unstable at low finite values of loop gain.

11.4   Basic Closed-loop Dynamics—Second-order System

(A) THE MODEL

A control system described by a second-order characteristic function is the next least complex after the first-order system. Since all basic second-order systems are described by the canonical form, the dynamic characteristics presented in Chap. 8 apply as well for the closed-loop second-order system as for the open-loop case. There are several arrangements by which loop components can together produce a second-order system, particularly:

1. *Position control of an object:* Two integrations are necessary to produce position from force, and hence a position-controlled process is minimally second order.
2. *First-order controlled process:* The velocity-control process of the previous section is itself first order, and if another loop component also has first-order dynamics (for example, the engine actuator), the overall closed-loop system is second order.

   Positional servomechanisms are extremely important in both the biological and the engineering contexts, and an abstract version is used as the example in this section, involving translational motion,

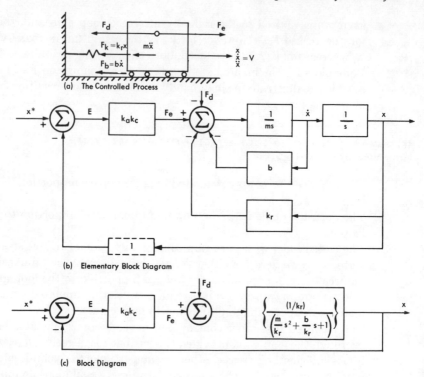

FIG. 11.11   *Basic position control—a second-order control system.*

with mass m, viscous friction b, and resisting spring k_r.   The basic
block diagram (Fig. 11.11) is essentially the same as Fig. 11.7 except
for the extra integration and except that position rather than velocity
provides the feedback.   Again, pure gain actuator k_a, controller k_c,
and feedback transducer $k_f = 1$ have been assumed, together with
disturbance force F_d.

(B) TRANSFER FUNCTION OF POSITION CONTROL SYSTEM

The relation between actuator force and output position is

$$F_e - F_d = m\ddot{x} + b\dot{x} + k_r x \tag{11.25}$$

or

$$x = \frac{1}{ms^2 + bs + k_r} (F_e - F_d) \tag{11.26}$$

By combining this with the other loop components (Fig. 11.11), the
closed-loop transfer function results:

$$x = \frac{1}{ms^2 + bs + k_r + k_a k_c} (k_a k_c x^* - F_d) \tag{11.27}$$

Comparison of Eqs. (11.26) and (11.27) shows that the same form of
closed-loop transfer relation results whether or not there is a resisting

spring force inherent in the controlled process, since the closed-loop spring effect equals the sum of the process spring k_r and the position feedback gains $k_a k_c$. We now obtain the usual operational form for Eq. (11.27) [and for Eq. (11.26)] by noting that the equivalent forward gain of the resistive spring is $1/k_r$ (Fig. 11.11c), so that the loop gain $k_l = k_a k_c/k_r$.

Closed-loop
$$x = \frac{1}{\dfrac{1}{1 + k_l}\dfrac{m}{k_r} s^2 + \dfrac{1}{1 + k_l}\dfrac{b}{k_r} s + 1} \left(\frac{k_l}{1 + k_l} x^* - \frac{1}{1 + k_l} \frac{F_d}{k_r} \right)$$
$$(11.28)$$

$$x = \frac{1}{(s/\omega_{nl})^2 + 2\zeta_l(s/\omega_{nl}) + 1} \left(\frac{k_l}{1 + k_l} x^* - \frac{1}{1 + k_l} \frac{F_d}{k_r} \right)$$
$$(11.29)$$

where w_{nl} and ζ_l are defined by Eqs. (11.33) in Sec. 11.4D.

Open-loop
$$x = \frac{1}{\dfrac{m}{k_r} s^2 + \dfrac{b}{k_r} s + 1} \left(x^* - \frac{F_d}{k_r} \right) \qquad (11.30)$$

$$x = \frac{1}{(s/\omega_n)^2 + 2\zeta(s/\omega_n) + 1} \left(x^* - \frac{F_d}{k_r} \right) \qquad (11.31)$$

where again $k_a k_c/k_r = 1$ has been assumed in order to calibrate the open loop. These open- and closed-loop transfer relations [Eqs. (11.31), (11.29)] for the second-order system should be compared with Eqs. (11.13) and (11.11) for the first-order case; apart from different parameters and order of equation, the formal effect of feedback is seen to be similar.

(C) TRANSFER FUNCTION OF AN ALTERNATIVE SECOND–ORDER SYSTEM

A rather similar second-order system arises if the engine dynamics are included as a first-order lag in the velocity-control system (Fig. 11.12). The following transfer relations apply:

$$V = \frac{1}{(s/\omega_{nl})^2 + 2\zeta_l(s/\omega_{nl}) + 1} \left(\frac{k_l}{1 + k_l} V^* - \frac{T_e s + 1}{1 + k_l} \frac{F_d}{b} \right) \qquad (11.32)$$

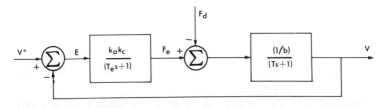

FIG. 11.12 *An alternative second-order control system—velocity control.*

where

$$\omega_{nl}{}^2 = \frac{1 + k_l}{TT_e} \qquad \frac{2\zeta_l}{\omega_{nl}} = \frac{T + T_c}{1 + k_l}$$

Note that because the disturbance signal F_d enters between the two first-order lags, its closed-loop transfer function has a first-order numerator, instead of the constant value which occurs in Eq. (11.29).

(D) EFFECT OF LOOP GAIN ON SYSTEM'S POLES

The effect of increasing loop gain is certainly to "speed up" the poles of the second-order system, but not so simply as in the first-order system because of the ability of second-order systems to oscillate. The effect can be neatly illustrated by following the locus of the poles as the loop gain is increased from zero, the open-loop condition. Some detailed but straightforward algebra is involved in obtaining the relation between open-loop ζ and ω_n parameters of Eq. (11.31) and those of the closed loop ζ_l and ω_{nl} [Eq. (11.29)]; thus, there results

$$\omega_{nl} = \omega_n (1 + k_l)^{\frac{1}{2}} = \left[(1 + k_l) \frac{k_r}{m} \right]^{\frac{1}{2}}$$

$$\zeta_l = \zeta (1 + k_l)^{-\frac{1}{2}} = \left(\frac{1}{1 + k_l} \frac{b^2}{4mk_r} \right)^{\frac{1}{2}} \qquad\qquad (\textit{11.33})$$

Hence also

$$\zeta_l \omega_{nl} = \zeta \omega_n = \frac{b}{2m}$$

The effect of increasing loop gain is therefore to increase the undamped natural frequency and decrease the damping ratio, each in a square-root relation, so that their product remains constant at a process-determined value of $b/2m$. Since the poles are real and separated when the damping ratio exceeds unity, the most general locus of closed-loop poles is given when k_l increases in a loop consisting of an initially overdamped system; the poles first come together at the critical damping condition for some value of k_l and then separate along a line parallel to the imaginary axis, becoming ever more oscillatory in nature as k_l is further increased (Fig. 11.13). The constancy of $\zeta_l \omega_{nl}$, however, means that the increasing oscillation of the transient responses occurs within the same exponential decay envelope (Fig. 8.8). Figure 11.13 illustrates these points for the example in which the open-loop process has separated time constants $T_1 = 2T_2$; the parameters for the previously used loop gain $k_l = 4$ are $\zeta_l = 0.475$ and $\omega_{nl} = 2.23\omega_n$.

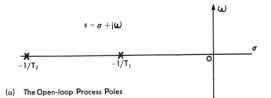

(a) The Open-loop Process Poles

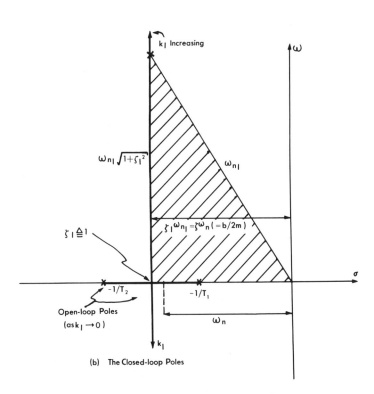

(b) The Closed-loop Poles

FIG. 11.13 **Loci of the closed-loop poles for second-order control system.**

(E) THE STEP RESPONSES

The closed- and open-loop responses to step inputs in reference position x^* and disturbance force F_d are obtainable directly from Table 8.2 in conjunction with Eqs. (11.29) and (11.31). As exemplified in Fig. 11.14, this finite-gain second-order closed loop also fails to achieve the ideal steady-state responses (refere. ʒ response of unity and disturbance response of zero). Dynamic response is faster, but in fact the possible improvement is limited because as k_l increases the response becomes unsatisfactorily oscillatory; in particular, although the undamped natural frequency ω_{nl} increases, the real part of the poles $\zeta_l\omega_{nl}$ remains constant (Fig. 11.13), and thus the response con-

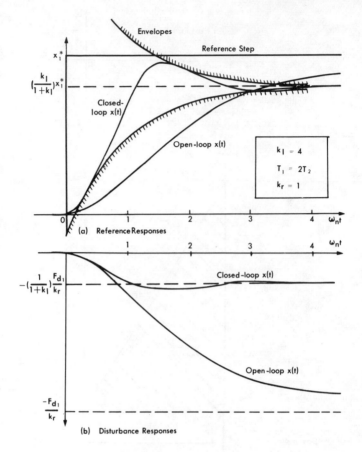

FIG. 11.14   Step responses of the second-order control system.

tinues to decay only within the envelope shown.   Actually there are
well-established techniques to overcome this problem, which are
considered in Sec. 11.5.

(F) FREQUENCY RESPONSE

The comparison of open- and closed-loop Bode plots is much as one
should suspect.   The log AR plots both fall off with a -2 slope at
high frequencies (Fig. 11.15), but in the intermediate-frequency
range there is considerable improvement.   For reference inputs the
system bandwidth increases by a factor of about 6 with $k_l = 4$,
approximately $\omega_n/2$ open loop, to $3\omega_n$ closed loop.   This is the correl-
ative advantage to the decrease in rise time evident in the step
response of Fig. 11.14.   The closed-loop phase is always less than
the open-loop phase, which is also a sign of better tracking dynamics.
The major feature of the disturbance response is of course the desired
extra attenuation in the low-frequency range.

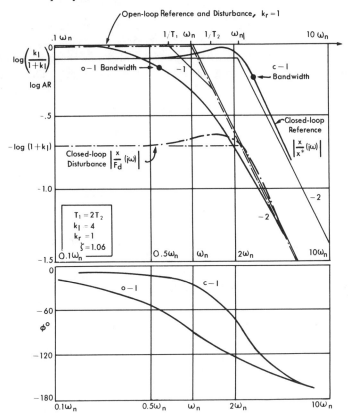

FIG. 11.15 Bode plots for second-order control system.

(G) SUMMARY

Pure gain control of a second-order process produces generally similar characteristics to that for first order, except that the system becomes progressively more oscillatory as loop gain is increased. In practice, both nature and the engineer augment pure gain control with more flexible techniques in order to overcome this difficulty.

11.5 Stability and Compensation of Linear Control Systems

The designer (synthesizer) of systems has a different motivation than the systems analyst. Thus the control engineer and the industrial manager are primarily concerned with synthesizing systems from available components, which will be optimal in some reasonable sense, while the biologist has classically been concerned with the analysis of existing systems. Unfortunately, however, almost no direct synthesis tools exist which allow a design to be computed automatically once a problem is specified. In practice, therefore, systems are typically designed by iterative analysis of promising (and often unpromising) procedures over varying parameter ranges.

Also in practice, the biological systems analyst must work similarly because he must design a mathematical model by successive comparisons of predicted with experimental data (analysis) in order to develop understanding of his complex systems. Therefore, the empirical methods developed by control-system designers should prove useful to biological analysts, and some are very briefly presented below. There is, however, much literature on the subject, for which the reader should consult the Bibliography.

(A) SYSTEM COMPLEXITY AND INSTABILITY

Although the idealized first-order control system cannot oscillate, and although the idealized second-order system's oscillations always decay, practical control systems can and often do become unstable if system parameters are not correctly adjusted, so that these simple models cannot be valid in extreme cases. Instability is used here in the limited sense that at least one pole temporarily moves into the right half-plane so that an oscillation starts to build up, for example:

1. The system "blows up" catastrophically; a predator-prey or disease phenomenon results in extinction of some species, individual organism, etc.
2. A stable limit cycle sets in due to inherent saturating nonlinearities, etc. (Parkinson's disease, hippus, clonus, etc.), or due to a relaxation-oscillation mechanism (for example, the cardiac-control cycle).
3. The runaway phenomenon itself produces system changes resulting in decay or stability (adaptation or learning being perhaps most obvious—for example, when my unpracticed daughter suffers increasing steering oscillations on her bicycle, she rapidly learns to damp them).

Stable limit cycles are of course widely used in oscillators for both physical and biological timing mechanisms, for example, pure-delay neuron rings in various species [10.13]. Many adaptive control systems are also designed to hunt around optimal performance (Sec. 15.3). Furthermore, certain nonlinear control systems exhibit intentional stable cycling, as, for example, in conventional house heating, and evidently to some extent in the sweating and vasoconstriction effects of the thermoregulatory control system. Having noted these limitations and exceptions, we can state that most biological and engineering control systems are required not to exhibit sustained oscillatory behavior. It is therefore relevant to consider briefly the mathematics of closed-loop-system instability.

(B) STABILITY AND FREQUENCY RESPONSE

In Sec. 2.9 the expression $H(s) = -1$ expressed mathematically the pictorially derived conclusion that sustained oscillations occur in a negative-feedback loop if the total loop has a dynamic gain of unity or greater when the phase lag is 180°. Effectively, then, there are

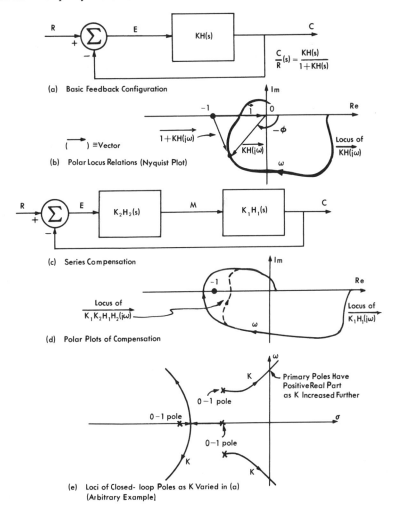

(a) Basic Feedback Configuration

(b) Polar Locus Relations (Nyquist Plot)

(c) Series Compensation

(d) Polar Plots of Compensation

(e) Loci of Closed-loop Poles as K Varied in (a)
(Arbitrary Example)

FIG. 11.16 *Control-system compensation and stability.*

two phase inverters which give rise to a total effect of positive feedback, so that the loop is self-regenerating when its gain is unity or greater. Actually this definition of stability is insufficient for the so-called *conditionally stable systems*, but their treatment is not justified at this point.

The analytical tools developed since Chap. 2 now permit a more analytical approach to the stability criterion. The nomenclature used is conventional (Fig. 11.16a) and separates the steady-state gain term K from the dynamic terms $H(s)$ involving polynomials in s; in fact, K is identically the loop gain k_l of previous sections. From Eq. (11.2) or by inspection of the figure the closed-loop transfer function is written

$$\frac{C}{R}(s) = \frac{KH(s)}{1 + KH(s)} \qquad (11.34)$$

In frequency response terms $KH(j\omega)$ and $1 + KH(j\omega)$ are readily definable vectors on the complex plane; thus, as shown on Fig. 11.16b, $KH(j\omega)$ is the basic frequency response vector from the origin, while the vector $1 + KH(j\omega)$ is the vectorial sum of 1 and $KH(j\omega)$ and therefore runs from -1 to the tip of the $KH(j\omega)$ vector. The magnitude and phase of $C/R(j\omega)$ are then given respectively by the quotient of magnitudes and phase difference of these vectors; that is, the closed-loop frequency response vector $C/R(j\omega)$ can graphically be derived from the open-loop vector $KH(j\omega)$. The instability condition now follows directly because as the $KH(j\omega)$ vector moves closer to the -1 point at some frequency, $1 + KH(j\omega)$ becomes smaller and hence $C/R(j\omega)$ becomes larger. When $KH(j\omega)$ passes through the -1 point the magnitude of $C/R(j\omega)$ becomes infinite at some frequency, corresponding to two undamped poles lying on the imaginary axis and hence to undamped nondecaying oscillations in the impulse response. So we can state that *if the -1 point lies on the right as one "walks around" the polar locus in the direction of increasing ω, the system has a right-half-plane pole and is actively unstable.* This definition is an elementary form of the *Nyquist stability criterion* which encompasses a broader range of conditions, such as conditional stability, at the expense of some complexity. Since it is not clear that biological systems normally incur these broader conditions, the full treatment is omitted.

One empirical design for control systems in the frequency domain has evolved from this result. It can be crudely stated as: Plot the locus of the fixed elements (process) of the system $K_1H_1(j\omega)$ (Fig. 11.16c), and then artfully insert compensation elements $K_2H_2(j\omega)$, often in series, so that the combined locus bends suitably away from the -1 point (Fig. 11.16d).

(C) STABILITY IN THE s PLANE

Another interpretation of Eq. (11.34) in pole-zero terms leads to the fruitful and powerful root-locus method. First it should be recalled that $H(s)$ is in general a ratio of two s polynomials which specify the poles (denominator) and zeros (numerator) of the open-loop system (K does not affect these). The poles of the closed loop, however, are those s values which provide roots of the denominator expression of Eq. (11.34), that is, the characteristic function

$$1 + KH(s) = 0 \qquad \text{or} \qquad KH(s) = -1 \qquad (11.35)$$

As shown in Sec. 11.7 a graphical construction in the complex plane is available to locate these closed-loop pole values of s and trace their loci as functions of K or other system parameters (Fig. 11.16e). Since typically some closed-loop poles tend to wander into the right half-plane as a parameter of interest, such as K, is increased to obtain faster responses, the system compensation problem in root-locus terms becomes one of inserting desirable pole-zero combinations to retard this inevitable onset of instability (and linear realizability).

(D) ANALYTICAL METHODS

There is also an analytical method for determining whether a system has right-half-plane poles which can be modified to give some useful quantitative information; it is called *Routh's test* and is presented in Sec. 11.7 for the interested student. There are other and more complex stability methods for nonlinear systems in the applied mathematics literature; for example, see Gibson in the Bibliography for a detailed account.

(E) DYNAMIC COMPENSATION TECHNIQUES

Since it is not clear that the compensation techniques of engineering control systems are freely used in living systems, we only mention a few of the conceptually more interesting examples.

Feedback of Derivatives. Referring to the position-control system and specifically to Fig. 11.13, Eqs. (11.33) show that our inability to decrease the settling time of the exponential decay envelope arose because b/m was fixed. However, since increasing the position-feedback gain $k_a k_c$ resulted in an increased closed-loop spring stiffness $k_a k_c + k_r$, and hence increased ω_{nl}, it should be intuitively clear that providing extra velocity feedback should complement b and hence move $\zeta_l \omega_{nl}$ to the left in Fig. 11.13. Furthermore, the normal viscous damping categorized by b is dissipative of power, whereas artificial velocity feedback can be implemented by such devices as d-c generators (tachometers) with essentially no power drain if the circuit is correctly designed. From a frequency response point of view, the important feature of increased damping is that the high resonant peak of a low damping ratio can be eliminated at the same time as bandwidth is increased.

In a similar way, higher derivatives, especially the second, can be fed back, although they are of more value in higher-order systems than the second. However, there is an early limit to the efficiency of derivative feedback, indeed of all compensation, in improving the bandwidth of high-order systems, especially when the controlled process has significant pure time delay with its consequent large phase lag. Even if significant improvement can be obtained by modifying the C/R transfer function, the designer had better carefully watch that his actuator can supply the resulting M/R peak requirements, and that it is economical to do so.

Lead-Lag and Proportional-plus-Derivative Compensation. Lead-lag networks have phase-leading properties for $\alpha > 1$ which make them of interest in compensating sluggish processes, because they can bend the open-loop locus "inside" the -1 point (Fig. 11.16d). The so-called proportional-plus-derivative (P + D) *controller* is widely used in chemical-engineering process control, but at high frequencies it operates like a lead-lag network (Sec. 9.6). In any case, the phase-leading property of both elements fulfills the same purpose as derivative feedback.

(F) STEADY–STATE–COMPENSATION TECHNIQUES; LAG–LEAD AND PROPORTIONAL–PLUS–INTEGRAL COMPENSATION

The compensation techniques of Sec. 11.5E are all aimed at improving the dynamic response of the closed loop. There is often an equally important need for reducing steady-state errors to zero, but it was noted that pure gain control could not ensure this for either reference or disturbance inputs. In fact, a follow-up device is needed which will continue to increase the gain as long as an error exists. Pure integral control is the ideal device and this can be realized without much difficulty:

$$H(s) = \frac{1}{T_i s} \qquad\qquad (11.36)$$

Unfortunately, this element is very destabilizing because it contributes 90° phase lag at all frequencies. Consequently proportional-plus-integral (P + I) control is another widely used process-control element because, although steady-state droop is still eliminated, the large phase-lag condition is restricted to a region of relatively low frequencies, where stability is normally less of a problem. The lag-lead element (lead-lag with $\alpha < 1$) is another approximation widely used by electrical control engineers, but this can only attenuate the steady-state error rather than eliminate it since its gain remains finite at zero frequency (Fig. 7.22).

(G) COMBINED COMPENSATION

In a complex system all promising compensation techniques will be tried. Typically, however, the electrical control engineer combines lead-lag and lag-lead compensation networks at resonant and low frequencies, respectively, at least for simple systems; correspondingly, the chemical process-control engineer has used three-term controllers (P + I + D) whose parameter settings are typically based on the experimentally determined transient or frequency responses of his controlled process. Goldman [11.2] discusses the body's glucose-control system in the context of these three control elements (as well as other nonlinear elements) and suggests that all three are present; in particular, he postulates that fat-depot storage constitutes the integral mode. Finally, in neural control the neuron's method of summing incoming action potentials is suggestive of an approximate integrator, although this occurs spatially as well as temporally.

▼

11.6 Frequency Response Techniques

In this section two frequency response techniques are briefly described which enable the closed-loop frequency response to be predicted from the open-loop frequency response plots; the first is a fast approximate method, and the second is more detailed.

(A) THE ASYMPTOTIC APPROXIMATION TECHNIQUE

This method uses only the log AR information of the open loop $|KH(j\omega)|$, but as we have noted, the phase information is not independent in minimum-phase systems. The relevant expression is Eq. (11.34), here expressed in frequency response form,

$$\frac{C}{R}(j\omega) = \frac{KH(j\omega)}{1 + KH(j\omega)} \qquad (11.37)$$

When $KH(j\omega)$ has small magnitude ($\ll 1$) in some ranges of frequency, the denominator approximates to unity magnitude and the closed-loop AR approximates to that of the open loop; if

$$\log |KH(j\omega)| \ll 0 \qquad \log \left|\frac{C}{R}(j\omega)\right| \approx \log |KH(j\omega)| \qquad (11.38)$$

When the loop gain K tends to zero, then the system's steady-state behavior is little modified by the feedback path, but Eq. (11.38) demonstrates also that dynamically the closed-loop system behaves as if it had no feedback when it is oscillated at frequencies to which the open-loop dynamics $KH(j\omega)$ are unresponsive. On the other hand, if

$$\log |KH(j\omega)| \gg 0 \qquad \log \left|\frac{C}{R}(j\omega)\right| \approx 0 \qquad (11.39)$$

Equation (11.39) shows that however large the open-loop static and dynamic gains at certain frequencies, the closed loop only approximates to a unity-gain following device. Equations (11.38) and (11.39) also allow asymptotic approximations to be superimposed easily on the open-loop log AR plot, and in particular to cut off all $KH(j\omega)$ peaks and follow $KH(j\omega)$ elsewhere, as shown in Fig. 11.17 for a hypothetical system. However, it should be noted that closed-loop resonant peaks are possible whenever the closed-loop asymptote changes slope by -2 logs/decade, for example, point B in Fig. 11.17, although there may have been no resonance at all in the open-loop plot.

The practical utility of this technique appears in the design problem of adjusting parameters until satisfactory closed-loop performance is achieved. For example, consider finding the k_l which is satisfactory for the second-order position servomechanism treated in Sec. 11.4. In the open-loop log AR plot, k_l appears only as an additive constant, and hence, varying k_l only moves the same plot up and down on the ordinate. However, as shown on Fig. 11.18, the cutoff at log AR = 0 applies invariantly for the closed-loop approximation, so that the latter curve is readily determined. Notice that as k_l is increased the closed-loop asymptotic breaks change from two successive -1 log/decade breaks to one -2 break, signifying the onset of closed-loop resonance. For $k_l = 10$ the true curve is shown and indeed has a considerable resonance peak. This true curve also shows that the low-frequency approximation is in error because it assumes unity gain, whereas the actual gain is $k_l/(1 + k_l)$; in consequence the estimate of resonant frequency is not quite right either. In practice, only one open-loop log AR plot need be drawn, since the effect of varying k_l can be equivalently and more easily achieved by moving the ordinate scale up and down. Hence, within known limitations a quick way is available for estimating closed-loop dynamics from open-loop frequency response information.

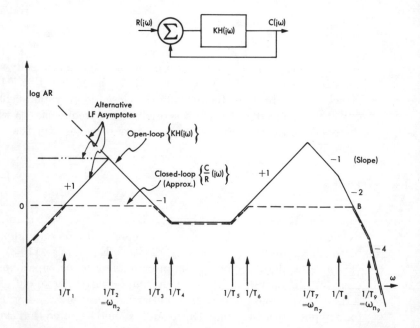

FIG. 11.17 *Asymptotic approximation technique—from open to closed loop.*

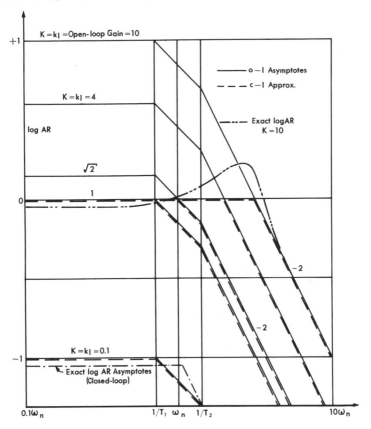

FIG. 11.18 *Approximation technique applied to second-order control system.*

(B) LOGARITHMIC POLAR PLOTS (NICHOLS CHARTS)

This method enables the complete Bode-plot data for the closed loop (Fig. 11.16a) to be found from the open-loop Bode-plot data. Moreover, various compensation schemes, including pure gain adjustment k_l, can be graphically tested for their effects on the closed-loop behavior. The logarithmic polar plot uses the open-loop log AR as ordinate, open-loop phase ϕ as abscissa, and frequency ω as parameter; corresponding closed-loop values of log AR and ϕ can then be interpolated directly from the superimposed closed-loop grid (Fig. 11.19).

The simplest use is for loop gain setting, in which case the open-loop curve is moved up or down until it is tangential to some desirable maximum value of closed-loop $\log AR_{max} \triangleq \log M_p$, where M_p is usually around $\sqrt{2}$ ($\log M_p = 0.15$). The allowable upward displacement equals the extra $\log k_l$ permitted in the loop. For example, the open-loop Bode data of Fig. 11.15 have been plotted on Fig. 11.19 both for $k_l = 1$, the basic curve, and for $k_l = 4$ by trans-

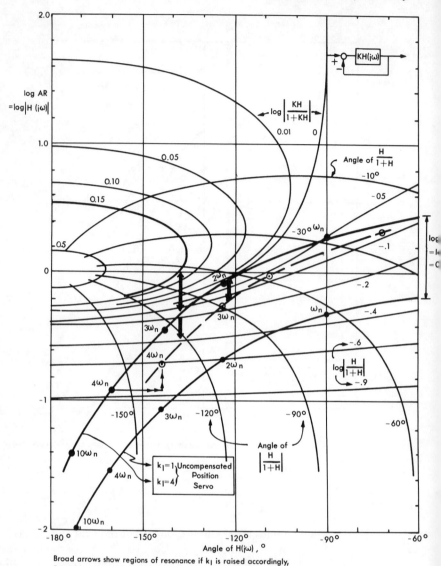

FIG. 11.19 Compensation on logarithmic polar plots—Nichols chart.

lating the curve upward an amount $\log 4 = 0.6$. The closed-loop
Bode data can then be read off, and the reader should check that the
data agree with the $|(x/x^*)(j\omega)|$ curve of Fig. 11.15.   In order to find
the extra k_l permitted before $\log M_p = 0.15$ is exceeded, the curve is
translated upward again until it is tangential with this $\log M_p$
contour. [One confusing detail arises, however, because the $\log M_p$
is on an absolute scale, whereas we should preferably only adjust for
this degree of resonance with respect to the closed-loop low-frequency
gain of $k_l/(1 + k_l)$ (see Fig. 11.15).   The result, using the absolute
scale, is that the total $\log k_l$ permitted is about 0.90 at a frequency

$\omega_{nl} \approx 2.8\omega_n$, and is not quite in agreement with the prediction of Eqs. (11.33). The reader should satisfy himself of these details.]

The incorporation of compensation for the process dynamics provides a less trivial use of the Nichols chart. Here we consider only the incorporation of a series lead-lag network as shown in Fig. 11.20, and we choose an arbitrary value of $\alpha = 2$ because its log AR and ϕ data are already shown on Figs. 7.22 and 7.23. The maximum phase lead of approximately 20° is achieved at $\omega = 1/(\sqrt{\alpha}\,T)$ and should be located at a frequency which will maximize the bandwidth of the overall closed loop within the stated M_p constraint. In order to keep this example simple, we place the ω of maximum phase lead at the resonant frequency which arose in the uncompensated loop with $k_l = 4$, namely $\omega = 2.23\omega_n$. Thus

$$T = \frac{1}{\sqrt{\alpha}\,\omega} = \frac{1}{\sqrt{2}\,(2.23)\omega_n} = \frac{0.316}{\omega_n} \tag{11.40}$$

In practice, lead-lag compensation is usually applied with greater α values, and to higher-order systems whose phase lag exceeds $-180°$, but it would be unnecessarily time-consuming to introduce a higher-order process just to demonstrate this point. The resultant plot is obtained by adding the log AR and ϕ data of the original system and the lead-lag network with the results shown in Fig. 11.19. The general effect is that the curve shifts to the right for any given frequency, away from the resonance area; for example, note the $4\omega_n$ point. In consequence, extra loop gain is permissible for the same closed-loop M_p value, which reduces steady-state errors, and resonance occurs at a higher frequency, which increases bandwidth and reduces rise time. In this case the resonance condition of the uncompensated $k_l = 4$ case is recovered by adding an extra logarithmic value of approximately 0.25, i.e., the new $k_l = 7$, and the resonant frequency is increased from nearly $2\omega_n$ to nearly $3\omega_n$. Alternatively, if a criterion of log $M_p = 0.15$ is chosen, the total permitted gain becomes about 14, compared with 8 in the uncompensated case; and the respective resonant frequencies become about $3.7\omega_n$ and $2.8\omega_n$ (see Fig. 11.19). Of course, this better performance must be paid for by higher manipulated variable requirements, but at least there is now no stability problem.

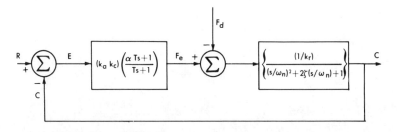

FIG. 11.20 Configuration for lead-lag compensation.

This extremely brief treatment has only been intended to show the motivation behind frequency response compensation and to give a glimpse of its techniques. Many details have been omitted which the interested reader can find in most control-engineering texts.

▼

11.7   *The Root-locus Method*

(A) THE METHOD

The root-locus method is a graphical method in the complex plane for describing and adjusting closed-loop poles and zeros by working from the open-loop poles and zeros. Equation (11.34), the basic expression for this method, shows that the closed-loop poles are determined from the roots of the characteristic function $1 + KH(s)$ [Eq. (11.35)], while the closed-loop zeros equal the open-loop zeros identically. (The last point is perhaps not obvious and is elaborated shortly.) Interest therefore centers on the characteristic function, which we now expand as a ratio of products of poles p_j and zeros z_i:

$$-1 = K \frac{(s + z_1) \cdots (s + z_i) \cdots (s + z_m)}{(s + p_1) \cdots (s + p_j) \cdots (s + p_n)} \qquad (11.41)$$

The problem now is to find those n points $s_1, \ldots, s_n$ on the s plane which solve the above equation for given p_j, z_i, and K, since these are in fact the closed-loop poles. Equation (11.41) makes available a simple graphical rule based on the fact that (-1) on the left-hand side is a vector of unit magnitude and of phase $(180 \pm l360)°$, where l is an integer. On the right-hand side, K has magnitude but zero phase, so that the factors of $H(s)$ must alone satisfy the angle condition. In Fig. 11.21 each of these factors is shown as a vector, from

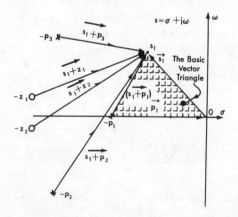

FIG. 11.21   Pole-zero vectors in the complex plane.

open-loop pole or zero to some arbitrary s value s_1. The rule for finding s values which are closed-loop poles is therefore: *The sum of angles of "zero" vectors minus the sum of angles of "pole" vectors must equal* 180° ± *l*360°. Actually this construction produces a set of loci rather than individual points, and the second condition, magnitude, is needed to specify a given point on the loci; thus, the loop gain K can be marked on the loci as a parameter. Numerically K equals the product of pole-vector magnitudes divided by zero-vector magnitudes, and is determined graphically in straightforward, if tedious, fashion. In order to make the point clearer that the closed-loop zeros identically equal the open-loop zeros, we expand Eq. (11.34) in the same fashion as Eq. (11.41):

$$\frac{C}{R}(s) = K \frac{(s + z_1) \cdots (s + z_i) \cdots (s + z_m)}{[(s + z_1) \cdots (s + z_i) \cdots (s + z_m)] + K[(s + p_1) \cdots (s + p_j) \cdots (s + p_n)]} \tag{11.42}$$

The numerator of the closed-loop transfer function is then identically that of the open-loop function, and therefore the zeros must be also.

(B) GAIN–ADJUSTMENT EXAMPLE

The method is first illustrated for simple closed-loop gain control of a process characterized by a first-order cascade:

$$H(s) = \frac{K}{(1 + Ts)^n} \equiv \frac{K}{T^2} \frac{1}{(s + 1/T)^2} \quad \text{for } n = 2 \tag{11.43}$$

The angle condition is readily checked by drawing vectors from the poles to any point on the loci. In regard to the magnitude condition, the closed-loop poles start at the open-loop poles for $K = 0$ and then move away toward infinity as K increases. For numerical calculations it should be noted that the required vectors appear in the right-hand-side version of Eq. (11.43) only. See Fig. 11.22.

When we consider $n = 3$ and 4 there must be an equal number of loci moving away from the multiple open-loop pole in each case [see Eq. (11.42)], and to satisfy the angle condition they must be as shown in Fig. 11.22. The striking difference from the previous loci is that in each case two loci move across into the right half-plane at some suitable values of loop gain and frequency. At crossover there are two undamped closed-loop poles, and then the presence of one or more damped poles does not stabilize the system. The necessary gains for this oscillatory condition are readily shown, trigonometrically or otherwise, to be 8 and 4, respectively, and the frequencies are those of the imaginary axis intercepts, namely, $\sqrt{3}/T$ and $1/T$, respectively. At this condition the poles for the third-order system can be

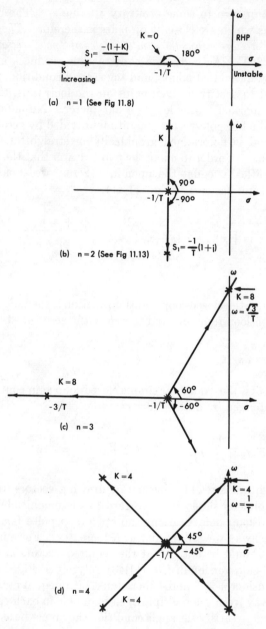

(a) n = 1 (See Fig 11.8)

(b) n = 2 (See Fig 11.13)

(c) n = 3

(d) n = 4

FIG. 11.22 Root loci for $H(s) = \dfrac{K}{(1 + Ts)^n}$.

written by inspection in transfer-function form but without the gain term:

$$\frac{C}{R}(s) = \frac{1}{\left(s + j\frac{\sqrt{3}}{T}\right)\left(s - j\frac{\sqrt{3}}{T}\right)\left(s + \frac{3}{T}\right)} \equiv \frac{1}{\left(s^2 + \frac{3}{T^2}\right)\left(s + \frac{3}{T}\right)}$$

$$\triangleq \frac{1}{(s^2 + \omega_n{}^2)\left(s + \frac{1}{T_1}\right)} \quad (11.44)$$

where $T_1 = T/3$, and $\omega_n{}^2 = 3/T^2$. The gain term is derived by combining the closed-loop effect of the loop gain, namely, $K/(1 + K) = \frac{8}{9}$, with the product of pole values [Eqs. (11.43), (11.44)], i.e.,

$$\frac{C}{R}(s) = \frac{K}{1 + K}\frac{\omega_n{}^2}{T_1}\frac{1}{(s^2 + \omega_n{}^2)(s + 1/T_1)} = \frac{\frac{8}{9}}{[(Ts)^2/3 + 1](Ts/3 + 1)}$$

The reader should check this result by using Eq. (11.2), with a 3 equal-time-constant plant and loop gain of 8.

(C) THE LEAD–LAG COMPENSATION PROBLEM

We now apply root-locus techniques to illustrate the lead-lag compensation example of Sec. 11.6. The original open-loop poles and the closed-loop poles for $k_l = 4$ with pure gain control are shown on Fig. 11.23a. The open-loop pole and zero of the lead-lag circuit are then added as shown in Fig. 11.23b, by noting that the value of the pole $-1/T$ is given from Eq. (11.40) as $-3.16\omega_n$ and that therefore the value of the zero $-1/\alpha T$ for $\alpha = 2$ is $-1.58\omega_n$. The important effect of the lead-lag pole and zero is to move the real part of the complex-conjugate loci back from $-1.06\omega_n$ to $-1.85\omega_n$ at high gain values, and hence almost to halve the time constant of the exponential decay envelope at reasonably high values of loop gain. Thus, as already noted in Fig. 11.19, a value of $k_l = 7$ is now permitted instead of 4 for the same damping ratio (≈ 0.5) of the conjugate pair of poles. The closed-loop transfer function also contains a real pole at approximately $-1.7\omega_n$, and the open-loop zero at $-1.58\omega_n$, but it is left to the reader to write this out, either from Eq. (11.2) or by inspection of the figure.

Wherever a pole and zero are very close together, their dynamic effects almost cancel each other. If the lead-lag zero had been placed precisely on top of the open-loop system's fast pole, the new system would have been a basic second-order system with poles at $-3.16\omega_n$ and at $-1/T_1 = -0.707\omega_n$. By increasing α the first of these poles could be made faster still, and hence the real part of the complex-conjugate loci could again be increased. Notice also that at $k_l = 7$ the real pole and zero of the closed loop almost cancel each other.

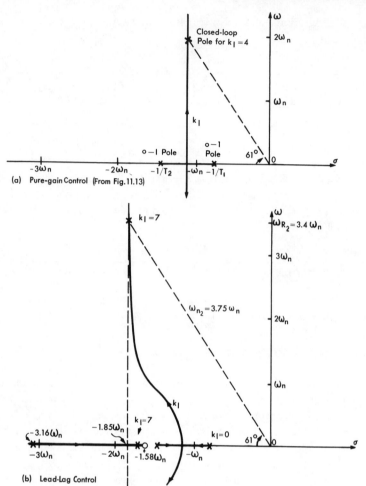

FIG. 11.23   Root-locus method for compensation.

(D) HELPFUL CONSTRUCTION RULES

The graphical technique described does involve considerable manual work, but fortunately this can be reduced considerably by using several auxiliary rules for root-loci construction which follow from the basic definition.   The auxiliary rules are largely self-explanatory, and it is left to the reader to apply them to the above and other examples.

Number of Root Loci.   There are as many loci as the order of the characteristic function [denominator of Eq. (11.42)] and generally this means as many as there are open-loop poles, namely n.

Termination of Loci.   One locus closes onto each open-loop zero as $K \to \infty$.   This leaves $n - m$ loci which must terminate at infinity.

Asymptotes of Loci. As K increases and the $n - m$ poles move toward infinity, all open-loop poles and zeros can be considered concentrated at the same point for computational purposes. In consequence the $n - m$ loci tend asymptotically to straight lines at angles defined by $(180 + l360)/(n - m)$, $l = 0, \ldots, n - m - 1$. These asymptotes intersect on the real axis at a point defined by

$$\frac{\sum\limits_{j=1}^{n} p_j - \sum\limits_{i=1}^{m} z_i}{n - m} \qquad (11.45)$$

Loci on Real Axis. Those sections of the real axis which lie to the left of an odd number of real poles and zeros constitute root loci. The points at which loci break away (in pairs) from the real axis can be found by writing the trigonometric sum of angle changes for an arbitrary point displaced an infinitesimal distance ϵ from the real axis, and requiring that this angle change be zero (angles from zeros counted opposite to those from poles).

Complex Poles and Zeros. Loci are symmetric about the real axis. The angles of departure from complex poles, or arrivals at complex zeros, can be found by evaluating the combined angle of the vectors from all other poles and zeros to a point infinitesimally distant from the pole or zero under study. The difference of this combined angle from $(180 + l360)°$ defines the desired angle of departure or arrival.

Conditions at Onset of Instability; Routh's Test. The patterns of root loci developed in Fig. 11.22 show that whenever the excess of open-loop poles over zeros is three or more, some loci eventually cross into the right half-plane as the gain is increased. Routh's test permits determination of the gain and frequency at the crossover by applying certain manipulations to the characteristic function, the denominator of Eq. (11.42). In general form, the characteristic function may be written

$$a_n s^n + \cdots + a_1 s + a_0$$

and from this a standard array is constructed:

s^n	a_n	a_{n-2}	a_{n-4}	$\cdots$
s^{n-1}	a_{n-1}	a_{n-3}	a_{n-5}	$\cdots$
s^{n-2}	b_1	b_2	b_3	$\cdots$
s^{n-3}	c_1	c_2	c_3	$\cdots$
.	.	.	.	
.	.	.	.	
.	.	.	.	
s^3	k_1	k_2		
s^2	l_1	l_2		
s^1	m_1			
s^0	n_1			

$$(11.46)$$

where

$$b_1 = \frac{a_{n-1}a_{n-2} - a_n a_{n-3}}{a_{n-1}} \qquad b_2 = \frac{a_{n-1}a_{n-4} - a_n a_{n-5}}{a_{n-1}} \qquad \cdots$$

$$c_1 = \frac{b_1 a_{n-3} - b_2 a_{n-1}}{b_1} \qquad \cdots$$

Routh's test asserts that the original function has as many right-half-plane roots (here, poles) as there are changes of sign in the first column of the array. Since we do not want positive roots we can find the permissible K range for stability by equating some relevant terms of the first column to zero. In particular the last two rows s^0 and s^1 each contain only one term, n_1 and m_1, which are functions of K, and their solution provides the K range. Furthermore, the frequency at the crossover condition may be found by writing the two terms of the s^2 row as

$$l_1 s^2 + l_2 = 0 \tag{11.47}$$

Routh's test is widely documented in the control literature, to which the reader is referred for deeper study.

We consider as an example the closed-loop gain control of a first-order cascade [Eq. (11.43)], with $n = 3$. The characteristic function is

$$(Ts)^3 + 3(Ts)^2 + 3Ts + 1 + K$$

and the array is

$$
\begin{array}{c|cc}
s^3 & T^3 & 3T \\
s^2 & 3T^2 & 1 + K \\
s^1 & \dfrac{8 - K}{3}\,T & \\
s^0 & 1 + K &
\end{array}
$$

In consequence, the stable range for K is

$$-1 < K < 8$$

and the frequency at crossover is given by

$$3[(Ts)^2 + 3] = 0$$

where $K = 8$ is inserted, that is,

$$s^2 = -\frac{3}{T} \qquad \text{or} \qquad s = \pm j\frac{3}{T}$$

which checks the results already found on Fig. 11.22c by trigonometric reasoning.

(E) SUMMARY

In summary, the root-locus technique is a powerful tool for closed-loop synthesis in linear systems, if the number of parameters to be varied is limited. After graphical manipulations have produced satisfactory closed-loop results, the transfer function can easily be written out, whence transient and frequency responses can be calculated exactly. Often the practiced analyst can almost visualize these responses from the pole-zero locations in reasonably simple cases.

11.8 Nonlinear Control

(A) GENERAL

It has been implied that linear control is not sensible in a practical world because significant cost is attached to both large energy consumption and high-effort actuators. Nonetheless, linear design remains popular in control engineering because the theory is generalized and tractable and because such designs work well enough in many practical cases. It is also true, however, that the control engineer, as well as nature, is learning to incorporate intentional nonlinearity into his control systems in order to make the overall system more effective and economic. At this stage a distinction should be drawn between two types of nonlinearity:

1. Inherent and often undesired nonlinearities which occur in the various system components, especially the process
2. Intentional nonlinearities designed into the controller of the system to achieve better performance in some sense

The first type of nonlinearity was treated briefly in Sec. 3.5. In Sec. 9.6 the biological nonlinearity of URS was introduced, and the question naturally arises whether this is an inherent, undesired type or an intentional, controller type. Probably the question has no clear answer, but it seems likely that URS must fit the second category, to some extent at least, since it provides much of that dynamic compensation which the system needs for stability, etc.

In this section we illustrate the intentional use of nonlinearity with the example of the *maximum-effort principle*, which seems significant in developing energetically efficient systems. To explain this, we recall that a linear system controller grades its output proportionally to the error signal (possibly also to its derivatives or integrals), so that this design therefore arises from an error-oriented viewpoint. The maximum-effort viewpoint, on the other hand, is energy-oriented and involves using all available effort to counteract error whenever corrective action is undertaken. One simple engineering example of such a maximum-effort system is the thermostatic control of house temperature, where the furnace is on-off controlled. Its block diagram and some hypothetical responses are sketched in Fig. 11.24. Its relay operation essentially converts the closed-loop dynamic equations into open-loop ones, which undergo discontinuous changes dependent upon a simple on-off logic function of the error (this latter

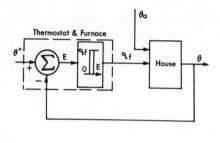

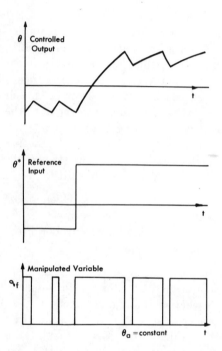

FIG. 11.24 *An on-off controlled system—house thermoregulation.*

function is not quite a pure relay due to hysteretic effects). The relay-type control is cheaper to instrument than proportional control, and since it can maintain the house temperature within acceptable limits, it is the obvious choice.

(B) MAXIMUM–EFFORT FORCE SERVOMECHANISM—THE HUMAN OPERATOR

The minimally second-order nature of the transfer relation between force as input and position as output has already been mentioned in Sec. 11.4; it has been noted, furthermore, that many limb and organ movements require position controlling—notably, of course, hands and eyes. In the engineering field many similar positioning devices are required, and there has recently been much work on the design of so-called *bang-bang systems* in which maximum positive actuation alternates with maximum negative actuation. This work has been particularly motivated by space-travel vehicles for which energy is obviously both expensive and limited, but at the same time investigators have found that the maximum-effort principle also operates to some extent in biological position control.

Consider first, then, the positioning of a massive object, without resistive damping or spring, given an actuator of maximum force $\pm F_M$ and the requirement of completing a step response in minimum time. First of all, it is intuitively clear that the maximum value of force available must be used at all times if the time is to be minimum. (Argue this out by assuming that it is not true.) Thus, the response will consist of full accelerating force until the object has moved halfway, followed by instantaneous reversal to full decelerating force; the resulting force, velocity, and displacement patterns are shown in Fig. 11.25. The following relations apply during acceleration, and reflected ones during deceleration:

$$\dot{x}(t) = \frac{1}{m} \int_0^t F \, dt = \frac{F_M t}{m} \qquad 0 < t < t_1$$

$$x(t) = \int_0^t \dot{x} \, dt = \frac{F_M}{m} \left(\frac{t^2}{2} \right) \qquad 0 < t < t_1$$

(**11.48**)

The plot of $\dot{x}(t)$ therefore consists of ramp segments, and that of $x(t)$ consists of parabolic segments.

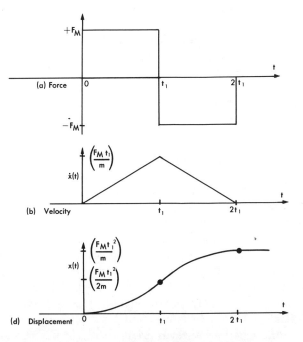

FIG. 11.25 Basic maximum-effort trajectory—inertial load only.

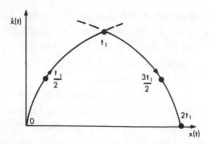

FIG. 11.26   Phase-plane trajectory of Fig. 11.25.

The *phase plane* of $\dot{x}(t)$ versus $x(t)$ provides a valuable method of showing such trajectories, and is widely used in the analysis of non-linear control systems (for example, [10.3]).   The trajectory for this simple case is readily obtained by eliminating t in Eqs. (11.48):

$$\dot{x}^2 = \frac{F_M}{2m}\, x \qquad 0 < t < t_1 \tag{11.49}$$

and comprises parabolic segments from the appropriate initial conditions (Fig. 11.26).

Several points should be made about this example:

1. This so-called *dead-beat position response* has no overshoot, nor does it decay exponentially into its steady-state value; instead, it arrives tangentially at the steady state with zero velocity, at which point the forcing function is cut off.   The dead-beat trajectory occupies minimum time for a given F_M, by definition.
2. The closed-loop control problem is one of *switching*, that is, of predicting the time t_1 at which the force should be reversed in order to achieve a desired displacement.   Naturally, this switching problem becomes more complex when other than pure inertia dynamics are involved.
3. Either a small dead zone with zero actuator output or a linear zone is needed if the system is not to limit-cycle.

Smith [11.3] has tested human subjects on this voluntary task of minimal-time movement of masses using the forearm and hand, both with and without spring and viscous forces.   The force-velocity-displacement patterns for pure inertia closely followed the ideal of Fig. 11.25 except that there were small turn-on and turn-off times during which muscle force built up and decayed, respectively. Smith found that overshoot and undershoot seldom occurred after preliminary learning, and that the response was essentially dead-beat.   Obviously, the results raise very interesting questions regarding the programming of switching time by the CNS.   For example, since the subject was merely told to move the mass as fast as possible, the magnitude chosen for the constant force F_M must derive from some appropriate performance criterion within the CNS because the F_M chosen was less than the maximum muscle force available.

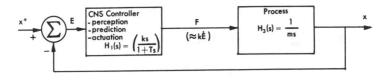

FIG. 11.27 *Highly simplified model of human-operator tracking system.*

When spring and damping forces were added to make the process correspond to the positional servomechanisms of Sec. 11.4, the subject modified his switching so as still to attempt a dead-beat response. Typically, however, a small overshoot or undershoot would result, which was subsequently corrected after a delay of about 200 msec.

In a similar study of human-operator characteristics for a visual-manual tracking task with small inertia only, Wilde and Westcott [11.4] find the same basic maximum-effort picture for the step response. Further interesting conclusions arise, however, from their experiment with a randomly varying input command. First, the idea of a sampling mode of operation and prediction is confirmed by the muscle-actuator program consisting of essentially constant force segments of about $\frac{1}{4}$ sec each. Second, and more pertinently for our present purpose, the human operator attempts to program his tracking velocity to be proportional to the error in position, although in detail, owing to the constant force segments, this can only be average velocity (achieved as an average over the sampled interval). Thus, in this task the CNS apparently acts essentially as a derivative controller (Fig. 11.27), at least within certain bandwidth and saturation limits. As a result the polar plot of frequency response stays away from the -1 point and its consequence of instability. Figure 11.28 shows Wilde and Westcott's experimentally derived polar plot, compared with that for the gain control of a pure inertia process which in principle passes along the negative real axis.

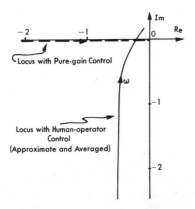

FIG. 11.28 *Approximate polar plots for Fig. 11.27.* (*Adapted from Wilde and Westcott.*)

12 The Special Challenge of Biological Control Systems

12.1 Introduction

Linear modeling was used in Chap. 11 only to demonstrate the principal effects of feedback control without excessive mathematical detail and with the generality permitted of linear systems. However, throughout the book it has been emphasized that real-life system analysis must normally include nonlinear phenomena, including both the inevitable "process" nonlinearities and the "desirable" nonlinearities of maximum-effort actuators, etc. In Chap. 3 the thermoregulatory control system for the body was presented in some detail, and since then much of the dynamic modeling in the book has referred to the homeostatic or regulatory system for the body. Consequently, the modeling examples of this chapter are chosen to emphasize the servomechanism (or tracking) aspect of biological systems. In particular, the following successively smaller but related systems are considered:

1. The human operator in manual tracking (Sec. 12.3)
2. The stabilization system for visual fixation (Sec. 12.2)
3. The skeletal-muscle servomechanism (Sec. 12.4)

In order to model these systems successfully it seems necessary to incorporate a number of special techniques which the control engineer now uses to some extent in his systems. These include non-linear controllers, sampled-data operation, disturbance feedback, prediction computing, and adaptation. However, the special challenge of these biological systems arises because such effects often occur together and also because the relevant performance criteria are not obvious.

The control of living systems comprising many individuals or species (ecologies) involves another major phenomenon, that of positive feedback due to the growth process. In addition, it becomes extremely difficult to specify the purpose, or performance criterion, of such a system or of its parts. However, a few comments can be made regarding such systems in order to show that they may be treated within the same general theoretical framework.

(A) PASSIVE VERSUS ACTIVE CONTROL; NATURAL SELECTION

Most of the biological models chosen have exemplified so-called *active control* because "causal" feedback paths and corrective actuators could be identified within a closed loop. However, the purely mathematical analysis of control does not inherently distinguish between active and passive feedback parameters, since typically their dynamic effects are complementary; in fact, it has been pertinently said, "Feedback is in the eyes of the beholder." Thus, although population models, such as those of simple growth, warring armies, and insect ecologies, have been termed *passive* in the sense that one cannot confidently attribute a causal motive to the feedback channels without incurring difficult teleological problems, nonetheless mutually causal (feedback) paths can easily be identified.

Population growth and decay processes are basic to ecologies at almost all system levels from cellular to communal, and it seems that in natural ecologies the respective birthrates are adjusted through feedback action according to their populations and environments. Thus, although the robin may predate with varying success in particular periods upon the earthworm population, the earthworm birthrate presumably varies as necessary to maintain some relatively stable population on the average [12.1]. This phenomenon is emphasized strikingly in such insect populations as that of the flour beetle, where typically there is a high birthrate of eggs, coupled with an almost zero probability that the eggs can become adult beetles since they are normally eaten by the beetle population itself unless a catastrophic population drop occurs. These are certainly examples of self-regulating feedback control from a system analyst's viewpoint.

There is another slightly different group of problems which we may define loosely as those subject to the *competitive exclusion principle;* that is, if a group of individuals from one species cohabit a given ecology, those subgroups suffering deleterious mutations will become completely excluded ([12.2], for example). This principle seems to apply in many zoological problems, but here we only mention one

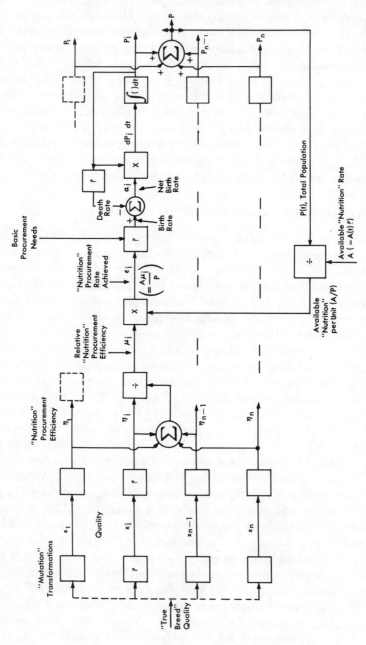

FIG. 12.1 Hypothetical information-flow diagram for competitive exclusion example. Shown in full for P_j only.

recent example having to do with cellular systems such as skin and liver. Curtis [12.3] suggests that, whereas both skin and liver cells undergo spontaneous mutations, the skin cells divide regularly and hence ultimately "breed out" the daughter generations of those which suffer deleterious mutations because they are competitively excluded in the struggle for existence. Liver cells, on the other hand, divide rarely and hence their deleterious mutations accumulate, to the increasing detriment of the organ's function. When one also includes cells which almost never divide, such as brain cells, a possible mechanism for aging emerges. Figure 12.1 is a hypothetical information-flow diagram to explain these processes in a generalized competitive exclusion situation. For simplicity we assume some "true-breed" quality, with regard to an appropriate survival characteristic such as ability to procure "nutrition." Different levels of this ability, and hence of relative efficiency, then result from differential *mutations*, using the latter term in the most general way. For a given stream, the actual nutrition rate achieved varies as the quotient of the relative efficiency and the available nutrition per unit. This rate may be assumed to affect the birthrate achieved and hence the population of this particular stream. The total population clearly results from summation over the streams. If a constant availability of nutrition is assumed, a negative-feedback path exists to modify selectively the achieved nutrition rate of each stream. Notice that the basic positive-feedback loop still exists internally for population growth of each stream except that the net birthrate is now not constant but modulated by each stream's relative efficiency. The net birthrate can therefore become negative for each less efficient stream, since there is only a relative, rather than an absolute, standard of efficiency for continued survival. Thus it should be clear that all mutant streams except that with the best relative efficiency will be eliminated in the steady state. In detail the functional relations of this process are not all known, but the important relations from the feedback point of view are as shown. There seem to be many biological examples of this kind, where positive-feedback inner loops are stabilized by negative-feedback outer loops ([12.4], for example), and functionally they belong without doubt to the class of feedback-control systems.

(B) GENERIC STRUCTURE OF CONTROL SYSTEMS

We now return to consideration of conventional biological servomechanisms, and therefore first review briefly their generic structure. In engineering control systems this structure comprises: process → feedback transducer → comparator → controller → actuator. Biological systems are not necessarily identical, however, and in particular the middle three components are often grouped in the feedback transducer, with the stabilizing feature of phase advance often being supplied by its rate-sensitive characteristic. Although there are multiple paths in the biological systems which we consider, especially those of neural transmission, there are often only one or at most a few input-output variables of interest. In other cases, however, there are strong cross couplings with other systems which must also be considered. Furthermore, a statistical treatment may become necessary for multipath channels when the signals are statistical in nature (Chap. 14). Finally, actuators (and often transducers also) in living systems operate unidirectionally, so that they are arranged in antagonistic pairs.

12.2   Stabilization System for Visual Fixation

(A) GENERAL

The visual system is vitally important for man's normal functioning. Its pupillary-light reflex has been considered in Chap. 2; in this section we consider its tracking servomechanism, in which the extraocular muscles rotate the eyeballs in their orbits so that the fovea of the retina may remain stationary upon a moving target of interest, or alternatively may select new targets. For present simplicity the target is assumed to maintain a constant radius, so that the coordinated actions of vergence and accommodation which are normally involved can be ignored. Vision, and hence perception, is achieved by stabilizing the retinal image, at least intermittently, for periods of about 180 msec. Actually, there are small rapid eyeball movements during this so-called *fixation period*, but it is not yet clear whether they are tolerable disturbances or desirable controlled movements [12.7]. Furthermore, if the retinal image is completely stabilized by artificial means, satisfactory perception rapidly ceases ([12.5], for example). These last two phenomena can be ignored in the present modeling, however, because interest centers on the gross target-tracking servomechanism itself. The following development largely follows [12.6] of G. Melvill Jones and the author, and [12.7] of Young with respect to the sampled-data model.

(B) THE SYSTEM TO BE ANALYZED

Eye movement in the skull is controlled by three pairs of extraocular muscles, allowing accurate movement in the three rotational degrees of freedom. However, eye movement in space can also be brought about by movement of the head on the body and of the body in space, so that there are three main sets of actuators whose complementary effects must obviously be coordinated. The complex neuromuscular-control system which performs this has four main feedback channels: visual, vestibular (semicircular canal), and two proprioceptive (eye in skull, skull on body). The tracking pattern of the eye in skull, and indeed of the skull on body in some species,

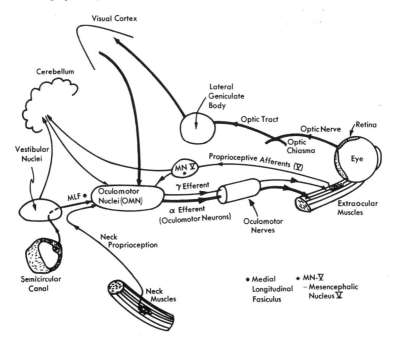

FIG. 12.2   Simplified anatomical schematic of visual-fixation system.
(*After Jones and Milsum.*)

comprises two basic modes, saccadic jumps and smooth tracking. One major problem of the analyst is to determine how these modes are modulated by the various control loops just mentioned.

(C) ANATOMICAL STRUCTURE

The anatomical structure for movements of the eye in skull is indicated in Fig. 12.2 in highly simplified form, and is briefly described below.

Actuation and Proprioceptive Feedback.   The system actuators are clearly the three pairs of extraocular muscles, and for simplicity we restrict our consideration to the lateral and medial recti which produce eye rotations essentially in the horizontal plane. These muscles have unusual physiological characteristics, all of which seem admirably designed for excellent dynamic response such as is needed for the saccadic jumps.   The muscles are innervated from the oculomotor nuclei by the oculomotor nerves, comprising both the large and fast-conducting α motoneurons, and the small and slowly conducting γ motoneurons.   In turn, afferent (feedback) information on muscle movement is provided by stretch receptors in the muscles themselves, their neurons apparently being carried in the Vth nerve to the corresponding mesencephalic nucleus whence they travel to the oculomotor nuclei to complete a basic loop for the muscle servomechanism.   Actually, there are other pathways of innervation, such as to the cerebellum, but these cannot be presently considered since their effects are not quantitatively known. This basic servomechanism seems to operate in both a velocity mode, as a smooth-pursuit tracker, and a position mode, as a target locator particularly in the saccadic mode. For simplicity and because of present ignorance, the skull-on-neck muscles and their proprioceptive feedback are omitted from this diagram, except for the proprioceptive feedback used in driving the eye-in-skull system. The latter pathway is via the upper cervical nerves and the spinal continuation of the medial longitudinal fasiculus and then presumably to the oculomotor nuclei. The analysis of such interacting eye and skull movements is aided by preparing an information-flow diagram (Fig. 12.3) which incorporates this spatial feedback and geometry.

Visual Feedback. The visual channel is normally considered the main feedback path. This originates in the retina, and of course relates the external environment, including particularly the current target of foveal vision, to the biological system. The visual information is carried neurally in the optic nerve, via the optic chiasma, optic tract, lateral geniculate body, and visual cortex to the oculomotor nuclei. However, it should be noted that the expression *visual information* hides a large ignorance. Thus, although we know that the retina must provide suitable error-sensitive information to drive the eye so that it may locate and lock on foveally to a target, this knowledge does not help us to specify how or where the raw retinal image is suitably data-processed. Fortunately, the problem is a similar cybernetics one, from this book's viewpoint, to that of the pupillary-light reflex-control system, and it is therefore ignored since the gross pattern is known.

Vestibular Feedback. The vestibular channel receives its information from the cupulae of the semicircular canals (Fig. 8.5) and from the otolith organs. The vestibular nerve carries the information to the vestibular nuclei in the brain stem, whence it proceeds via the medial longitudinal fasiculus to the oculomotor nuclei. The otolith organs are not further considered since they are probably mainly concerned with supplying a signal to orient the eyes with respect to gravitational and translational accelerations. On the other hand, the semicircular canals provide without doubt the main dynamic drive to compensate the eyes for the short, sharp, angular head movements of everyday life. This is an extremely important disturbance feed-forward function which considerably increases the system bandwidth, as shown below.

Overall System. Visual, vestibular, and two proprioceptive feedback channels are the main sources of information with which the oculomotor nuclei synthesize the actuating drive signals for eye-in-skull control. These movements must, however, be considered within the complete geometric picture of the system's operation.

(D) THE SPATIAL PICTURE

The geometric framework of the visual stabilization problem plays an important part in our understanding of it. The inset "uniocular" sketch in Fig. 12.3 defines the various horizontal-plane angles of target θ_t, eye in space θ_e, eye in skull $\theta_e - \theta_s$, skull on body $\theta_s - \theta_b$, and body in space θ_b. The spatial summation point Σ_2 in the main figure emphasizes that the eye position in space θ_e results from the sum of three "biological" angles. Their complementary actions are illustrated, for example, in the task of fixating a new target at an angle from the previous one; typically the eye in skull undertakes a saccadic movement onto the target, which is then followed more slowly by a follow-up of skull on neck to recenter the eye in skull, and usually, if the target condition persists, by body in space to line up eye, skull, and body essentially on target. In this development,

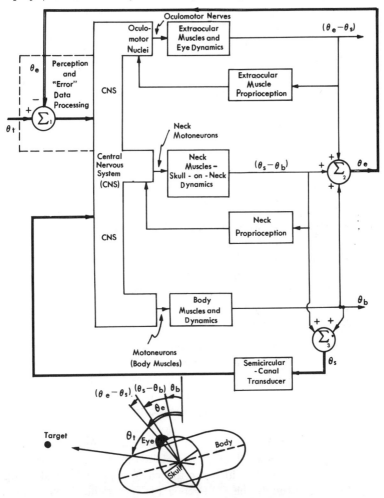

FIG. 12.3 Spatial configuration of visual-fixation system. (*From Jones and Milsum.*)

however, we are especially concerned with dynamic conditions, which are more typically directed toward utilizing eye-in-skull movements for rapid compensation against these other motions.

The visual feedback of θ_e is shown on Fig. 12.3 as providing one main closed loop, and θ_t is clearly the reference input of this system. Unfortunately, the dynamic response of the smooth-pursuit mode in the visual loop is insufficient to stabilize the eye against the angular velocities induced in the head from the disturbance of normally encountered body motions, and it is in this context that the semicircular canal significantly increases the system's disturbance response characteristics. Apparently such improvement has been very important for evolutionary survival, since the semicircular canal provides a good example of the exact "engineering" which nature occasionally evolves. The dynamics of this transducer seem to be unusually linear and they are also bidirectional owing to a d-c discharge frequency when the cupula is undeflected. The basic transfer function between skull acceleration $\ddot{\psi}$ and cupula deflection ϕ is second order [Eq. (8.22)], but when the Bode frequency response plots are numerically constructed for the system's hydrodynamic and other parameters (Fig. 7.2), it becomes clear that the intended role is that of an angular-velocity transducer, and the equivalent transfer function is therefore given by Eqs. (8.24) and (8.25), here combined:

$$\frac{\phi}{\Omega} = \frac{k_c s}{(T_1 s + 1)(T_2 s + 1)} \tag{12.1}$$

Two important points should be emphasized with regard to establishing this transfer function:

1. A basic second-order system could operate as either an acceleration, a velocity, or a displacement meter, depending only upon the numerical relations between the parameters.

2. A transducer can only be designed for a specific task within some frequency range; it is therefore necessary to combine the numerical knowledge of system parameters with the nature of environmental inputs if the system's function is to be correctly inferred.

In this particular case the inference is based on the fact that the gain and phase are essentially constant (at k_c and 0, respectively) over the naturally encountered frequency range of disturbance input, when the input is the skull's angular velocity. This is certainly the desirable frequency response of any basic transducer, namely, $y/x = k_c$.

This transducer provides a second major feedback path for the system, as shown in Fig. 12.3. There remain the two proprioceptive feedback channels which are drawn to suggest that although they do pass through the CNS they really provide local feedback for monitoring the muscle servomechanisms. This spatial picture reveals that when the skull is subjected to an angular disturbance velocity, then stabilization of the retinal image could be achieved in principle by (1) using skull-on-body muscles to "null" the skull movement—the classical regulating action—in this case, to provide a stable platform; (2) using eye-in-skull muscles to compensate the angular velocity of the disturbed skull with a suitable angular velocity of the eye in the skull; such a task involves less inertia than (1); (3) moving the body in space; or (4) combining the above. It is well established that large animals such as man use the eye-in-skull compensation path (2) extensively. However, without "going into the loop" to measure skull-on-neck muscle forces the analyst can never know what proportion of the potential disturbance is normally nulled by (1), because such proportion never appears as a vestibular signal. It is interesting that some small animals such as pigeons exhibit "head nystagmus" when walking (saccadic forward and smooth backward compensation movements). It is true that this is desirable because of their side-facing eyes, but in addition it may exemplify the preferential use of head rather than eye movement as the inertia of the head decreases, with presumably decreased angular controllability in the smaller eyes.

(E) THE GENERAL PATTERN OF EYE MOVEMENT

The spatial picture of Fig. 12.3 is essential to an understanding of how the overall system *could* operate. Obviously, there are too many degrees of freedom to consider exhaustively here, and indeed experimental data is insufficient for generalized model building. However, the greatest importance in everyday living and the fastest dynamics are undoubtedly evidenced by the eye-in-skull movements under control of the visual and vestibular feedback loops. Extraocular muscle proprioception is considered part of the muscle servomechanism rather than a basic feedback, and it is therefore not emphasized until the next section, while eye-in-skull response to skull-on-body proprioception is presently ignored for lack of suitable data.

The general patterns of ocular response comprise the saccadic-jump and the smooth-pursuit modes which are basic to the present theme, and some small amplitude modes which are presumably not.

Saccadic Mode. The reader may readily observe the pure saccadic mode by watching a friend's eyes as he scans a visual field from side to side with head fixed. In this case the eye normally holds on to each successive scene, θ_e fixed, for 180 msec or more, before jumping rapidly and accurately to a new target, and so on (Fig. 12.4). A suitable experimental setup for evaluating this mode is to present target motion to a seated subject by a spot of light on a screen. The movement of eye in skull can be measured by a variety of techniques, including one similar to that for the pupil ([12.7], for example).

Smooth-pursuit Mode. The smooth-pursuit mode seems to have at least three separate submodes of interest:

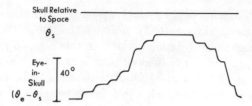

FIG. 12.4 Pure saccadic mode of eye movement—skull stationary. (*From Jones and Milsum.*)

1. *Predictable mode:* If a predictably moving target such as a pendulum is watched with head stationary, in general the eyes track smoothly, with neither saccadic jumps of position nor jumps in velocity. (Actually there may be a few saccades in practice, as shown in Fig. 12.5*a.*) In this mode periodic waveforms such as sinusoidal and square waves are followed essentially "exactly,"

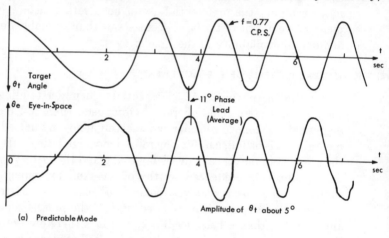

(a) Predictable Mode

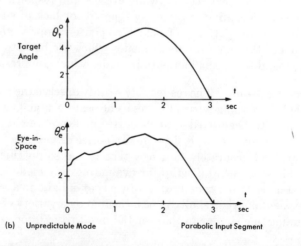

(b) Unpredictable Mode   Parabolic Input Segment

FIG. 12.5 The smooth-pursuit mode of eye movement—skull stationary. (*From Young.*)

except for the rather unusual feature, reported by some investigators (for example, Young, [12.7]), that the eye actually shows phase lead relative to the target and slight amplification of its magnitude, as illustrated by the frequency response plot of Fig. 7.6. Young's mathematical model for this pattern-recognizing and prediction submode is presented in Sec. 12.2*G*.

2. *Unpredictable mode:* When the target motion becomes unpredictable, such as by combining nonharmonic sinusoids or by generating a random signal, the tracking system reverts to using the visual loop for generation of its primary drive signal in a "watch-and-follow" type of action. Young proposes a sampled-data mathematical model for this mode which is presented in Sec. 12.2*H*. An interesting feature of this model is that constant-velocity segments are apparently generated for the next discrete interval following "computation" of the recent tracking-error rate (Fig. 12.5*b*), which is the idealized law of operation postulated for manual tracking in Sec. 11.8. A sampled-data model is also postulated for the saccadic mode, with the two modes of position and velocity tracking being used in complementary fashion. In this regard the unpredictable sampled-data smooth-pursuit mode has a tracking-velocity limit of about 25°/sec, above which the saccadic mode must be utilized for large corrective jumps. The frequency response curves for unpredictable inputs show considerable deterioration in gain and phase at all frequencies relative to the predictable curves (Fig. 7.6) (except for a local resonance around 2 cps due to the sampled-data mode). It should be noted that even when the unpredictable input waveform is synthesized as a sum of sinusoids, the output waveform is not sinusoidal in form, since it results from a set of constant-velocity and saccadic-jump segments (Fig. 12.6). In consequence, frequency response plots represent considerable idealization.

3. *Vestibularly driven mode:* The dynamic shortcoming of the system so far is that the eyes are not carried in an inertially stable platform; thus, for example, if the head suffers a 5° amplitude oscillation during such normal activities as running at 5 cps, its peak angular velocity would be about 150°/sec. The intended use of the semicircular canals in this context is then to drive the eyes in a compensatory direction so that the original performance of the undisturbed tracking system just described may be reestablished. Figure 12.7*a* illustrates that when the eye fixates a stationary target, while the head is rotated sinusoidally at about 0.5 cps, perfect compensatory movement of the eye in skull is achieved, so that the eye remains stationary in space. Figure 12.7*b* illustrates a combination of saccadic and pursuit modes, in which the eye evaluates the visual field while the head is again oscillated. The essential feature is that after the eye makes saccadic jumps to each new target as in Fig. 12.4, it then maintains fixation by smooth pursuit in each sample interval. Thus when the head movement is added to this so-called

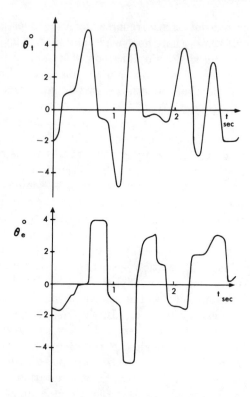

FIG. 12.6 Combined tracking of unpredictable signal—skull station-ary. *(From Young.)*

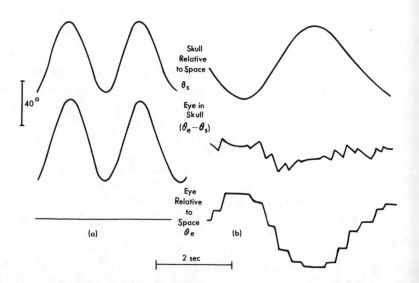

FIG. 12.7 Saccadic mode with vestibular compensation—skull rotat-ing. *(From Jones and Milsum.)*

nystagmoid trace of eye-in-skull movement, Fig. 12.4 is indeed reestablished as regards eye movement relative to space.

Combined Saccadic and Smooth-pursuit Modes. In general the eye-response pattern comprises both saccadic and smooth-pursuit modes, as shown, for example, in Figs. 12.6 and 12.7, with the smooth-pursuit action being in any of its three submodes. It seems probable that a graded modification of the frequency response plots of Fig. 7.6 occurs as a signal becomes less and less predictable, which would imply that the actions of the submodes can be combined.

The oculomotor subsystem (the inner $\theta_e - \theta_s$ loop in Fig. 12.3) must be able to generate both the saccadic jumps and the graded smooth-pursuit movements, either separately or together. The saccadic motion is very rapid, with a 10° saccade typically lasting somewhat less than 60 msec. Westheimer [12.8] found that the saccadic curve could be fitted reasonably well by the step response of a basic second-order transfer function, with eye inertia, eye damping, and antagonistic muscle spring as the appropriate constants, and with the agonist muscle providing a step input. His parameters are $\omega_n = 120$ rad/sec and ζ is 0.7, for which the step response settles with little overshoot in about $t = 7/\omega_n \approx 60$ msec (see Fig. 8.11). Westheimer's model was open-loop because the presence of proprioceptive feedback in the extraocular muscles was not then generally accepted. More recently Vossius [12.9] has proposed a feedback model incorporating proportional plus approximately derivative proprioceptive feedback, but excluding antagonistic muscle spring. In other recent work, Robinson [12.10] postulates that the eyeball dynamics have a much lower natural frequency, of about 6 rad/sec, and that the short time of saccades is achieved by a preprogrammed pattern of neural firing which initially "overdrives" the system massively. Two points should be made now with regard to the eye's rather special muscle servomechanism:

1. The oculomotor nerve impulses to the agonist muscle for initiation of the saccade are of brief duration, and there is no evidence for any subsequent switch to the antagonist muscle for deceleration.
2. The initial acceleration achieved is very high, but the required muscle torque could be reduced if less inertia than that of the whole eyeball were involved. This would be the case if relative motion could occur between the corneal lens complex and the humor (aqueous and vitreous; Fig. 2.11) so that the humor was initially left behind. Indeed, there is some circumstantial evidence for this from the subjectively observable, relatively slow movement of one's own "floating shadows" following a saccadic flick.

(G) THE PREDICTABLE MODEL

We return now to model the smooth-pursuit tracking mode when the input is predictable, and in particular when it is sinusoidal. Young [12.7] was able to fit the log AR data of Fig. 7.6 by a fourth-order transfer function (Fig. 12.8):

$$H_1(s) = \frac{1.08}{\left(\dfrac{s}{\omega_0} + 1\right)^2 \left[\left(\dfrac{s}{\omega_0}\right)^2 + 2\zeta \dfrac{s}{\omega_0} + 1\right]} \tag{12.2}$$

where ω_0 is 9.4 rad/sec and ζ is 0.35. However, the experimental phase curve does not match well with that of this transfer function, and in particular is always less severe. Since such incompatibilities between gain and phase curves do not occur if the system is minimum

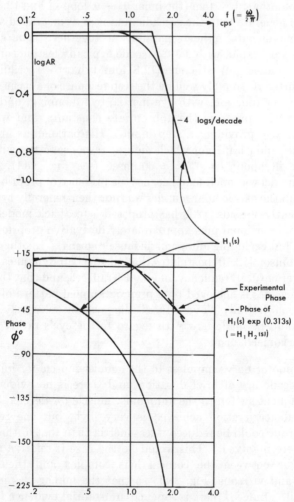

FIG. 12.8 Transfer-function model for the predictive mode. (From Young.)

phase and linear, it seems probable that the CNS provides in addition a predictive element in the control: $H_2(s) = e^{+Ts}$. By setting $T = 0.313$ sec, Young obtained excellent agreement between experimental and model phase curves. This implies that the CNS analyzes the pattern of the incoming signal, presumably by comparison with past records, and then provides sufficient prediction to cancel out the various loop dynamics. Once the pattern is recognized, open-loop control is presumably implemented, with occasional feedback monitoring of present performance, which could account for the occasional saccades of Fig. 12.5a. Figure 12.11 presents a suggested form of the overall functional tracking system, including this channel.

Some questions arise, naturally enough. For example another entirely smooth-pursuit submode is that driven by the semicircular canal, but this transducer supplies continuous "analog" information to the system, which is therefore suitable for producing continuously graded responses. On the other hand, one would superficially expect a predicting computer in the CNS to operate in a discrete fashion and hence to drive through the velocity and position sampled-data modes. Indeed, there is the further puzzling point that if one closes one's eyes after following a sine wave visually, and voluntarily continues to track by visual "mentalization," the eyes immediately switch to the saccadic mode with an overall movement rather like that shown in Fig. 12.4, although the subject has no knowledge of this result. Obviously, then, the visual monitoring is still functionally important. Young speculates that the forward predicting of the signal by about 0.3 sec could be accomplished by obtaining the autocorrelation function of the input, which is a statistical subject taken up in Sec. 14.3. Another speculation is that the repetitive waveform of the input signal "reverberates" around neuronal rings, so that an output can be tapped off with the appropriate time advance. However, these ideas have little neurophysiological basis as yet.

This section should have demonstrated that some rather simple mathematical model making, based on the tools we have built up, produces quantitative "answers" which immediately open up new questions for the physiologist. This represents an entirely different level of understanding than could be obtained by the reasonably obvious qualitative fact (or ex post facto, at least) that man can anticipate when he recognizes an input-signal pattern.

(H) THE SAMPLED-DATA MODEL

Young has been able to match the experimental behavior of the eye in tracking unpredictable signals with skull stationary, for both the saccadic and smooth pursuit modes, by using two separate but complementary sampled-data systems. The following development essentially follows his work [12.7]. Such a development is also relevant for the manual-tracking task of Sec. 11.8. A sampled-data system is one in which new information on the variables, in at least one point of the closed loop, becomes available only at discrete intervals of time.   It is not linear in the most general sense, because the output sequence following an input sample cannot be affected until another input is available one sample interval later.   However, superposition does apply at the sample times if the other elements are linear.   Typically, the use of sampled data produces destabilizing effects in a control loop due to the time delay between successive samples, but this disadvantage is often more than nullified in the overall system because of the desirable data processing and hence control-loop compensation which can be performed in the sampled-data section, and because of the ease with which one computer-controller can be "time-shared" among several tasks.   There is an extensive engineering literature on sampled-data and computer-controlled systems (see Bibliography).

The experimental evidence for the sampled-data model can best be studied from certain revealing transient inputs, much as was done in Sec. 2.8 for the pupil-control system:

Pulse Response.   The response of the eye to a short pulse of target position is shown in Fig. 12.9a and compared with the mathematical response of a simple sampled-data model.   The first observation is that there is no response until after a somewhat variable delay of about 180 msec.   At this point a saccadic jump occurs to the place where the target was pulsed.   This saccade is usually very accurate, but if it is not, a small corrective saccade occurs after another delay period.   It should be noted that the input has already returned to zero before the response occurs, so that when the system again samples its performance, it is still in error and hence starts another corrective response down the refractory delay line, which finally is effected two sample periods after the pulse was first initiated.   As exercises the reader should construct the response to a pulse longer than the sample interval, and the response to a step input. Although Young provides even stronger evidence, the above case should be sufficient to exemplify that once the sample has been taken and the response entrained, no corrective action can be undertaken until after another sample period when the response is subsequently perceived to be in error.   This is a necessary property of sampled-data systems, but its existence is not a sufficient condition to prove the presence of a sampled-data system.   Unfortunately, the scope of the present book does not warrant the mathematical development of sampled-data control theory, for at this stage it seems important only to introduce the major modeling possibilities.

Ramp Response.   We now present the evidence for a smooth-pursuit sampled-data system that is distinct from but complementary to the sampled-data model for the saccadic mode.   The simplest input which exhibits this mode is the ramp, as shown in Fig. 12.9b, and

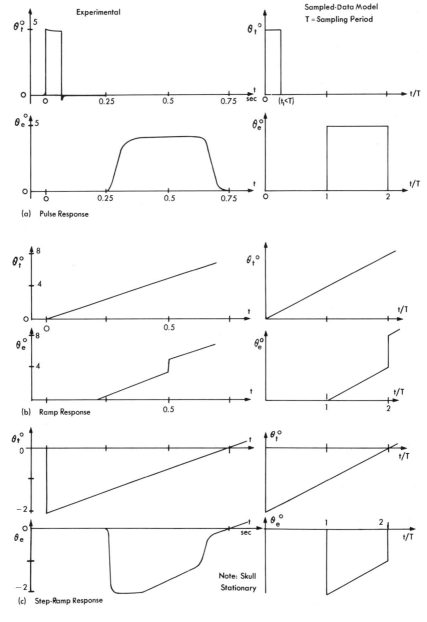

FIG. 12.9   Sampled-data models for saccadic and smooth-pursuit unpredictable modes.   *(From Young.)*

again there is a delay period without response, after which the eye starts to track with a constant-velocity segment.   For a ramp input in the velocity range 3 to 10°/sec this segment is then merely continued, while at faster ramp rates the saccadic mode occasionally switches in, apparently to reduce the position error, and finally, above about 30°/sec, there is normally no smooth-pursuit mode left.

By referring back to Fig. 12.5 for a parabolic input, one sees that the velocity segments do indeed adjust themselves to varying input rates, although there are some errors in estimating velocities, and some saccades also. The question naturally arises of how the retinal error rate is estimated at each sample period in order to give a suitable command to the smooth-pursuit mode for the ensuing segment, but again there is little neurophysiological knowledge to answer this question. Clearly, however, the corresponding numerical computation on a digital computer is estimating the derivative by subtracting the previous value from the present value and dividing by the interval (Sec. 10.6). This process inevitably involves a sampling delay also, at least when performed in a straightforward manner.

Step-ramp Response. In order to establish the independence of these two sampled-data modes the so-called *step-ramp input* and its response are useful (Fig. 12.9c). A saccadic jump generally occurs after one delay period, and in addition the system sets off with a correct-velocity segment, which is of opposite sign to the saccade. One or more saccades then occur later to correct for any accumulated position error. Since the system response could obviously be improved by collaboration between the two modes, it would seem that they do in fact work independently.

A possible block diagram for the sampled-data model which emerges is shown in Fig. 12.10. The position-tracking or saccadic mode is called a *zero-order hold system* in sampled-data work because it is controlled by the retinal position error of the target as held over from the previous sampling instant. On the other hand, the velocity-tracking mode is called a *first-order hold system* because it is controlled by an estimate of the retinal error velocity, where the latter is computed by comparing a previously held error sample with the most recent sample, as shown on the figure. The block diagram also contains certain experimentally determined dead-zone and velocity-saturation limits. Finally, it should be noted that since the model's data appear in pulsed form after the sampler, they must be integrated somewhere $(1/s)$ in order to provide an appropriate continuing neural command. The neurophysiology is not known, however, and hence no direct verification of the model is possible. With certain simplifications Young does analyze this combined system using sampled-data theory, and one simplification particularly worth noting is that the eyeball dynamics may be neglected because the transient response is completed before each new sample is taken. One interesting feature not originally included in Young's analytical computations is the effect of a statistically variable sampling interval. Its inclusion certainly complicates the calculations considerably, but some work has been done on the general sub-

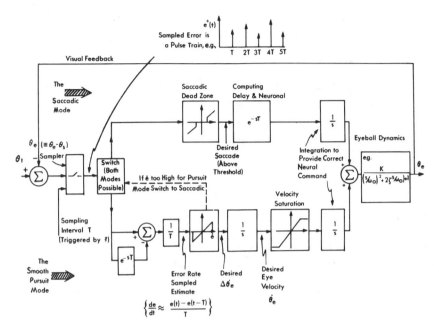

FIG. 12.10   Possible functional organization of sampled-data visual tracking system.   (After Young.)

ject.   Once again it appears that complete analysis of a biological control system requires a statistical treatment.

An extensive theory has been developed for the analysis of sampled-data systems, especially if all components other than the sampler are linear.   Usually the so-called *z-transform method* is used to simplify matters, but we cannot introduce its theory here. Obviously, sampled-data systems can be programmed easily for digital-computer analysis, and hence arbitrary nonlinearities can also be included.

(I) THE COMPLETE OCULOMOTOR SYSTEM

Figure 12.11 represents a highly conjectural attempt to synthesize the complete oculomotor system, which may help to pose some neurophysiological questions.   A new input variable, the angle of the next target, has been added to account for the system's good ability to jump saccadically to a new target (for example, Fig. 12.4). This information presumably only feeds to the saccadic-control block, however.   The three smooth-pursuit modes previously discussed are also shown as separate blocks with appropriate driving inputs.   As regards the actuation of the extraocular muscles them-

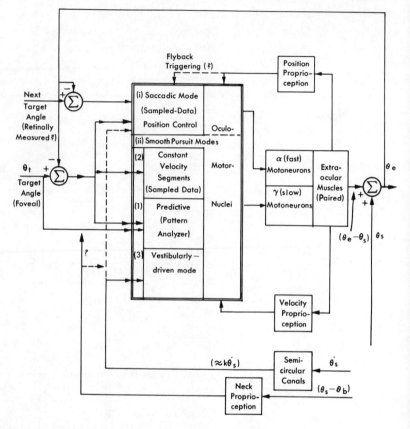

FIG. 12.11   Conjectural synthesis of oculomotor control.

selves, it is tempting to speculate that separate position and velocity proprioceptive units are concerned with position and velocity control, respectively, but whether or not these also operate through fast and slow motoneurons, respectively, is not at all clear. Proprioceptive feedback of eye position has been shown to the saccadic mode since the eye angle presumably plays a part in triggering flyback saccades. An afferent path to there from the semicircular canals has also been added because of evidence that the mean eye position during sinusoidal oscillations of the head is proportional to the head's angular velocity [12.11]. Finally, because of complete ignorance the neck's proprioceptive feedback has not been connected to any particular functional block.

(J) SUMMARY

The patterns of eye movements turn out to be very complex, especially because of the multiple, spatially complementary modes of operation which are possible. In this situation, thoughtful dynamic experiments combined with mathematical model building have enabled significant advances in understanding to be achieved. This experience suggests that the overall system can indeed be modeled mathematically, but in view of its complexity and of the need for much new neurophysiological data it does represent a special challenge to the systems analyst. The example has also served as a useful vehicle to introduce very briefly the functional nature of sampled-data systems, to present quantitative evidence for predictive operation when the tracking waveform can be recognized, and to integrate the previously deduced transfer function of the semicircular canal purposefully into a larger system.

12.3 The Human Operator

In view of the sampled-data and prediction modeling developed in the previous section, it is worthwhile to return to the manual-coordination task of Sec. 11.8 for some further study. The first big difference to be noted between the tasks is that of quantitative numbers, namely, the available torque-inertia ratios. As we have seen, the eyeball dynamic system has, in effect, a natural frequency of about 20 cps for saccadic jumps, with a settling time of about 60 msec. Since new sampled programs are initiated only every 180 msec or so, one can essentially ignore the eyeball dynamics in studying the performances of both the position and velocity sampled-data systems. The eye musculature and its effective inertia are rather specially designed to achieve this excellent response, but the torque-inertia ratios of skeletal-muscle–limb systems are inevitably much lower. Thus, for example, the typical tremor frequencies in Parkinson's disease, clonus, etc., are almost 1 decade lower, namely, around 4 cps. Naturally, limbs can be moved very fast when the muscles are highly stimulated, but such rapid movements are expensive metabolically and are therefore normally reserved for emergency situations. Thus, it is interesting to note that in skilled fast actions, such as playing the piano at maximum speed (about 20 cps), the movements are apparently uncontrolled in the proprioceptive and visual feedback sense, because no recall seems possible once the muscle action has been initiated. As one knows from experience, training is essential for good performance in such tasks, which therefore presumably involves developing an open-loop preprogram of muscle activation.

Returning now to the manual-tracking task with unpredictable visual input, it was stated in Sec. 11.8 that the operator apparently prepares his program for successive segments by making the average velocity of the segment proportional to the predicted error of the upcoming interval [11.4]. This raises two interesting and correlated differences between the manual-tracking and the eye-movement models:

1.   The manual-tracking system does not attempt to produce saccades, nor even constant-velocity segments, because its actuator output is force-limited (and therefore acceleration-limited) within the available interval.   However, it does the next-best thing and programs a maximum-effort segment ($\pm$ forces), producing typically a velocity triangle within each interval (Figs. 11.25, 12.12).   Interestingly enough, Wilde finds that operators do not like to "coast" in any interval, even where a posteriori examination of the record would show this to have been desirable.   A further feature is that only a few rather constant ratios of force were selected for the successive segments, for example, $3:6:11$ in their experiment.

2.   The operator often attempts an error prediction on which to base the next velocity triangle, presumably because the performance is still not good enough by some subconscious standard.   Certainly in no other way can the stability and dynamic performance achieved by the eye's sampled-data smooth-pursuit mode be approached, because in that system a desired constant-velocity segment can be achieved almost instantaneously relative to the sample time.   Wilde and Westcott were not able to resolve clearly from their data whether the operator attempted a constant pure prediction, and if so, how much; however, when prediction is attempted, it apparently varies all the way up to the sampling period of about 180 msec.   In view of the eye's graded change in visual-tracking prediction from 313-msec pure prediction for sinusoidal inputs to watch-and-follow tracking with unpredictable inputs, it seems plausible that the prediction attempted in the manual-tracking case should depend upon the predictability perceived in the signal, and typically this would vary subjectively with time.   The problem is further complicated by the fact that if too much prediction is attempted and proves faulty, instability is just as likely to result as if too little prediction were attempted.   For example, this type of instability occasionally results when two hurried people meet in a corridor and incorrectly predict how they should pass.

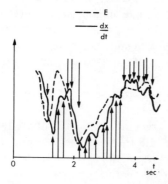

FIG. 12.12   *Manual tracking of random input.   Arrows indicate suspected beginnings of sample periods and their "velocity triangles."* (*From Wilde and Westcott.*)

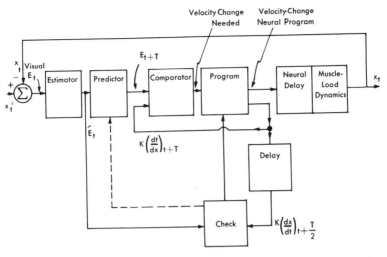

FIG. 12.13 *Proposed model for human operator—unpredictable compensatory tracking.* *(From Wilde and Westcott.)*

Wilde and Westcott postulate the following model for the human operator, in a manual-tracking task where the positional error only is displayed for him [E_t in Fig. 12.13 rather than $x^*(t)$ and $x(t)$], the so-called *compensatory tracking*. The operator first filters this perceived error E_t in some manner which experience of the waveform has shown to be beneficial. This best estimate $\hat{E}_t$ is then passed on to a predictor, whose prediction time is probably correlated with the confidence available in the waveform estimation. The predicted error E_{t+T} is then compared with the previously programmed error (which equals the velocity triangle currently being effected) and any difference is integrated and otherwise modified in the program block to set the new muscle motoneuron program into operation. The system is similar in form to the sampled-data smooth-pursuit mode of Fig. 12.10 except for the estimation and prediction blocks, and except for the less ambitious velocity program. The authors suggest in addition a checking feature by which the current error estimate is compared with the program which achieved it, by delaying the program one half-period on the average, and in consequence the program can be modified in some as yet unspecified way. This is an adaptive feature and could with equal value be fed back to modify the estimator-prediction blocks. Clearly, the prediction of the eye system generally performs the same function, although it must be emphasized that such a block diagram as Fig. 12.13 can only represent the analyst's conjectures at a particular time about how the system *could* operate.

In recent years there has been a tremendous amount of experimental work in determining human-operator characteristics, because these must be known reasonably well in order to design complex man-machine systems which are optimal or at least stable—for example, modern aircraft. Both *ergonomics* and *human engineering* are terms frequently used to describe this area (see, for example [12.12], for the state of the art). Earlier work has largely been concerned with simple linear or quasi-linear analogs of the human operator, and one example of a second-order transfer function with time delay has been introduced in Sec. 8.11. Although there is undoubtedly little psychophysiological basis for linear models, they can be useful in some circumstances, especially when adaptively set up so that any modifications are continuously identified, a situation which is discussed for the above model in Sec. 15.4.

12.4   The Skeletal-muscle Servomechanism

(A) GENERAL

Muscles are the basic actuators for closed-loop control of limb and eye movements, but in turn they and their neurons also constitute servomechanisms, with the specific task of power amplification. In this section the status of neuromuscular system modeling is reviewed from a control-theory viewpoint, but it is not claimed that completely satisfactory attention is given to all aspects of the voluminous neurophysiological literature. The work on this system perhaps typifies the difficulty of a systems approach in biology, namely, that the required quantitative data on internal system variables from revealing dynamic experiments are very hard to obtain. Thus, although the essential anatomy has been established, it seems that only a beginning has been made in understanding how the system works in the full dynamic sense. In particular, although each neuromuscular system has the deceptively simple output of a single limb's position, each such system involves multiple local-feedback loops, quite apart from multiple integrating pathways through higher nervous centers, so that purely descriptive data concerning these pathways are quite insufficient for dynamic analysis of the system. Fortunately, neurophysiologists have been attacking this dynamic problem with increasing vigor in recent years.

(B) ANATOMY

The known essential anatomical pathways are outlined schematically in Fig. 12.14. The muscle itself consists of extrafusal contracting fibers which are organized into motor units each of which is "fired" by one motoneuron from the spinal chord, called an α *motoneuron* because of its relatively large diameter. The fibers are attached onto the relevant bone through tendons which exhibit a finite spring stiffness, and in which are embedded Golgi tendon organs. These organs are mechanoreceptors, or proprioceptors, sensitive to the

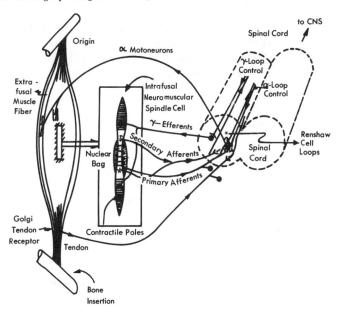

FIG. 12.14   Schematic of skeletal-muscle control system.

total muscle tension since they are placed in series with it.   The
tendon organs provide inhibitory feedback to the α motoneurons but
they have a high threshold level and do not constitute the basic
position-feedback element for this servomechanism.   Instead, this
latter proprioceptive function is performed by specialized structures
called *neuromuscular spindles*, which consist largely of intrafusal
fibers having contractile poles and noncontractile centers.   The
poles contract under stimulation by the γ-*efferent neurons*, so-called
because of their relatively small diameter.   The noncontractile
center contains at least two types of afferent, or feedback-receptor,
endings, the annulo-spiral end organs which give rise to the primary
afferents, and the flower-spray endings which give rise to the secondary afferents, both being sensitive to some function of the spindle's
stretch.   For completeness, joint proprioceptors and even skin
receptors should be mentioned.   However, there is little quantitative
knowledge of how they subserve the overall system except that joint
proprioceptors especially may be involved in learned, skillful motor
actions.   For more details of the preceding and following material
there is extensive reference material (for example, [3.1], [3.8], and the
Bibliography).
    The primary afferents and the tendon afferents are both large
(group 1), with diameters of 12 to 20 μ, and they consequently have
a fast conduction velocity of 70 to 120 m/sec; the secondary
afferents are small (group II), with diameters of 5 to 12 μ and a
conduction velocity of 30 to 70 m/sec.   The spindle cell's primary
afferents have an excitatory effect on the α motoneurons travel-

ing via the spinal chord in a fast monosynaptic reflex arc, whereas the tendon organs have an inhibitory effect on a disynaptic connection with the α motoneurons. The spindle cell's secondary afferents conduct more slowly and through a multisynaptic connection. A local feedback loop also exists in the spinal chord in that the α motoneuron branches as the recurrent axon collateral and feeds back through the so-called *Renshaw cell* to the local motoneuronal pool [12.13]. This feedback effect is inhibitory and in general is thought to regulate the firing behavior of the motoneuronal pool for a given muscle.

Since muscles cannot push, every movable limb has an agonist-antagonist pair of muscles which are reciprocally innervated. With normal muscle tone some tension is maintained in each muscle, and hence a controlled limb torque can be obtained by increasing tension in one muscle and reducing it correspondingly in the other. Although this method can provide fine control as muscle tone is increased, it is obviously expensive metabolically to maintain a large tone. Since animals are laterally symmetric, there are crossed lateral innervations also; for example, if one leg is raised by the classic avoidance reflex to a painful stimulus, the tension in the contra-lateral antigravity muscle is simultaneously increased reflexly so that the animal may remain supported. However, the last cross coupling provides an unnecessary addition for present purposes, while the reciprocal innervation phenomenon can be implicitly included in an analysis using only one muscle, by assuming that the muscle can push as well as pull.

(C) INFORMATION–FLOW DIAGRAM

Figure 12.15 attempts to present the information-flow diagram of the system, but uncertainites arise even at this relatively qualitative stage, for example: Does the Golgi tendon organ respond only, or even principally, to an input of tension? Does the recurrent-axon-collateral-plus-Renshaw-cell path feed back to its own α motoneuron? Actually this last question especially may imply too narrow a viewpoint since limb positioning results from the coordinated action of multiple neuronal paths in parallel. In any case, the diagram conforms generally with current ideas (for example, [12.14] through [12.18]).

This servomechanism can apparently be driven by either of two reference command inputs from higher nervous centers, the so-called α- and γ-*efferent pathways*. The basic closed loop or reflex arc which implements this consists of spindle-cell primary afferents $\rightarrow \alpha$ motoneurons $\rightarrow$ muscle fibers. In detail, however, there are also less well understood closed paths through higher centers. In addition, the disturbance-compensation feature of the closed loop is available to counteract any external torques which may cause undesired limb movements. It seems generally agreed that the γ-efferent path acts essentially as a position-command signal, at least for accurately controlled movements in which the slower action of the γ route is acceptable. The neuromuscular spindle fulfills (at least partially) the functions of feedback transducer, comparator, and controller. In particular, the comparator action arises from the multivariable nature of the spindle's response in the parameter-changing mode already discussed in Sec. 2.7. Thus, the firing frequency of the spindle's primary afferents in the steady state is approximately proportional to muscle extension over the physiological range (Fig. 12.16), but this characteristic is displaced increasingly to the left when the frequency of the γ-efferent command is increased. The actual process involved is that the spindle's contractile poles are activated rather as are ordinary muscle fibers by the γ-efferent impulses, so that in effect a variable bias is placed on the spindle afferents. The α-efferent command route from higher centers activates the muscle's α motoneurons directly, without passing through the spindle system. This route produces a rapid response because of fast conduction and few synapses, but it is presumably at the cost of less accurate control since feedback does not act as directly in this case. Hence, it seems likely that the α route is normally used for both sudden and fast movements, especially those of the skilled task and the avoidance reflex.

Disturbance torque enters after the muscle dynamics on Fig. 12.15, and therefore produces its undesired effect of limb deflection very rapidly. However, the closed-loop effect is then brought into play relatively slowly to restore the desired limb position. The effects of the other feedback paths should also be mentioned. The high threshold of the Golgi tendon organ as a receptor of force seems to imply that this pathway provides an overload protection for the system. The sudden relaxing of the ankle when the foot lands

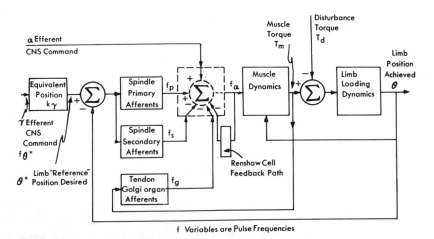

f Variables are Pulse Frequencies

FIG. 12.15 Information-flow diagram of skeletal-muscle control system.

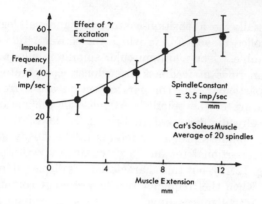

FIG. 12.16   *Effect of* γ *efferent upon spindle-cell primary afferent frequency.*   (*After Granit.*)

awkwardly after slipping, the so-called *clasp-knife reflex*, is caused by this tendon reflex.   Much less is known about the functions of the spindle's secondary afferents, although their arc is multisynaptic and inhibitory upon α motoneurons.   The Renshaw-cell feedback loop seems to regulate and load-share among the motoneuronal pool, but again the detailed dynamic functioning is not known.

Finally, the poorly understood integrative pathways from the various receptor organs through higher levels of the CNS should be mentioned.   Such pathways can have important dynamic effects on the system and its stability, as evidenced by such clinical phenomena as Parkinson's disease and decerebrate rigidity.   However, many normal actions can probably be analyzed initially on the assumption that such pathways play no important dynamic part, partly because the transmission time of such pathways is certainly longer.

(D) MUSCLE–LOAD DYNAMICS

The muscle is an actuator producing a torque on the limb (Fig. 12.15). The limb's dynamic conditions at any moment depend upon the difference between this driving torque and any externally applied disturbance torque.   This net torque must then be fed through the limb and load dynamics in order to establish the limb's velocity and position in the usual way.   Unfortunately, however, the analysis is made somewhat complicated because the muscle tension available to produce limb torque depends upon the stretch position and velocity, which in turn depend upon the dynamics of the mechanical load. In particular, the available force reduces as the stretch velocity increases, but this is a "droop" characteristic universal to all actuators, for example, the car-engine characteristic of Fig. 2.4*a*.

Many data have been amassed on muscle tension-length curves. However, they do not directly provide the dynamic model needed for our block diagram since they represent steady-state results with tetanic (*i.e.*, maximum-effective) stimulation, and are of either the isometric (constant-length) or isotonic (constant-force) type. In order, therefore, to extract such a model we must consider briefly the gross dynamic response characteristics of muscle contraction. Most current models assume that muscle can be considered to be composed of an active contractile component together with series and parallel elastic components. Two possible configurations are shown in Fig. 12.17 [12.20 to 12.22]. Unfortunately, Jewell and Wilkie [12.21] were not able to infer confidently from their data either the precise physiological sites for these elements or a strongly preferred configuration. Certainly the tendon provides much of the series stiffness, but probably some is also distributed along the muscle. Other models [12.23, 12.17] propose that the purely passive mechanical properties must involve a damping element dependent upon the velocity of shortening, as shown dashed in Fig. 12.17. However, in this development we follow Wilkie [12.20], who sufficiently modified Hill's energy-based equation for muscle contraction that the muscle can work into any specified loading system of Fig. 12.15. His equation for the tetanized whole muscle is

$$\frac{dx}{dt} = \frac{b(F_0 - F_m)}{F_m + a} - \frac{d}{dt}f_1(F_m) \qquad (12.3)$$

where dx/dt = contraction velocity $\sim d\theta/dt$
F_m = external muscle force $\sim T_m$
$f_1(F_m)$ = stress-strain curve of series elastic component
F_0 = isometric tension (tetanic)
a, b = experimentally evaluated constants

This equation can be put into block-diagram form, but it does not yet represent a satisfactory transfer equation because it is valid for tetanic stimulation only. Therefore, since there do not yet seem to be satisfactory experimental data on normal operation with graded stimulation rates of the α motoneurons f_α, Stark et al. linearly

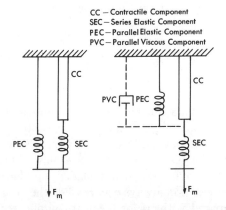

CC — Contractile Component
SEC — Series Elastic Component
PEC — Parallel Elastic Component
PVC — Parallel Viscous Component

FIG. 12.17   Muscle analogs.   (*From Jewell and Wilkie.*)

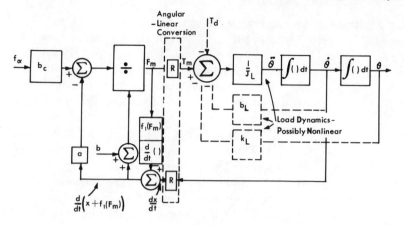

FIG. 12.18   Muscle-load dynamics.

interpolated the muscle force as a function of this f_α [12.18].   On the
other hand, they did not provide for the $f_1(F_m)$ term of Eq. (12.3).
Furthermore, Hammond [12.17] in his proposed model did not allow
for external loading because his experimental data were largely concerned with velocity-forcing inputs.   Thus, for present exemplary
purposes and in the hope of provoking relevant experimental data,
we simply modify Eq. (12.3) by assuming that the isometric tension
at firing frequency f_α is less than the isometric tetanic tension F_0 in
the ratio of actual and tetanic impulse frequencies:

$$F_\alpha = F_0 \frac{f_\alpha}{f_0} = c f_\alpha$$

where c = constant
$\quad$ f_0 = tetanic impulse frequency
$\quad$ F_α = isometric force with firing frequency f_α

This reduced value of tension should replace F_0 in Eq. (12.3), which
therefore becomes, after reorganizing for input-output purposes,

$$F_m = \frac{(bc)f_\alpha - a \dfrac{d}{dt}\,[x + f_1(F_m)]}{b + \dfrac{d}{dt}\,[x + f_1(F_m)]} \tag{12.4}$$

The block diagram of Fig. 12.18 illustrates this relation and also
clearly shows how the limb-loading dynamics inextricably couple
with the muscle-actuator dynamics.   Thus, for example, the muscle
tensions in the heavy manual task of Sec. 11.8 are quite different
from the muscle tensions in moving the arm unloaded, for the same
velocity pattern.   This interaction always occurs when the actuator
is not overpowering compared with its load, but it complicates the
task of obtaining suitable experimental data for modeling purposes.

(E) RECEPTOR DYNAMICS

In general one cannot yet satisfactorily infer dynamic models for the spindle and tendon receptors, despite a fairly impressive amount of experimental data—for example, on the adapting nature of the response to constant stimuli, with its corresponding phase lead in the frequency response characteristic. This subject is discussed more generally and fully in Chap. 13; here we only aim for a representative transfer characteristic. Lippold et al. [12.24] have investigated the frequency response of muscle spindle in the cat, and find that phase lead approaches 90° as frequency increases, at least up to about 20 cps (Fig. 12.19a). They did not fit a linear transfer function since the gain and phase curves are not readily compatible, and furthermore the phase-lag curve varied if the amplitude of

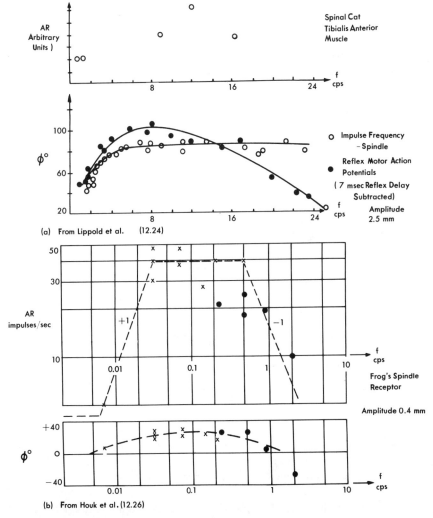

(a) From Lippold et al. (12.24)

(b) From Houk et al. (12.26)

FIG. 12.19 Frequency responses of spindle receptors. [(a) *From Lippold et al.* (b) *From Houk et al.*]

oscillation was changed.   Davey and Roberts [12.25] have fitted the
constant-velocity (ramp) response of a frog's muscle spindle with
three exponentials to obtain a transfer function, but they do not
quote numbers.   More recently, Houk et al. [12.26], also using frog
muscle (extensor longitus digitus IV), performed frequency response
tests using a preparation which averaged the responses of several
spindle cells.   The sinusoidal stretch was superimposed on a steady
tension which neither saturated the receptor nor gave rise to silent
periods, and the resulting frequency response (Fig. 12.19*b*) shows a
phase-leading characteristic of about 20° over a 1-decade range
around a center frequency of 0.1 cps.   From these data Houk
postulates a transfer function, which in turn does provide reasonable
correlation between experimental and predicted step responses:

$$H(s) = \frac{27(30s + 1)}{(5s + 1)(0.35s + 1)} \tag{12.5}$$

(F) POSITION AND VELOCITY SERVOMECHANISMS

A few attempts to model the overall muscle servomechanism quantitatively have been reported [12.17, 12.18], and certain established
qualitative features of this system can be summarized.   First, the
spindle receptors provide some rate-sensitive action which helps to
nullify phase lags elsewhere in the system, and hence to ensure a
stable servomechanism for postural and voluntary control.   Thus,
for example, both Lippold et al. (Fig. 12.19*a*) and Partridge and
Glaser [12.16] find in their experiments on cats that the reflexly
induced tensions resulting from externally imposed sinusoidal oscillations of muscle length also exhibit some phase lead with respect to
the length-change input in a certain frequency range.   Of course, as
we have seen from consideration of the human-operator characteristics, this predictive effect is often advantageous.   Secondly, the
spindle's secondary afferents, Golgi tendon afferents, and Renshaw-
cell feedback paths all provide inhibition of various types, which
cannot yet be dynamically modeled.

Quantitative information cannot be given on the α and γ control
loops in the higher nervous centers (Fig. 12.14).   However, it has
been postulated that the cerebellum plays an important part in
deciding how a particular muscle task is to be carried out [12.15],
because of evidence that when the cerebellum is cooled it is possible
to reroute some particular reflex contraction from the "length-servo"
γ route through to the direct α route [12.27].   The general current
thinking is that the γ-efferent path provides length-servo action for
accurately controlled movements, whereas the α path controls the
reflex and the skilled rapid movements.   In any case, the two paths
are probably used cooperatively in many normal operations.   Furthermore, the nature of the response depends upon the input signal
as well; for example, in the case of the eye saccade this is typically
a short burst of impulses down the α motoneurons.   On the other

hand, what is the stimulation pattern during the constant-force acceleration segments of manual tracking, of constant-velocity smooth eye pursuit, or of the smoothly varying velocity in predictable eye and manual-tracking tasks? In this regard it is worth noting that the eye receives its velocity-compensation signal for head movements from the semicircular canal, which presumably provides a neural signal proportional to angular velocity. The simplest system would be realized if this "desired velocity" signal were used by the muscles in its raw "velocity" form; in turn, the data-processing requirements would then be minimized if the feedback signal were that of "achieved velocity" directly. Finally, this implies the need for velocity-sensitive spindle receptors, although positional information is also needed, of course, for saccadic control, etc.

(G) THE CHALLENGE

Neuromuscular systems challenge the control engineer because it is probable that his well-proved techniques of system analysis for the corresponding physical servomechanisms may not be fully applicable. This inapplicability would exist if it were found impossible to obtain the adequate experimental dynamic data on which present engineering methods depend. The difficulties arise, as we have noted, in obtaining data to characterize at least the following components:

1. *Actuator:* The neural-input–force-output characteristics of muscle.
2. *Receptors:* The linear, or more probably nonlinear, dynamic characteristics of the various receptors.
3. *Information channels:* Adequate characterization of the effective information in the various multiple neuron channels is needed; in particular, for example, what information-rate changes are brought about by Renshaw-cell feedback on the motoneuronal pool?
4. *Multiple pathways:* The relative dynamic effects of the various afferent pathways in Fig. 12.15 need to be satisfactorily established, as do, very importantly, the relative dynamic effects of the various hierarchical integrating pathways in higher centers (Fig. 12.14). Owing to the complexity and uncertainty of the paths, the latter problem will probably require the development of new cybernetic techniques. It has, of course, important clinical implications for such diseases as Parkinson's disease and clonus.

For the neurophysiologist the corresponding challenge is to ascertain what small part of these idealized data needs can ever be satisfied, and then to consider the directions in which developments of control theory would most aid his task of understanding this system. It is interesting to note that one mutual task of physiologist and engineer which undoubtedly promises rapid advances for this area is the design of flexible, efficient, and cheap prostheses which work from bioelectric signals, both as artificial limbs and as power boosters for natural limbs involved in otherwise impossible tasks such as maneuvering under high accelerative forces.

13 Biological Receptors

13.1 Introduction

In previous chapters we have referred to the retinal photochemo-receptors, cutaneous thermoreceptors, the cochlear mechanoreceptor of sound, the semicircular canals as mechanoreceptors of angular velocity, and the various neuromuscular proprioceptors. There are of course many other external and internal sensory receptors in the body, each providing feedback for either homeostatic regulators or tracking servomechanisms; some relevant examples are: baro-receptors of arterial blood pressure in the carotid sinus for cardio-vascular control; chemoreceptors of O_2 tension in the carotid glomus for respiratory control; chemoreceptors of CO_2 by structures in the medulla; osmosensitive structures for osmotic pressures in the brain stem, for water and electrolyte control; cutaneous pain and touch receptors (mechanoreceptors); electroreceptors of fish for "radar"; olfactory and taste chemoreceptors.

In this chapter some basic material on receptors is reviewed (Sec. 13.2); analytical modeling is presented for one particular mechano-receptor (Sec. 13.3); and finally a feedback model for sensory reception is presented (Sec. 13.4).

Deterministic input-output relations are implicitly assumed in the analysis of this chapter, but, it is interesting to note, this is not true of the detailed picture. Thus, even when the known input signal is of constant value, the impulse frequency of a single receptor afferent exhibits some irregularity. The fact that such irregularity tends to average out over either an ensemble of receptors or a period of time provides the basis for the deterministic models used so far. In consequence, however, receptors do seem to provide one of the noise generators in biological systems, but the treatment of statistical signals is deferred until Chap. 14.

13.2   Receptor Characteristics

In this section we review briefly the functional nature of receptor characteristics, but the reader is referred to such texts as [3.1] and [3.8], and to the Bibliography, for greater detail. In addition [13.11] presents models for the relevant pulse-modulation process.

(A) STIMULATION MECHANISM

In general each receptor is most sensitive to some particular form of energy (heat, light, sound, etc.) although it may also respond to other forms (especially mechanical deformation and heat). The form to which it is most sensitive is called its *adequate stimulus*. The response characteristic of receptors is that impulses (spikes) are propagated along the afferent nerve—the so-called *all-or-none nature of discharge*. The *doctrine of specific nerve energies* states that if any receptor discharges as a result of stimulation by another form than its adequate stimulus, the CNS perceives it nevertheless as the "intended" stimulus—for example, "seeing stars" when one's eye is hit. The biochemical nature of energy conversion and modulation into impulses is not completely understood, but the general pattern seems to be that there are two phases to the process.

1.   The first phase is that of converting from the adequate stimulus to a common mechanism which can produce impulses; this mechanism produces a *generator potential* of depolarization and obviously seems analogous to the depolarization necessary to produce impulses in the ordinary neuron. Generator potentials have been measured in recent years and appear to have dynamics which correlate reasonably well with observed impulse frequency as a result of inputting stimuli of varying intensity. Furthermore, the generator-potential concept provides a simple basis to explain the action of those receptors in the CNS which may not signal their output directly as spike potentials, for example, hypothalamic thermoreception. If the potentials were generated in appropriate controlling neurons of the thermoregulatory system, their action would be that of a nonlinear comparator by biasing cellular sensitivity (cf. Fig. 12.16).

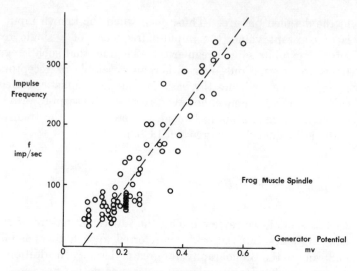

FIG. 13.1   Receptor characteristics—generator potential versus impulse frequency. (*Replotted from Katz.*)

2. The second phase involves converting from the generator potential to impulse frequency and is common to all neural receptors, as opposed to the differing conversions necessary from adequate stimuli to generator potential.   (In this respect, however, it should be noted that the hair cell has been utilized in several evolved configurations so that it can respond to different energy inputs— for example, in the cochlea responding to sound, and in the semicircular canals responding to angular velocity.)   In at least one recorded case (from frog's muscle spindle), this conversion is essentially linear in the steady state (Fig. 13.1).   The broken line is a best-fit regression line drawn through the observation points, and illustrates from another viewpoint why statistical considerations are often necessary in biological systems.

With regard to the neuronal impulse discharge, it is important to note that no information content in the mathematical sense appears to be carried by the form of the spike discharge itself, except that insofar as there is spatial ordering of afferent neurons in the CNS, the latter has available information about the site and nature of the peripheral receptor initiating the spikes.   Neuronal information transfer therefore depends only upon recognizing the presence of spikes when they occur.   In consequence, a relatively high threshold level for recognition of a spike can be utilized and considerable degeneration of spike shape can be tolerated before information loss occurs, just as in a digital computer.   Presumably this threshold decision is implemented by the success or otherwise of the axonal spike in releasing a pulse of transmitter substance at the synapse after traversing the axon.   In receptor and efferent paths it thus seems reasonable to assume that information is conveyed by the time between pulses, in particular by the pulse frequency (the inverse of the time interval). However, in the CNS this may not always be true since information can evidently be transmitted by changing the impulse time pattern while retaining the same overall frequency (Sec. 14.4).

(B) RESPONSE TO STIMULUS INTENSITY

The Weber-Fechner generalization, that receptor output varies as the logarithm of the stimulus intensity, was introduced for the retina in Sec. 2.6. Since there is some evidence for a linear relation between generator potential and impulse frequency, presumably this logarithmic relation is implemented somewhere in the first stage of receptor conversion. For illustration, Fig. 13.2 is a classical result from Matthews [13.2] showing that the impulse frequency of the frog's muscle spindle does vary approximately as the logarithm of the applied load after adaptation. This particular law is considered further in Sec. 13.4.

There is a practical upper limit in impulse frequency of about 500 impulses/sec from any single receptor due to the refractory period, an effect equivalent to saturation. Since there is an absolute lower limit in frequency of zero, and a practical limit of around 2 impulses/sec due to the noise level, neuronal "forgetting," etc., the output of a single receptor has a limited dynamic range of the order of 100. However, this range can cover a very large range of the input stimulus because of the logarithmic sensitivity relation. Nature's further technique for extending the response range is to recruit more receptors as the stimulus intensity increases. Precisely how this is achieved in a well-coordinated fashion is not known, but the variable threshold characteristics of the receptors at a given site are probably instrumental, and it is also possible that the efferent innervation of receptors may be involved in this task. This small-fiber feedback

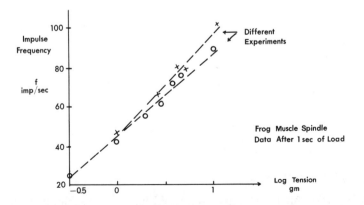

FIG. 13.2 Receptor characteristics—impulse frequency versus applied load. (*Replotted from Matthews.*)

from the CNS into the receptor organ (efferent path) has been established in a number of organs, including the retina and cochlea, while the γ-efferent pathway to the spindle cells seems anatomically analogous. The detailed functioning of the efferent pathways is not known, but it seems likely that they are responsible for setting the threshold levels of the various receptor organs so that the CNS need not be burdened with the data processing of information in which it is not currently interested. (Wooldridge discusses this question in pertinent comparison with computer technology [10.13].) For example, the efferent pathways to eye and ear [13.3, 13.4] probably have a similar functional purpose to that of the muscle's γ efferents, although the latter's task is perhaps more dynamic.

(C) SPONTANEOUS DISCHARGE

Many receptors exhibit spontaneous discharge, that is, a maintained impulse frequency even when the relevant adequate stimulus is at zero level, and this we also call a d-c level of impulse frequency. Discussion of this phenomenon centers around two ideas:

Centrifugal Control. It is known that the efferent innervation of many receptors permits selective filtering of environmental sensory information. In addition, such information is known to be very important to central arousal, which may be considered a CNS function that prevents us from falling asleep. Thus this efferent innervation, or centrifugal control, can subserve the organism's awakeness by controlling the spontaneous discharge frequencies of the various receptors. See, for example, Granit and Wooldridge.

Bidirectionality. Unless a receptor exhibits spontaneous discharge it cannot provide bidirectional proportional feedback. The semicircular canal is one example of a receptor which is apparently linearly and bidirectionally sensitive to its stimulus. The muscle spindle can also show this effect after being set to some desired length by the γ-efferent system. In other cases, two receptor organs are used to provide bidirectional information, for example, hot and cold thermoreceptors.

There do not seem to be many receptors which are bidirectionally rate-sensitive, possibly because there has been no great evolutionary advantage conferred by this mode. Thus, in many reflex systems (such as pupillary control and pain avoidance), the prime need is only to minimize exposure to an undesirably increased level of stimulation.

(D) ADAPTATION; RATE SENSITIVITY

If a receptor is subjected to a rapid increase of stimulus from some equilibrated level to another steady level, the impulse discharge frequency typically shows an almost instantaneous jump to a high

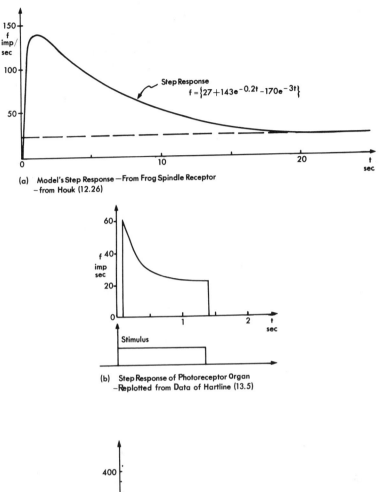

(a) Model's Step Response—From Frog Spindle Receptor
— from Houk (12.26)

Step Response
$$f = \{27 + 143e^{-0.2t} - 170e^{-3t}\}$$

(b) Step Response of Photoreceptor Organ
— Replotted from Data of Hartline (13.5)

Stimulus

(c) Step Response of Cockroach Mechano Receptor
— from Chapman & Smith (13.6)

FIG. 13.3   Receptor adaptation characteristics. [(*a*) *From Houk.*   (*b*) *Replotted from data of Hartline.*   (*c*) *From Chapman and Smith.*]

value followed by decay to some new level (Fig. 2.2).   This phenomenon has been termed *adaptation* in physiology, but *rate sensitivity* seems a more pertinent title for control-oriented terminology.

The term *slow adaptation* is used to describe those receptors which maintain a response long after the rate-sensitive response has died away.   Since it persists after perhaps several hours, one should presumably consider it a proportional response.   Slow adaptation therefore appears to be a misnomer, and we call it *proportional-plus-rate sensitivity* instead.

In general those receptors concerned with regulation of the *milieu interne* and with neuromuscular servomechanisms must maintain a proportional response, but the decay time of their rate-sensitive mode varies considerably.   Figure 13.3a shows the step response of Houk's model [12.26] for the frog's spindle receptor whose frequency response is shown in Fig. 12.19.   Figure 13.3b shows adaptation in a photoreceptor organ [13.5], and here the time constant of decay is obviously much shorter.

The term *fast adaptation* is used in physiology to describe those receptors whose response is essentially equilibrated within a short interval, say about 1 sec.   Such receptors are particularly the exploratory ones such as touch for which it is desirable to forget past impressions rapidly in order to be sensitive to new ones, although typically the old response does not die away completely.   For example, Fig. 13.3c shows the step response of a cockroach's mechanoreceptor on the large tactile spine of the femur [13.6]; it is quantitatively modeled in Sec. 13.3.

This concludes our qualitative review of the functional characteristics of receptors.   We now consider some transfer-function models for such receptors.

13.3   Transfer-function Models of Receptors

A linear model for the frog's muscle spindle receptor has already been presented in Sec. 12.4 and its step response in Fig. 13.3a. Other models have also been developed, of which one interesting example is now described.

Chapman and Smith [13.6] performed both transient and frequency response tests on the cockroach mechanoreceptor and were able to model their results by a rather unusual form of transfer function, which had been earlier reported by Landgren for the cat's carotid sinus baroreceptors [13.7].   This transfer function arises

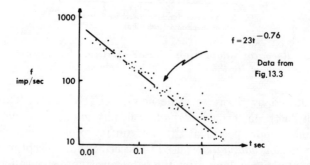

FIG. 13.4 Logarithmic plot of cockroach mechanoreceptor. (*From Chapman and Smith.*)

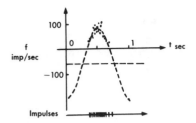

FIG. 13.5 Frequency response data of cockroach mechanoreceptor. *(From Chapman and Smith.)*

because the step response data (Fig. 13.3c) are well fitted by a straight line on log-log coordinates (Fig. 13.4), which implies a power-law relationship rather than the classical exponential decay:

$$f = Bt^{-\beta} \quad \text{impulses/sec} \tag{13.1}$$

where in this case β is 0.76 and B is 23.   Analysis of the frequency response data was complicated by the fact that the receptor was almost always quiet during some period of the cycle (Fig. 13.5). Therefore the mean value and fundamental frequency component were found from a Fourier series analysis, and the amplitude of this fundamental sinusoid was used to evaluate the AR.   The resulting Bode plot data from experiment and model are shown in Fig. 13.6; the data of the model are obtained as follows: The step response of Eq. (13.1) may be manipulated to give the mechanoreceptor's transfer function

$$H(s) = B[\Gamma(1 - k)]s^{\beta} \tag{13.2}$$

where $\Gamma(1 - k)$ is the gamma function, a constant dependent upon the parameter $1 - k$ (for example, [3.13]).   This transfer function

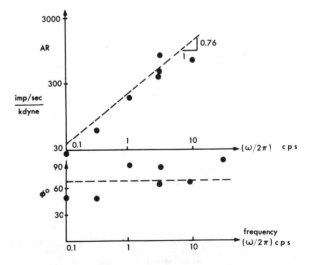

FIG. 13.6 Bode plots for cockroach mechanoreceptor.   *(From Chapman and Smith.)*

contains a nonintegral power of s and in the present case represents "0.76 of a differentiation" upon the input. The model's frequency response form follows fairly directly from Eq. (13.2) as

$$H(j\omega) = B[\Gamma(1 - k)](j\omega)^\beta$$
$$= \{B[\Gamma(1 - k)]\omega^\beta\}e^{j\beta\pi/2} \tag{13.3}$$

by recalling that $j = e^{j\pi/2}$ (Fig. 7.10). The phase ϕ is specified by the exponent of Eq. (13.3), and in this case $\phi = 68°$. This phase is constant at all frequencies, but the log AR increases linearly with log ω, having a slope of $+0.76$ log/decade. Figure 13.6 shows that these predicted curves fit the pseudoexperimental data reasonably well over the range of 0.1 to 30 cps. It should be noted that a conventional linear transfer function was earlier derived by extracting exponentials from the step response for this mechanoreceptor [13.8]. Three exponentials were involved, but the results do not seem to agree well with the frequency response results just presented (see Prob. 13.1).

We now leave the problem of receptor transfer functions, noting that there is evidently need for much further experimental work on these devices, using a suitably broad dynamic range of transient and other input stimulations. Also, it should be noted that the presentation of linear models here does not necessarily contradict the validity of the URS modeling of Sec. 9.6, because many of the relevant step responses were taken in the "increase" direction only. Furthermore, when the transfer function has been found by frequency response techniques, the waveform has not usually been symmetrical.

13.4 Receptors and Perceived Intensity

We now consider an interesting generalized feedback model of receptors and their resultant sensory perception. The Weber-Fechner law states that there is a logarithmic relation between stimulus intensity I and output neural frequency f_R:

$$f_R = k_R \log (I - I_0) + a_R \tag{13.4}$$

where k_R, I_0, and a_R are constants. On the other hand, Stevens has presented much experimental psychophysical evidence that perceived intensity varies as a power law of general form [13.9]

$$\psi = a(I - I_0)^\beta \tag{13.5}$$

where the power β varies between 0.33 and 3.5 for various modalities (sight, hearing, and touch, for example).

MacKay [13.10] has recently proposed a feedback model which renders these two laws theoretically compatible and at the same time raises some important neurophysiological questions. His basis is that

> . . . perception is regarded as an internal, outwardly directed, adaptive or "matching" response to stimulation, generated within the organizing system, that determines the "current state of readiness" of the organism. On this view the perceived intensity of a stimulus should reflect, not the frequency of impulses from the receptor organ, but the magnitude of internal "organizing activity" evoked to match, or in some other sense "counter-balance," that frequency.

Clearly, such a model might have relevance to the γ-efferent or "centrifugal control" systems for afferent discharge rates which are mentioned in Sec. 13.2. On the other hand, the functional nature of the model does not depend on whether the feedback components occur within the CNS, near the receptors, or in both places. The postulated information-flow diagram, shown in Fig. 13.7, in addition to the basic receptor obeying Fechner's law [Eq. (13.4)], comprises two extra elements. For perceiving the intensity of the stimulus a more central element O is postulated which acts as an "effort generator" or "organizer" and produces a matching frequency

$$f_O = k_0 \log (\psi - \psi_0) + a_0 \qquad (13.6)$$

where ψ is the internal activity of O which suffices to match f_O against the receptor frequency f_R, given certain characteristics of the third element—the comparator C—and where k_0, ψ_0, and a_0 are other constants. The final element C operates upon the two frequencies, thus providing a corrective path to modify ψ until matching is achieved. A suitable comparator action can be specified by

$$f_c = \phi(f_R - bf_O - k_c) \qquad (13.7)$$

where ϕ = monotonic function
b = relative weighting of frequencies in C
k_c = some "constant"-firing-rate parameter of C

MacKay postulates that a neural comparator could achieve the negative weighting of f_O by utilizing inhibitory synaptic connection for this path, in contrast to normal excitatory connection for the f_R path. This effect is discussed in Sec. 2.7.

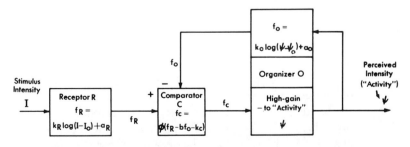

FIG. 13.7 Model for perceived intensity. (*Adapted from MacKay.*)

The reader should readily recognize the equivalence between the configurations of Fig. 13.7 and the operational amplifier (Fig. 10.8). Thus, if the loop gain of this system is high, or if the loop contains an integrator (Sec. 11.5), f_O and f_R can be made to match even when the "mismatch" error signal f_c is arbitrarily small. In particular, equilibrium is obtained when

$$f_R = bf_O + k_c \qquad\qquad (13.8)$$

and incorporating Eqs. (13.4) and (13.6) in Eq. (13.8) yields

$$\psi - \psi_0 = a(I - I_0)^\beta$$

where

$$\beta = \frac{k_R}{bk_0}$$

$$\log a = \frac{a_R - ba_0 - k_c}{bk_0} \qquad\qquad (13.9)$$

This equation shows that the "perceived intensity" of organizer O does indeed vary as a rather general power law of the stimulus intensity, as postulated in Eq. (13.5). The constant β is determined by the relative gains of the two neural signals before comparison. The constant a is a measure of the various biases and is decreased as the coupling gain bk_0 increases.

It is of interest to examine several further aspects of this model.

(A) THE "JUST-NOTICEABLE DIFFERENCE" AND FECHNER'S LAW

Let us suppose that the organizer O can perceive a change in stimulus when the frequency f_c changes by Δf. From any previous equilibrium this requires that f_R change by this same amount Δf (if $\phi = 1$), which in turn requires that the stimulus intensity I change by an amount ΔI proportional to $I - I_0$ before a just-noticeable difference can be perceived. Since this requirement that

$$\frac{\Delta I}{I - I_0} = \text{constant} (= \Delta f)$$

gives rise to the logarithmic law, the model is mathematically and physiologically compatible with Fechner's observations.

(B) HIGH-GAIN LOOP AND THE LOOP FUNCTIONS

One basic advantage of the closed-loop configuration is that the actual nature of the function ϕ is relatively unimportant in obtaining the matching condition of Eq. (13.8), provided that the forward gain is high. However, in order to obtain exactly the generalized power law of Eq. (13.9), a differencing relation between f_R and f_O is needed. In the same way, the presence of negative feedback ensures that the frequency functions f_R and f_O need not be exactly logarithmic functions, as in Eqs. (13.4) and (13.6), in order for Stevens' law still to hold. In fact, it still holds exactly when f_R and f_O are power laws under certain parameter conditions (Prob. 13.2).

(C) THE ORGANIZER OR EFFORT GENERATOR O

As MacKay points out, we should not expect O to have its "activity" ψ represented by a neural firing rate because its output firing rate f_O must be related to the logarithm of ψ. Rather, one would expect to find O representing an analogous quality to receptor intensity, such as ". . . the intensity of some metabolic or other physical disturbance known to be logarithmically related to firing frequency. . . . " [13.10].

(D) THE HIGH–GAIN CONTROLLER

Finally the problem arises of how the error signal f_c is processed to produce the activity ψ. If the relation were exponential (antilogarithmic), a linear overall loop would be obtained, but this does not seem likely; if the relation were essentially that of integration (that is, if any f_c produced an accumulating increase of activity, such as by continuing release of biochemical material), the loop would be nonlinear, but the high-gain feature would certainly be achieved in the steady state.

This completes our brief consideration of the deterministic characteristics of biological receptors. The data limitations so exposed emphasize the need for considerably more experimentation, preferably utilizing dynamic inputs which are designed to reveal nonlinearities, inherent statistical processes, etc.

14 Statistical Signals

14.1 Introduction

The analytical methods treated so far in this book have implicitly assumed deterministic input-output relations for the various components of dynamic systems. Thus, a complete muscle responds essentially deterministically to a given motor nerve signal for any specified conditions of applied tension, length, etc., while equally the body's sweating rate depends deterministically upon analogous variables. This determinacy, however, does require using such overall analog variables as the temperature, pressure, and length of relatively large lumps of material. If, instead, we examine the functioning of individual microscopic units, such as, especially, neurons and receptors, then indeterminacy appears. For example, even when a receptor is subject to a known constant value of its adequate stimulus, the instantaneous impulse frequency of its afferent neuron exhibits some degree of irregularity, although the time-averaged frequency is deterministically related to the appropriate input variable. It is for this reason, of course, that we have so far been able to ignore the problem of statistical signals. We should also note that multiple parallel-afferent neurons are involved in a given control task, and that the phenomenon of recruitment, by which extra neurons are brought into effect, must be treated statistically.

Thus the receptors introduce some randomness (or *noise*) into the biological systems which they innervate. The CNS also introduces neural noise through its complex integrating and cross coupling effects, for example, in pupil tremor. However, it is not clear whether this CNS noise is independently generated within the CNS or whether it derives from imperfect filtering by the CNS of the body's various afferent signals.

The development of the material in this chapter is as follows: Section 14.2 is concerned with discrete statistical signals, where *discrete* specifies signals which are not continuous functions of time, nor even necessarily of continuously variable amplitude. For this development, the discrete nature of information transmission at the neuromuscular junction is used as an exemplary vehicle. The analysis leads directly to the important standard probability density functions (PDFs) of statistics, in particular the exponential, the Poisson, and the Gaussian. Some basic properties of such signals and their distributions are then briefly reviewed.

Such discrete signals are obtained neurophysiologically by micro-electrode techniques, which can provide records of the firing of single cells. Alternatively, larger internal probes can be used to obtain records from groups of cells, such as the electromyograms (EMGs) of muscle, electrocardiograms (ECGs) of heart, and electro-encephalograms (EEGs) of the CNS. Large surface electrodes can be used more conveniently for the same purpose. Since both larger electrode types average the impulses of many neurons, the resulting signals have a continuous waveform. A similar treatment of continuous-waveform statistics is made in Sec. 14.3, and this leads naturally to consideration of correlation functions and frequency spectra.

Averaging and correlating computers have become important biological tools because they can increase the signal-noise ratio of a signal which is embedded in general CNS or other noise (Sec. 14.4). Essentially, this is achieved by averaging out the noise content. Such computers are particularly useful when the process is performed on-line so that the experimenter may modify his experimental program as necessary while he proceeds. The problem of desirable length of averaging time is also briefly reviewed (Sec. 14.3); first of all, the inevitably limited storage capacity of computers and the need to obtain results quickly provide an upper limit to the averaging time; secondly, however, it is possible that the biological parameters are themselves time-varying, so that an extended averaging could produce meaningless results. In such cases a trade-off in data length usually results, as between increased statistical variability of short records and decreased biological significance of long ones.

A statistical approach to CNS work is essential, but it is not quite so basic for the autonomous and somatic systems to which we have limited ourselves in this book. Nonetheless, there are many aspects of our systems analysis which can benefit from such an approach, and since this particular biological subject is in its infancy, we can expect its applications and value to increase rapidly.

14.2   Discrete Statistical Signals

(A) CHEMORECEPTORS, SYNAPSES, AND NEUROMUSCULAR JUNCTIONS

In the carotid and aortic bodies there are chemoreceptors that measure oxygen tension and that fire irregularly at all frequencies.   Statistical tests show that their successive firing intervals are random although the overall frequency does indeed depend upon the stimulus intensity on the average.   What, then, do we mean by the term *random?*   In detail the functioning of this receptor involves release of granules of catecholamines, and it is these individual releases which are thought to be unpredictable and independent of each other and are therefore termed *random.*   Rather similar behavior seems to exist at neural synapses, although both excitatory and inhibitory responses can be produced upon the downstream neuron, dependent upon the type of synapse; in both cases the action is effected by the release of transmitter substances from the upstream axon.   The neuromuscular junction provides a third, rather similar process, which has been very well studied, and which we now turn to for our introduction to discrete statistical signals.   It should be repeated that *discrete* is defined as applying to a variable which occurs either at discrete times or with discrete magnitudes; in this case, the pulse-like events considered are discrete in time.

Each motoneuron may innervate several muscle fibers, and the combination is called a *motor unit.*   In detail the fiber is innervated at a special neural zone called the *end plate* (Fig. 14.1), the axon diameter being approximately one-tenth that of the muscle fiber (of the order of 100 μ).   There is considerable evidence for a chemical form of transmission because in normal healthy muscle the arrival of

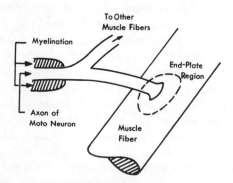

FIG. 14.1   Simplified schematic of neuromuscular junction.

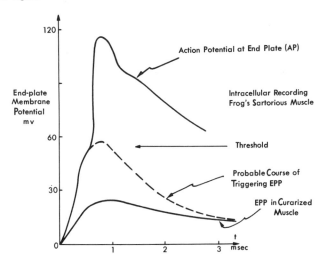

FIG. 14.2   End-plate recordings of AP and EPP.   *(After Fatt and Katz.)*

an action potential down the motoneuron triggers the release of acetylcholine (ACh) which diffuses across to the motor end plate.   The arrival of the ACh initiates events resulting in a depolarization wave called the *end-plate potential* (EPP), which normally exceeds the muscle's threshold level so that an action potential is propagated in both directions along the muscle fiber (Fig. 14.2).   This action potential is similar to the neuron's except that it generates the unit twitch activity of the muscle rather than an information transmission.   In Fig. 14.2 the normal EPP curve is only estimated since it is blanked out by the action potential which follows; however, when the neuromuscular junction is curarized to block transmission of the action potential, a full EPP waveform can be recorded, as shown from [14.1].

In 1952 Fatt and Katz reported upon a further phenomenon, spontaneous and apparently random "miniature" potential disturbances at the end plate [14.2].   These resemble the EPP in their

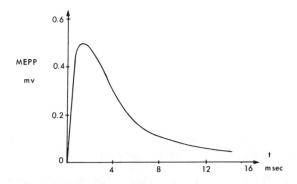

FIG. 14.3   Typical MEPP.

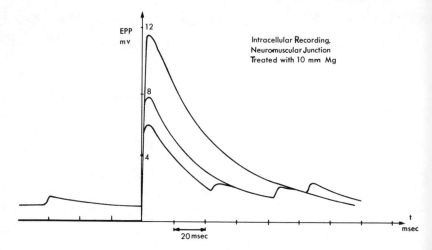

FIG. 14.4   Superimposed EPP and MEPP records.   (*From DelCastillo and Katz.*)

waveform and time constant, but are of much lower amplitude; they are consequently called *miniature end-plate potentials* (MEPPs) (Fig. 14.3). Actually these MEPPs vary rather randomly both in amplitude and in release time, thus providing the need for a statistical treatment. A summary is now presented of the working theory which has been constructed by Fatt, Katz, et al. [14.1 to 14.6] to explain these neuromuscular events. ACh is continually synthesized in the motoneuron endings, but it is in bound form and is only released from the terminal portion when suitably triggered. The units of ACh in this bound form are called granules or quanta, each of which may contain thousands or more molecules of ACh. The arrival of an action potential produces a high probability of their being discharged, but in addition individual quanta are discharged randomly from time to time. The triggering mechanism for this latter discharge is not well understood, although a "thermal-noise" excitation has been considered [14.2] (that is, random fluctuation of the resting membrane potential due to internal thermal agitation of the ions).

By utilizing certain techniques which accentuate the MEPP relative to the EPP [14.3], it has been concluded that the normal EPP is composed of a finite integral number of superimposed MEPPs whose synchronous release is triggered by the arrival of the action potential; for example, the introduction of magnesium ions reduces the overall EPP but does not prevent the release of individual MEPPs (Fig. 14.4).

(B) STATISTICAL CHARACTERIZATION OF THE EPP AND MEPP DATA

Since the MEPP have randomly varying amplitudes (maximum excursion from the base line), a reasonably obvious classification

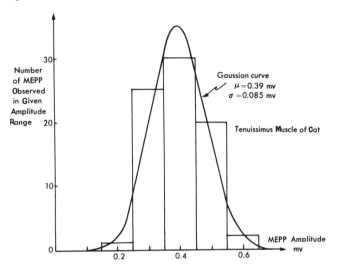

FIG. 14.5 Histogram of MEPP amplitudes. (*From Boyd and Martin.*)

scheme is to plot the respective number of events which fall into each of a set of amplitude ranges (slots); for example, 0.1-mv slots around 0.1, 0.2, 0.3 mv, etc., as in Fig. 14.5 for the tenuissimus muscle of the cat.   Many familiar examples of this type of classification exist— for example, the height and weight distributions of a population— and because of their stepwise nature such empirical plots are called *histograms*.   Obviously the average value of the event (average amplitude in Fig. 14.5) occurs somewhere within the total range, and furthermore, the spread of the histogram about this average may be expected to be significant in describing the data.   These topics are taken up in the mathematical modeling of Sec. 14.2C, and we merely note that the data of Fig. 14.5 do seem to be fitted adequately by the well-known Gaussian distribution curve.

The MEPPs also exhibit variable times between their successive spontaneous discharges, and these data may similarly be plotted in histogram form, with the time interval as the (stepwise) abscissa (Fig. 14.6 from [14.5]).   These data are well fitted by the exponential PDF, which is shown in Sec. 14.2C to result when the events are independent of each other's occurrence.   It was stated earlier that the EPP is composed of a finite, integral number of superimposed MEPPs.   This conclusion is justified by further statistical analysis of MEPP and EPP data, involving the Poisson PDF, also presented in Sec. 14.2C.

Figures 14.5 and 14.6 show that the histograms are subject to some irregularities as compared with the smooth curves from theoretical models.   This is due to either insufficient data for averaging purposes, changes in statistical parameters (a nonstationary process), or choice of the wrong mathematical model.   Statistical methods are available for estimating the *confidence* one can have that the mathe-

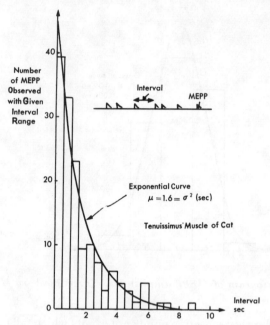

FIG. 14.6 Histogram of MEPP interval spacings. *(From Boyd and Martin.)*

matical model selected does indeed fit the experimental data (Sec. 14.2*D*).

(C) SOME MATHEMATICAL MODELS FOR STATISTICAL DATA

The Probability Density Function. Two histograms have been presented of the number of events (event frequency) versus the event's measure of interest (amplitude, time interval). Presuming that the range of measure includes all possible events, there is unity probability (namely, complete certainty) that the histogram contains them all. (Strictly, of course, this may not always be true of experimental data, since there may be improbable events which did not occur within the necessarily short recording time of practical experiments.) If one divides the number of events $n(x_k)$ in each slot x_k by the total number of events N, the histogram becomes one of relative frequencies, or probability of occurrence $P(x_k)$:

$$P(x_k) = \frac{n(x_k)}{N} \tag{14.1}$$

where x_k represents the kth slot. The requirement that some event shall certainly result yields

$$\sum_{k=1}^{K} P(x_k) = 1 \tag{14.2}$$

In some physical situations there are only a finite number of possible events (for example, in tossing one and two dice), so that simple computations yield the relative frequency or probability plots of Fig. 14.7a. (The reader can check by counting the possible outcomes.) Since no meaning attaches to noninteger events in this case, a bar chart rather than a histogram is used. The histogram form is relevant when the actual variable (e.g., height) is continuously graded but when its data are classified for convenience into a discrete set of slots (for example, ±0.25 in. in measuring height distribution). If this slot width is made progressively smaller, one passes in the limit to a PDF, which is usually a smooth curve. A

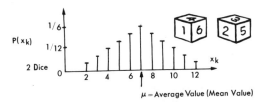

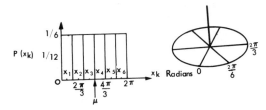

(a) Bar Probability Functions for Dice Tossing

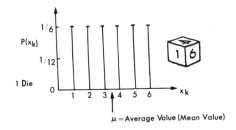

(b) Histogram Probability Function for "6 – slot" Roulette Wheel ("Rectangular")

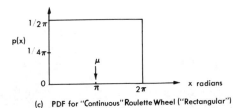

(c) PDF for "Continuous" Roulette Wheel ("Rectangular")

FIG. 14.7 *Discrete and continuous probability functions.*

PDF is defined so that there is a small finite probability $p(x)\,dx$ of the event's lying between $x - dx/2$ and $x + dx/2$ but zero probability of its being exactly x, while the area under the complete $p(x)$ curve must by definition represent complete certainty,

$$\int_{-\infty}^{+\infty} p(x)\,dx = 1 \tag{14.3}$$

Comparing Eqs. (14.2) and (14.3), it is clear that the following conversion holds between histogram and PDF:

$$P(x) = p(x)\,dx \tag{14.4}$$

The roulette wheel provides an easy analog for the single dice-tossing situation, since the angle from some arbitrary zero is a continuous variable in the range from 0 to 2π rad. If the periphery is first marked off in six slots, a *rectangular* histogram results (Fig. 14.7b), of $P(x_k) = \frac{1}{6}$ probability for each slot (provided the wheel is unbiased); if, however, the slot width is made infinitely small, a continuous PDF results which is also rectangular. The magnitude of this PDF must then be given by $p(x) = \frac{1}{2}\pi$, where x is counted in radians (Fig. 14.7c), in order that Eq. (14.3) shall be satisfied.

As suggested in Figs. 14.5 and 14.6, a common problem is to determine whether a particular PDF seems to be a good model for given experimental data. For comparative purposes, the ordinate scale of the PDF must then be converted from a probability density scale to an event frequency scale (or vice versa). To outline the process, consider the typical experimental data of Fig. 14.8 for event frequencies in stopping a roulette wheel. The PDF presumed a priori to govern the process is the rectangular one of Fig. 14.7c, but now this function must be shown compatibly in predicted event frequencies. If, then, the relative frequencies of events $n(x_k)/N$ are divided by the slot width [that is, combining Eqs. (14.1) and (14.4)], the averaged

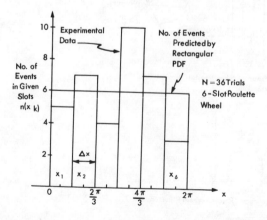

FIG. 14.8   Conversion from PDF to event frequency.

PDF in the slot results:

$$p(x_k) = \frac{P(x_k)}{\Delta x} = \frac{n(x_k)}{N \, \Delta x}$$

For our present purpose, however, the conversion is made the other way round:

$$n(x_k) = N \, \Delta x \, p(x_k) \tag{14.5}$$

Since the number of trials N equals 36, Eq. (14.5) then yields, for the predicted event frequencies,

$$n(x_k) = 36 \, \frac{2\pi}{6} \, \frac{1}{2\pi} = 6 \tag{14.6}$$

It should also be noted that in this particular case one can work more directly from the histogram probability function of Fig. 14.7b to predict that each slot should contain $NP(x_k)$, that is, $36 \times \frac{1}{6} = 6$ events.

With the mathematical model plotted in compatible fashion, there now remains the question of whether the data are likely to have come from a process governed by the mathematical model in question. There is a chi-square test for this which is deferred until Sec. 14.2D.

Moments of the PDF. In Sec. 9.5 the first two moments of the impulse response are defined [Eqs. (9.15), (9.16)]. The analogy between the impulse response of dynamic systems and the PDF of statistical processes is then pointed out with particular regard to the Gaussian function. Now, from the viewpoint of statistics, the first two moments of the PDF are also important statistical measures. Furthermore, there is an additional requirement that the zeroth moment or area under the PDF curve shall be unity [Eq. (14.3)], whereas the impulse response is not limited to a unity gain constant.

Mean μ: The first moment μ defines the mean or average value of the PDF,

$$\mu = \int_{-\infty}^{+\infty} x p(x) \, dx \tag{14.7}$$

and of the frequency histogram, by using Eq. (14.1) and defining x_k as the value of the variable at midrange of the kth slot:

$$\mu = \sum_{k=1}^{K} x_k P(x_k) = \frac{1}{N} \sum_{k=1}^{K} x_k n(x_k) \tag{14.8}$$

The relevant means have been computed and marked on Figs. 14.5 to 14.7 using these rules.

The *mode* and *median* are other, somewhat similar first-order measures, which are presented in elementary textbooks on statistics (see Bibliography).

Standard deviation σ and variance σ^2:   The standard deviation σ is the square root of the variance, which in turn is defined as the second moment with respect to the mean of the PDF or of the frequency histogram:

$$\sigma^2 = \int_{-\infty}^{+\infty} (x - \mu)^2 p(x)\, dx \qquad\qquad (14.9)$$

and

$$\sigma^2 = \frac{1}{N} \sum_{k=1}^{K} (x_k - \mu)^2 n(x_k) \qquad\qquad (14.10)$$

The standard deviation measures the dispersion of the PDF about its mean value.   It corresponds to the mean dispersion time T_s in the dynamic analogy, while μ corresponds to the mean delay time T_m.

Higher-order moments are necessary to characterize a statistical distribution more fully, but these first two often suffice in practice. The Gaussian PDF is an important example which is completely specified by these two moments alone, and which we now consider.

The Gaussian PDF.   The Gaussian has particular importance as a mathematical model of statistical processes because it arises so ubiquitously.   Furthermore, the *central limit theorem* (see Bibliography) predicts that the overall PDF of the sum of n variables tends toward the Gaussian as n tends toward infinity, as long as the n variables are independent of each other.   The reader may demonstrate this trend by computing the probability functions for the dice-tossing situation when the number of dice is increased beyond two. A further point is that the individual variables generally need not be statistical in nature for this result to apply.   Thus, we noted in Sec. 7.1 that an unpredictable, apparently random signal arises when a number of nonharmonic sine waves are added together, and indeed its PDF does tend toward the Gaussian as the number of waves is increased.

Further evidence is also available from the statistical-dynamic analogy by using the result from Sec. 9.5, where it was shown that the impulse response of a cascade of first-order systems tends toward the Gaussian (Fig. 9.7).   The impulse response of a basic first-order system is the same exponential decay as that characterizing the exponential PDF shown in Fig. 14.6 (and discussed below), so that the combined PDF resulting from the sum of n variables each having an exponential PDF tends to the Gaussian as n increases.

The Gaussian PDF is defined as

$$p(x) = \frac{1}{\sigma\sqrt{2\pi}} \exp\left[-\frac{1}{2}\left(\frac{x - \mu}{\sigma}\right)^2 \right] \qquad\qquad (14.11)$$

and is identical with the Gaussian impulse response [Eq. (9.25)].   It is extensively tabulated in the literature [3.13, etc.] for a normalized variable,

$$\tau = \frac{x - \mu}{\sigma}$$

as

$$p(\tau) = \frac{1}{\sqrt{2\pi}} e^{-(1/2)\tau^2} \qquad (14.12)$$

Its major characteristics are shown in Fig. 14.9, which in turn is identical with Fig. 9.8 if the x scale is used. It should be noted that in the unnormalized form the required ordinate scale for Fig. 14.9 is $\sigma p(x)$, so that the total area under the PDF may remain unity.

The Gaussian PDF is closely approximated by experimental data from many areas—for example, distribution of such properties as weight and height among a population of individuals; thermal and other noise; and measurement errors. Thus, Fig. 14.5 shows that a Gaussian PDF fits MEPP amplitude data well, although one would wish for more slots to establish this with confidence. Note that in order to convert from the tabulated values of probability density to the density of number of events one must use Eq. (14.5), and also

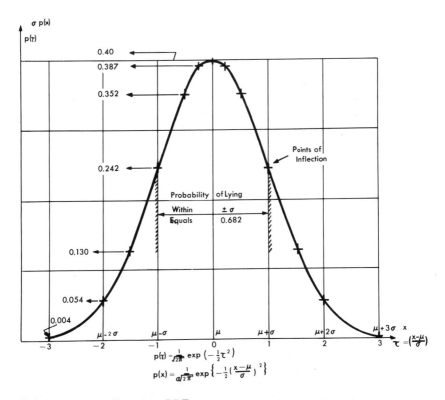

FIG. 14.9 The Gaussian PDF.

that the $p(x)$ value to be used [Eq. (14.11)] requires the tabulated value of $p(\tau)$ to be divided by σ.   For example, at the condition

$$x = \mu \qquad \tau = 0 \qquad \text{and} \qquad p(\tau = 0) = 0.399$$

and with

$$\sigma = 0.085 \qquad N = 78 \qquad \Delta x = 0.1$$

there results

$$n(x = \mu) = \frac{p(\tau = 0)}{\sigma} N \, \Delta x = 36.7$$

The Exponential PDF.   The exponential PDF (Fig. 14.10) is defined as

$$p(x) = \frac{1}{\mu} e^{-x/\mu} \qquad\qquad\qquad\qquad (14.13)$$

where μ is the mean value.   The standard deviation is not independent of the mean value as in the Gaussian case and is readily shown to equal

$$\sigma = \sqrt{\mu}$$

In Fig. 14.6 this exponential PDF is shown to characterize reasonably well the frequency distribution of time intervals existing between successive MEPP.   In general, it seems to characterize the distribution of times between events of a considerable number of random processes, for example, the intervals between pulses in a radioactive material, and the intervals between arrivals at (and departures from) various servicing facilities (such as gas stations).   It can be shown that this probability function arises when the events in question are independent of each other, and indeed this has been the postulate behind the triggering of MEPP.   Of course the parameter μ can be altered by changing the parameters of the statistical process, so that,

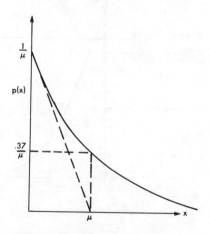

FIG. 14.10   The exponential PDF.

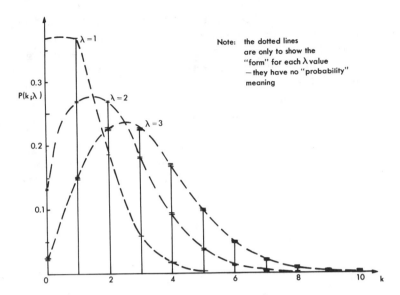

FIG. 14.11 The Poisson distribution—bar chart.

for example, the mean time between successive MEPP decreases when the temperature of the neuromuscular junction is raised [14.5].

In many cases one is interested in the probability function that k events occur in a given interval, and the exponential PDF provides a special case of this because it can also be viewed as specifying the probability that $k = 0$ events occur in an interval. The more general distribution is called the *Poisson*, and to it we now turn.

The Poisson Distribution. The Poisson distribution specifies the probability that exactly k events occur within a given time or space interval, when each event is random and independent as in the exponential case. The length of interval chosen does not enter directly as a parameter. This distribution is only concerned with an integral number of possible results and is therefore shown as a bar chart rather than as a PDF (Fig. 14.11). The required mathematical form is (for proof see [14.7], for example):

$$P(k,\lambda) = \frac{e^{-\lambda}\lambda^k}{k!} \qquad (14.14)$$

where λ is the mean number of events in the stated interval and k is the number of events in the stated interval.

The probability that no events occur in the given interval is therefore

$$P(0,\lambda) = e^{-\lambda} \qquad (14.15)$$

Since λ is the mean number of events in the stated interval, and since events are random, $\lambda = x/\mu$, where x and μ are, respectively, the interval time and the mean time between events of the exponential distribution. In consequence, Eq. (14.15) becomes an exponential PDF of the variable x which is identically Eq. (14.13).

The Poisson distribution has proved to be widely applicable, along with the Gaussian and binomial distributions. Besides the examples noted under the exponential PDF, the Poisson distribution has been used to show that "flying-bomb" hits on London during World War II were randomly distributed, and similarly that chromosome interchanges are randomly induced by X rays. Figure 14.11 shows the bar-chart distribution of Poisson probabilities for values of λ, the mean number of events per interval, equal to 1, 2, and 3. To exemplify its application in the flying-bomb case, the total area is first divided into a number of subareas ("intervals"). Dividing the total bomb hits (N events) by the number of subareas yields the mean number of events λ in each interval, and the Poisson-predicted number of subareas sustaining exactly k hits is then found by multiplying the corresponding probability of Fig. 14.11 by N. Since the predicted values agree well with the observed values, one can conclude that the bombs fell randomly in the London area.

We now return to the neuromuscular junction, to show that the release of ACh quanta does satisfy the Poisson distribution, thereby signifying randomness. The picture becomes a little more complicated in this case, however. Recall that in a healthy junction there are a large number m of quantal units available on the average to respond to motoneuronal stimulation, and further that this mean number can be reduced to the order of unity by withdrawing calcium or adding magnesium, so that each EPP results from a small finite number of quantal units (each of which separately would constitute an MEPP). It is with these conditions that the following statistical analysis is performed [14.3]: Clearly, the mean number of responding units is

$$m = \frac{\text{mean amplitude of EPP}}{\text{mean amplitude of MEPP}} \tag{14.16}$$

as illustrated by Fig. 14.4 (although a relatively high m of the order of 10 resulted in that particular experiment because neuromuscular blocking was small). If the release during motoneuronal stimulation is still random, although with a much higher probability, the distribution should be Poisson, with k being the number of quantal units released during the small stimulation interval. However, a Poisson process implies a finite probability that no units will respond, the so-called *failure response* (Fig. 14.11), a probability that is specified by Eq. (14.15). An appropriate experiment is therefore to stimulate the motoneuron a large number of times and observe the number of times that EPP fails to result. These experimental data then yield a second estimate of m when Eq. (14.15) is rearranged as

$$m \ (=\lambda) = \log_e \frac{\text{number of motoneuron impulses}}{\text{number of failures of EPP response}} \tag{14.17}$$

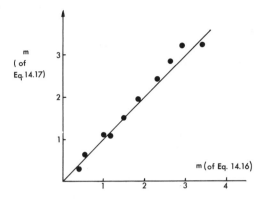

FIG. 14.12 Correlation of experimental methods for determining *m*.
(From Boyd and Martin.)

Excellent correlation is found [14.3, 14.6] between these two estimates of *m* (Fig. 14.12).   However, when one compares the probabilities for other values of *k* with those predicted by the Poisson
distribution, a complication arises because the mean amplitude of
the quantal unit's MEPP is not constant but rather is distributed in
the Gaussian fashion of Fig. 14.5.   Thus, although the EPP results
from an integral number of MEPP, its amplitude distribution is
essentially a continuous function of potential, rather than a discontinuous distribution at integral multiples of the MEPP mean
amplitude.   As shown in Fig. 14.13 from [14.6], this difficulty is
overcome by spreading the Poisson probability of each integral
number of quanta in the Gaussian PDF shape of Fig. 14.5.   Some
individual Gaussian curves are indicated on the figure, using the
relevant data from Fig. 14.5 for μ and σ, and it is seen that the pre-

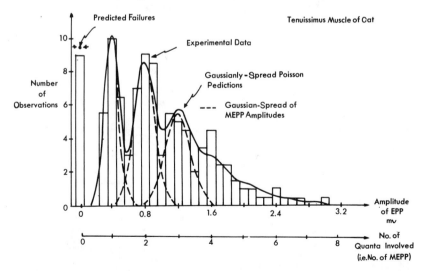

FIG. 14.13 Histogram of observed EPP amplitudes. *(From Boyd and
Martin.)*

dictions of the combined curve are in good agreement with the numbers of events observed.

On Other Probability Distributions. In the preceding paragraphs the events at the neuromuscular junction were used to introduce several important statistical distributions. Another important and basic distribution is the binomial (for example, coin tossing), which has been mentioned implicitly through the dice-throwing experiment; for a single coin the binomial distribution results, but in the general case of multiple coins the distributions are termed multinomial. The Poisson can be shown to approximate the distribution from a coin-tossing experiment when either the number of coins is increased (similar to Fig. 14.7a) or the number of tosses of a single coin is increased. Furthermore, as the value of the λ parameter increases, the Poisson distribution becomes increasingly symmetrical and in the limit becomes identical with the Gaussian distribution.

(D) CONFIDENCE IN STATISTICAL MODELS

Section 14.2C considered the modeling of experimental data by abstract probability functions, with an implicit assumption that an infinite population of data is available. However, a universal practical problem is that one can only sample from the population to obtain a limited quantity of data, either because the statistical parameters are changing or because one cannot wait long enough. In consequence, the data sample can only provide estimators of the population's distribution, and the analyst needs some quantitative measure of confidence in any probability models which he fits to such data. This subject is reviewed very briefly below.

The simplest version of this problem is that of fitting straight lines, parabolas, etc., to experimental relations between variables (for example, Figs. 12.16, 12.19, 13.1, 13.2, 13.4, and 14.12). Often this can be done by eye, especially after replotting on log-log or semilog paper (for example, Figs. 10.5 and 13.4), but the analytical method of best-fitting by least squares is a standard way of increasing accuracy. It is also called the *regression line technique* (see Bibliography or, for example, [14.8]).

When the data points are rather scattered but some possible functional relation is indicated (for example, Fig. 13.1), the problem is not only to find the best fit for the postulated relation but also to evaluate whether it carries any confidence. The *correlation coefficient* is a measure which is useful in this regard. Essentially, it is found by summing the products of all xy points, and in suitably normalized form has a value of ± 1 when the points lie in a straight line. Then, as the points tend to a random scatter, the correlation coefficient tends to zero.

We now take up the more complex problem of estimating the goodness of fit between experimental data and a proposed probability distribution. The χ^2 (chi-square) test is a standard method, depending upon a probability distribution of the same name. Consider first the comparison between observed and expected frequencies in the die-tossing experiment (Fig. 14.8). It seems reasonable that as the average deviation of experimental from expected frequencies increases, the model fits less well, and a square-of-deviation measure is in fact used which automatically penalizes positive and negative deviation equally. Hence a number called the χ^2 value is computed from the formula

$$\chi^2 = \sum_{k=1}^{K} \frac{[n_e(x_k) - n_0(x_k)]^2}{n_e(x_k)} \qquad (14.18)$$

where $n_e(x_k)$ is the expected frequency in slot x_k and $n_0(x_k)$ is the observed frequency in slot x_k.

We must now judge whether this χ^2 value seems reasonable in the given circumstances; that is, can we have confidence in the hypothesis that the expected frequencies are as assumed? For example, with the die we test the hypothesis that it is unbiased. In this case each x_k should be equally probable, and the value of χ^2 is $3\frac{2}{6}$. All possible outcomes of the 36 trials and their relative frequencies of occurrence can be computed using the multinomial distribution on this a priori assumption of no bias. Hence the frequency distribution of all possible χ^2 values can also be worked out, as a result of which we can judge how probable the given outcome is. Fortunately, this has already been done, and summarized by tabulation of an approximating PDF called the χ^2 *distribution*. However, its analytical form is too complex to merit detailing here, except to note that it is characterized by one parameter ν called the *degrees of freedom* (which normally equals 1 less than the number K of slots being tested). The approximation is usually satisfactory, provided that

$$\begin{aligned} \nu &\geq 5 \\ n_e(x_k) &\geq 5 \qquad \text{for all } k \end{aligned} \qquad (14.19)$$

In goodness-of-fit problems it is usual to assume that the result is "not significant" unless the χ^2 value is greater than that found at the 0.05 probability level of the χ^2 distribution for the given degrees of freedom; that is, if the experimental value exceeds that value which should only occur 1 time in 20 on the average, we conclude that the fit is bad and therefore that the data do not arise from the postulated probability model. In the example cited, the χ^2 value tabulated for the 5 percent level is 11.1; hence we conclude that the die is probably unbiased—at least until further data prove us wrong.

In this development an a priori knowledge of the probabilities has been assumed, but the more general problem is that of fitting a PDF model to frequency histogram data (for example, Figs. 14.5 and 14.6), where the parameters of the PDF—for example, μ and σ—

can only be estimated a posteriori from the experimental data. In this case the χ^2 test still applies provided that one degree of freedom is subtracted for each parameter so estimated. A further difference is that the expected frequency of occurrence for each slot must be estimated from the continuous PDF function. Problems 14.5 and 14.7 illustrate these points, but for deeper study the reader should consult the literature.

14.3   Continuous Statistical Signals

(A) GENERAL

It is common experience that one's hand exhibits a small tremor or fluctuation of position when it is held out at arm's length. Such a

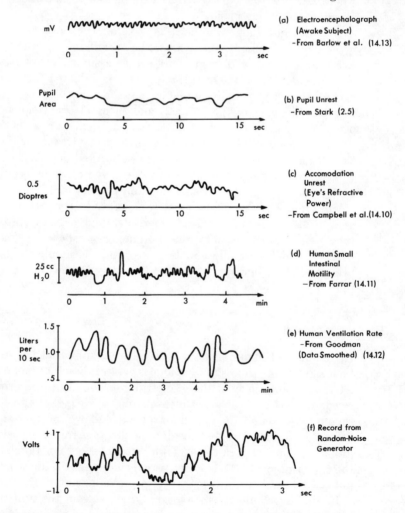

FIG. 14.14   *Some continuous statistical signals.*   [(*a*) *From Barlow et al.* (*b*) *From Stark.*   (*c*) *From Campbell et al.*   (*d*) *From Farrar.*   (*e*) *From Goodman* (*data smoothed*).]

small movement may be ignored during some actions, such as rapid arm swings, but becomes very important during others, such as threading a needle. Furthermore, in certain pathological cases, such as Parkinson's disease, the amplitude of tremor can increase to an intolerable level. There are many other interesting examples of such statistical variations in biological variables, of which a few are illustrated in Fig. 14.14, namely:

a. Electroencephalograph of awake subject [14.13]
b. Area of pupil during constant intensity illumination [2.5]
c. Variations of eye accommodation during constant target conditions [14.10]
d. Motility of small intestine [14.11]
e. Ventilation of resting human [14.12]

Continuity and Filtering. Physical variables such as those illustrated in Fig. 14.14 must be continuous in waveform because they result from applying effort in the generalized sense of Chap. 4 (for example, muscle tension), and these variations cannot occur instantaneously. Furthermore, any change in the variable usually involves variation in the energy stored which again increases the smoothness of the record. However, it should be noted that all measurements inevitably display statistical variations when sufficiently high accuracy is required. Determinacy only arises when this statistical variability is sufficiently small to be ignored in the given situation.

Bandwidth. In order to relate discrete and continuous waveform variables, recall first that the basis of the elementary discrete statistics approach is that each event is independent of past events. One result is that little continuity is likely if one draws a smooth curve through the time record of successive events, for example, dice tosses. In turn these rapid changes imply the need for high-frequency components in a Fourier analysis, so that one can consider a discrete statistical signal as essentially unlimited in bandwidth. On the other hand, continuous statistical signals are always bandwidth-limited according to the dynamic characteristics of the relevant generating and/or filtering processes.

PDF. The PDF is one basic measure of discrete and continuous signals alike. In the continuous case it is found experimentally by evaluating the proportion of time during which the signal falls into each of a sufficient number of slots. Continuous statistical signals may often be fruitfully considered as the result of a large number of more or less independent discrete events; for example, the continuous EEG potentials are presumably related to the multitudinous almost-discrete neural action potentials in the relevant cortical area. Consequently, the central limit theorem should apply, and in practice such continuous variables usually do have a Gaussian PDF. (Actually, since the events are not completely independent, the theorem is not always rigorously applicable.)

Frequency Content and Correlation. The fact that the continuous signals discussed are not completely independent of their past history necessitates the introduction of a further statistical measure in addition to the PDF which proved adequate for characterizing independent discrete signals. This second measure may be expressed either in the frequency domain, in which case it is called the *power spectral density* (PSD), or in the time domain, where it is called the *autocorrelation function* (ACF). These measures are introduced in Secs. 14.3*B* and *C*, respectively.

Stationarity. In order to calculate statistical measures satisfactorily, a long record must be processed and averaged, which raises the problem of stationarity. Briefly stated, a process is stationary if its statistical measures do not change with time. However, since a long record is needed to obtain reasonably constant estimates of these measures even when the process is stationary, inaccurate estimates will be obtained if the process is nonstationary. In practice the analyst is frequently faced with nonstationary processes and so must exercise great care; for example, the human ventilation rate (Fig. 14.14*e*) is modified by physical exercise, so that if the record were taken during variations of exercise a nonstationary process would be acting and the statistical measures obtained from such data (mean rate, variance, frequency content, etc.) could prove meaningless.

(B) POWER SPECTRAL DENSITY $\Phi(j\omega)$

The concept of PSD follows as a natural extension from the concept of harmonic content discussed in Sec. 7.1, and for this reason the subject is often called *generalized harmonic analysis*. As an example, consider the amplitude spectrum for the harmonic content of some repeatable waveform such as a violin string (Fig. 7.4). Experimentally this amplitude spectrum could be obtained by measuring the responses of a series of tuning forks each tuned to only one of the harmonic frequencies. If, however, the string's fundamental frequency were rapidly but irregularly modified by changing tension or finger position, then clearly the individual harmonics would each spread over a local range of frequencies, so that many more individual oscillators could be excited. In the limit there would be a continuous spread of amplitude, on the average, over a wide band of frequencies, rather like the Bode plot of frequency response (for example, Fig. 7.5 for communication channels).

The PSD is the mathematical measure used to define this generalized case where the waveform can no longer be broken down into (or synthesized from) a discrete set of harmonics. Furthermore, the waveform is constantly changing in an unpredictable manner, so that the continuous distribution in frequency is only an average measure, just as is the PDF of amplitude. The conventional units of the PSD are those of power per unit frequency; the total power in the signal therefore equals the area under the curve of all frequencies. Thus there is zero power at any precise value of frequency but some small finite power in each small band of frequency $\Delta\omega$, just as there is a small finite probability that a statistical variable described by a PDF lies in a small amplitude slot Δx. Conventionally the power unit is amplitude-squared, following from the electrical case in which the power dissipation numerically equals volts squared x^2 when voltage x is placed across a 1-ohm resistor (Table 4.2). With a statistical variable this becomes mean, or averaged, power, in the sense of mean square amplitude $\overline{x^2}$, and by the above definition equals the integral of the PSD with respect to frequency (Fig. 14.15a):

$$\overline{x(t)^2} = \frac{1}{2\pi} \int_{-\infty}^{+\infty} \Phi(j\omega)\, d\omega \qquad (14.20)$$

where $x(t)$ is the magnitude of the statistical signal and $\Phi(j\omega)$ is the PSD as function of radian frequency. Note that for mathematical convenience the PSD is normally defined to include negative as well as positive frequencies, although there is no physical meaning to this concept.

PSD of "White Noise." If the PSD were constant over all frequencies it would be termed *white*, in analogy with white light. However, the power would then be infinite [Eq. (14.20)], which is a certain guarantee that such a waveform cannot exist. In fact, the continued availability of power as infinite frequencies are approached would allow the waveform to make essentially instantaneous changes in direction, whereas energy considerations assure us this is not possible. Thus, all realizable spectra can be expected to "cut off" at sufficiently high frequency, but the white-noise concept is nevertheless very useful, especially in linear filtering considerations (Sec. 14.3F). The waveform of white noise can be viewed as exhibiting complete independence between conditions at infinitesimally separated intervals of time, and as such it is clearly analogous to the discrete waveform in which each event is independent of previous ones.

Measurements of PSD. The PSD can be measured in practice by the method implied above; that is, the power in each small frequency slot is found as an average value after passing the record through a narrow bandpass filter having an AR Bode plot such as shown on Fig. 14.15a. Measurement of the whole spectrum is achieved either by successively moving the center frequency of the filter until the whole bandwidth is scanned, or by setting up an array of such filters as suggested by the tuning-fork example. Alternatively, the spectrum may be evaluated by Fourier-transforming another measure, the ACF, which is introduced in Sec. 14.3C. There are many conceptual and practical problems relating to the details of these computations (see, for example [14.9, 14.12, 14.14, 14.15]).

Examples of PSD. Most of the PSDs corresponding to the sample records of Fig. 14.14 are shown in the cited references. For example, curve 1 of Fig. 14.16a presents the PSD of an EEG such as Fig. 14.14a, showing the expected concentration of power around the α-rhythm frequency of 10 cps; curve 2a shows the PSD for an eye-accommodation record such as in Fig. 14.14c with an entrance pupil of 7 mm, while curve 2b is for a pupil of 1 mm. The wide pupil shows a distinct reproducible power concentration around a resonant frequency of 2.1 cps, which disappears for the narrow pupil, while the very-low-frequency power correspondingly increases. In a somewhat similar way the α rhythm disappears during other conditions than that of alertness with eyes closed. These changes in spectral

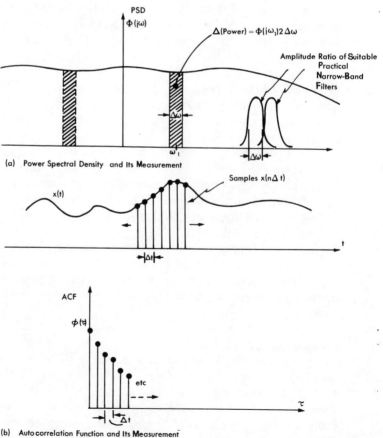

FIG. 14.15   Power spectral density and autocorrelation function.

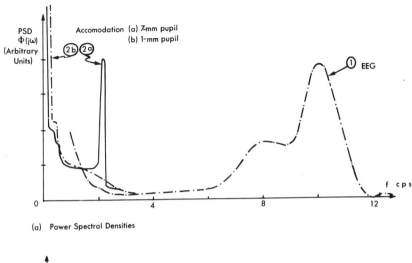

(a) Power Spectral Densities

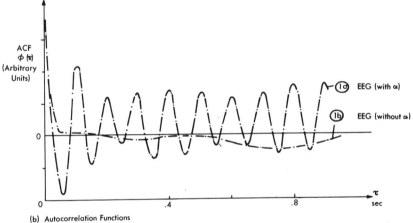

(b) Autocorrelation Functions

FIG. 14.16 Some experimental PSD and ACF. *(Curves 2 from Campbell et al. Curves 1 from Barlow et al.)*

characteristics clearly imply causal mechanisms, and although PSD techniques cannot provide all the desired answers, they do at least allow a start to be made by eliminating much of the uncertainty otherwise prevailing because of a fluctuating waveform.

As a further interesting example, Fig. 14.17 shows the striking qualitative similarity among the spectra of four different motor systems in a steady environment [14.10].   The systems are: pupil diameter during steady illumination of the retina; accommodation during steady viewing of near target; finger displacement during steady pointing; and eyeball position during steady fixation between saccadic movements.   It should be noted that these are plotted on log-log scales, just as for the Bode plot of AR.

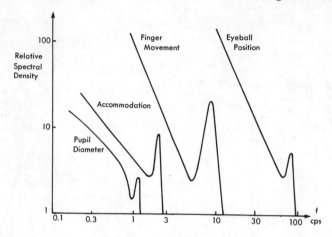

FIG. 14.17   PSD of some motor systems.   *(From Campbell et al.)*

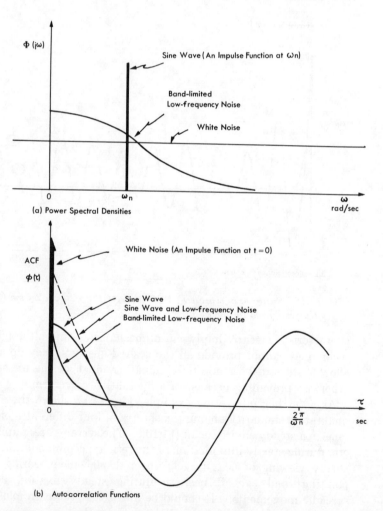

FIG. 14.18   Some theoretical PSD and ACF.

Some Analytical PSDs. It is useful to present together the spectra for certain special waveforms. The sine wave is of course completely predictable, and since its power is concentrated at one frequency, the PSD must consist of an impulse function at the appropriate frequency ω_n (Fig. 14.18a). Conversely, white noise has a constant PSD at all frequencies, as already pointed out. The figure also shows the general shape of the PSD for a bandwidth-limited signal such as the filtered output of a random-noise generator (Fig. 14.14f). Many other shapes are possible depending upon the generating process and its subsequent filtering.

(C) AUTOCORRELATION FUNCTION $\phi(\tau)$

In continuous waveforms the value of the signal $x(t)$ at any time t normally depends to some extent upon the values of the signal $x(t - \tau)$ at earlier points in time. This dependency usually decreases as the time shift τ increases, and is expressed statistically by the ACF. Before considering this function, however, the concept of finding the *average, mean,* or *expected value* of a continuous function of time should be mentioned. This measure is identical with the mean value defined by Eq. (14.7), but it is now defined as the result of averaging a continuous signal $x(t)$ over a time interval T:

$$E[x(t)] \equiv \overline{x(t)} \triangleq \lim_{T \to \infty} \left[\frac{1}{T} \int_0^T x(t)\, dt \right] \qquad (14.21)$$

where $E[\ ]$ is read "expected value of $[\ ]$" and $\overline{x(t)}$ is read "mean or average value of $x(t)$." Clearly, this averaging process can only yield meaningful results if the underlying statistical process is not changing during T, a condition termed stationarity.

The ACF $\phi(\tau)$ of a variable (or signal) is defined as the expected or mean value of the product of $x(t)$ and its time-shifted value $x(t - \tau)$:

$$\phi(\tau) = E[x(t)x(t - \tau)] = \overline{x(t)x(t - \tau)} \qquad [\equiv \overline{x(t)x(t + \tau)}] \qquad (14.22)$$

$$\phi(\tau) \triangleq \lim_{T \to \infty} \left[\frac{1}{T} \int_0^T x(t)x(t - \tau)\, dt \right] \qquad (14.23)$$

The ACF is a continuous function of its independent variable τ, and is symmetric about $\tau = 0$ [because of the equivalence of the two mean products shown in Eq. (14.22)]. Also, if $\tau = 0$ in Eq. (14.22) the value of the ACF for zero time shift is identically the mean square value of the signal

$$\phi(0) = \overline{x(t)^2} \qquad (14.24)$$

Consequently $\phi(0)$ numerically equals the area under the PSD. Also, it is intuitively clear and can be shown that for all τ,

$$\phi(\tau) \leq \phi(0) \qquad (14.25)$$

Actually, the ACF and the PSD are related as a Fourier-transform pair, which is essentially a two-sided Laplace transform with s

restricted to the imaginary part $j\omega$:

$$\Phi(j\omega) = \int_{-\infty}^{+\infty} \phi(\tau)e^{-j\omega\tau} \, d\tau$$

$$\phi(\tau) = \frac{1}{2\pi} \int_{-\infty}^{+\infty} \Phi(j\omega)e^{+j\omega\tau} \, d\omega \qquad\qquad (14.26)$$

Clearly, then, these two measures contain the same mathematical information, but in practice each often accentuates different aspects of a particular waveform since one is in the time domain and the other in the frequency domain. Thus the ACF plots of Fig. 14.18b, in comparison with the PSD plots of Fig. 14.18a, show that compression in the time domain results in expansion in the frequency domain, and vice versa; for example, the ACF of a sine wave is also a sinusoid, whereas its PSD is an impulse function. On the other hand, the ACF of white noise is an impulse function since successive samples are independent of each other, whereas the PSD is constant out to infinite frequencies. Note also from the figure that if a sine wave is corrupted with low-frequency noise, the latter appears only at the small-τ end of the ACF, so that the signal is effectively filtered of undesired noise.

Measurement of ACF. The defining operation of Eq. (14.23) can be directly instrumented by electronic analog equipment, comprising time-delay unit, multiplier, and averager. It can also be performed on a digital computer after first converting the continuous analog signal into a set of sample values (Fig. 14.15b). A number of detail problems arise, as now mentioned, but for greater detail the reader should consult the literature (for example, [14.9, 14.12, 14.14, 14.15]).

1. The speed of sampling must be fast enough to retain the information contained in the signal due to the highest frequency present in the PSD. In principle, two samples per cycle of the highest significant frequency suffice, but in practice, a higher rate is often used.
2. There should normally be at least several hundred samples available for the computation. The sample period Δt sets the shortest time shift τ available, while the longest, τ_m, should be restricted to about 10 percent of the overall record length.
3. In finding the PSD by digital computation, the most economical method is via the ACF [Eq. (14.23)] and its Fourier transformation [Eq. (14.26)].

Examples of ACF. Figure 14.16b shows the ACF of the awake subject's EEG, corresponding to the PSD already discussed. The strongly defined α rhythm appears in the ACF as a wave of the same frequency, although the amplitude varies irregularly. On the other hand, when the α rhythm is absent the ACF decays essentially to zero as τ increases.

(D) RECORD LENGTH, STATIONARITY, AND CONFIDENCE

If the signal does not have stationary characteristics, one cannot meaningfully compute averages from long lengths of record. On the other hand, if the averaging computation is performed on a short length of record considerable variability can be expected in the result, since this result itself is an estimate having its own statistical distribution. Confidence in the estimate is then determined by the variance of this distribution or by its square root, the standard deviation. A trade-off situation therefore applies between record length and confidence, which is now briefly discussed using the example of PSD measurement. However, it should also be noted that some biological signals may be sufficiently nonstationary that it is impossible to estimate their characteristics with sufficient confidence.

In general the PSD is computed for each of a set of slots having a bandwidth W (equaling the $\Delta\omega$ of Fig. 14.15a) which primarily determines the resolving power of the PSD analysis. The PSD estimates for each slot may generally be assumed independent, in which case a chi-square distribution applies [14.14], for which the degrees of freedom are

$$\nu = 2WT \hspace{3cm} (14.27)$$

where T is the record length. Confidence curves for the PSD estimate $\hat{\Phi}$ can now be constructed, as shown in Fig. 14.19 for 80 and 90 percent limits. The resulting requirement on record length is surprisingly severe to the intuition. For example, one could reasonably want to resolve an EEG into slots of 0.2 cps, but Eq. (14.27) shows that in order to produce 100 degrees of freedom a 500-sec length of record must be processed. In turn these degrees of freedom

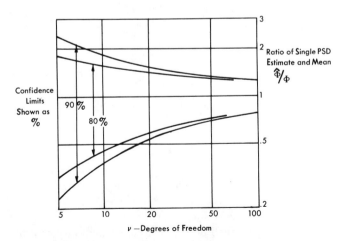

FIG. 14.19 PSD estimation as function of confidence and degrees of freedom.

would still only provide 90 percent confidence that any single estimate of Φ would lie within the range $0.8 \rightarrow 1.25$ of the mean Φ. From such requirements it becomes clearer why nonstationarity may render impossible the accurate measurement of some spectra.

If the analysis is performed by sampling at intervals Δt from a record of length T, and if the maximum time shift used in the calculation of the ACF is τ_m, the degrees of freedom may equivalently be shown to be

$$\nu = 2 \left(\frac{T}{\tau_m} - \frac{1}{3} \right) \qquad\qquad (14.28)$$

$$\nu \approx 2 \frac{T}{\tau_m} \quad \text{if} \quad \frac{T}{\tau_m} \gg 1 \qquad\qquad (14.29)$$

The term $\frac{1}{3}$ arises because as τ is increased, fewer products become available in the averaging technique implied by the discrete equivalent of Eq. (14.23). However, its effect remains negligible if the condition $\tau_m \leq 0.1T$, already mentioned, is respected. Comparison of Eqs. (14.27) and (14.29) shows that the maximum time shift of the ACF equals the reciprocal of the resolving bandwidth available for the PSD.

(E) SUMMARY OF PSD AND ACF

A continuous statistical signal is partly characterized by its PDF, which tends to be Gaussian in many practical situations. However, because of the inevitable "smoothness" of continuous signals, another measure must be introduced, which actually is only the second of a possible hierarchy of PDFs. The PDF is the first of these, while the PSD and ACF are both derived from the second distribution and represent the smoothness characteristic in the frequency and time domains respectively. The latter two functions constitute a Fourier-transform pair. Computation may be done by both analog and digital methods, but the practical requirement of finite record length and/or the presence of nonstationarity mean that the computation can only yield estimates of the true PSD and ACF. In practice, much of the use of PSD and ACF functions arises in the filtering of signals through dynamic systems (see Sec. 14.3F).

(F) LINEAR FILTERING

The term *linear filtering* refers as well to the input-output relations involved when statistical signals are processed by linear dynamic systems as to those involved when the signals are determinate. Actually, an output may also be statistical in nature as a result of the dynamic system itself acting as a noise generator and adding its effect to the ongoing input-output process, but in this section we shall concentrate upon the linear filtering process. *Optimal linear filtering* provides a mathematical method for designing optimal linear control systems when subjected to reference and disturbance signals of given spectral densities. The theory assumes that minimization of the mean square error is a valid criterion of optimal performance, and furthermore that the signals are described by the Gaussian PDF. Within these restrictions this theory has been valuable in providing an analytical design method, and, for example, it has been applied with particular success to the design of radar gun-aiming systems. It should be noted also that, in this application, prediction of future values of a random variable is implemented on a best-estimate basis, which is a function that, as we have already noted, the human operator performs. There are many limitations to optimal linear filtering, however, because it cannot be applied generally to the design of systems which incorporate nonlinear controllers to advantage.

This section is largely concerned with the important effects of filtering on spectral density and correlation functions, but a few comments should first be made concerning filtering effects upon the PDF:

1. A statistical signal of almost any PDF tends toward the Gaussian after linear filtering. Any signal already Gaussian remains so after linear filtering.
2. If a statistical signal of Gaussian PDF is inputted into a nonlinear dynamic system (that is, a filter), the PDF of the output is generally distorted from the Gaussian.
3. The various computations concerning spectral densities and correlation functions are not affected by the nature of the PDF, since these two types of measure are essentially independent.

Frequency-domain Relations. The idealized white spectrum was introduced earlier, and although it is unrealizable in practice, considering any given PSD as the result of linear filtering by an appropriate dynamic element whose input is white noise is often conceptually useful. For example, the low-frequency noise whose PSD and ACF are shown in Fig. 14.18 could typically result from passing white noise through a filter with the basic first-order transfer function $H(s) = k/(Ts + 1)$. Fortunately, these relations can be expressed analytically, as examined below.

The output spectrum $\Phi_{yy}(j\omega)$ resulting from linear filtering, with known transfer function $H(j\omega)$ of a known input spectrum $\Phi_{xx}(j\omega)$, can be shown to be defined by the relation

$$\Phi_{yy}(j\omega) = |H(j\omega)|^2 \, \Phi_{xx}(j\omega) \qquad\qquad (14.30)$$

This formula has a simple graphical interpretation because $|H(j\omega)|$ is of course shown by the Bode plot of AR, which must therefore be squared for use in Eq. (14.30) (Fig. 14.20). In consequence the value of the output spectrum at any given frequency ω is the product

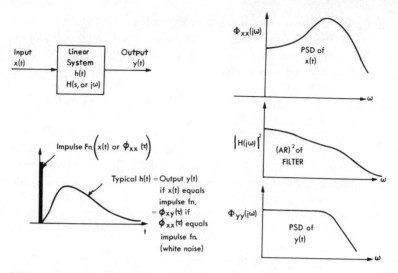

FIG. 14.20   Linear filtering relations.

of the corresponding input spectral value and the square of the system's AR at this frequency; in other words, each spectral component is processed independently by the system as if the others did not exist, which results from the superposition property of linear systems.   Figure 14.20 presents an example in which a peaked spectrum is converted by an appropriate filter into one much "flatter" out to some higher frequency; that is, the spectrum is whitened.   This process has been used, for example, in designing random-noise generators for analog computers, since it is usually desirable to maintain a flat spectrum out to frequencies which are high compared with the natural frequencies into which the noise will be inputted. As previously noted, such band-limited white noise is then for all practical purposes equivalent to ideal white noise.   The actual record of Fig. 14.14*f* results from filtering such band-limited white noise through several first-order lag systems.

The spectral calculation specified by Eq. (14.30) can be carried out either analytically, or graphically by "counting squares."   In addition, it is often necessary to find the mean square value of the resulting output signal, the total area under the PSD, and this is accomplished by substituting Eq. (14.30) in Eq. (14.20):

$$\overline{y(t)^2} = \frac{1}{2\pi} \int_{-\infty}^{+\infty} |H(j\omega)|^2 \Phi_{xx}(j\omega)\, d\omega \tag{14.31}$$

Fortunately, the actual integration procedure can be omitted because special tables of integrals are available (for example, [14.16]) which specify the mean square value as a function of the parameters of $H(j\omega)$ and $\Phi_{xx}(j\omega)$ (see also Prob. 14.10).

Equations (14.30) and (14.31) work only with the (squared) magnitude characteristic of the filter's transfer function because the overall mean square value results from summing an infinite set of component mean square values, and in this summation the relative phases are unimportant. For a complete dynamic description, however, the phase effect must also be included. In this case the input-output relation is

$$\Phi_{xy}(j\omega) = H(j\omega)\Phi_{xx}(j\omega) \tag{14.32}$$

where the new term on the left-hand side is termed the *cross-power spectral density*. However, it is now easier to transfer from the frequency domain into the time domain, where the equivalent relation to Eq. (14.32) is expressed in correlation functions. They are connected through the Fourier or Laplace transform as already noted.

Time-domain Convolution with Statistical Signals. Recall first that the convolution integral [Eq. (6.51)]

$$y(t) = \int_{-\infty}^{t} h(t - \sigma)x(\sigma) \, d\sigma$$

where σ is a dummy time variable of integration, is the time-domain equivalent of the transfer-function form

$$y(s) = H(s)x(s)$$

If the former equation is multiplied on both sides by $x(t - \tau)$ and averaged, there results

$$\phi_{xy}(\tau) = \int_{-\infty}^{\tau} h(\tau - \sigma)\phi_{xx}(\sigma) \, d\sigma \tag{14.33}$$

where $\phi_{xy}(\tau)$ is called the *cross-correlation function* (CCF) between variables x and y. By direct use of the Laplace-transform pair, one may then write from Eq. (14.33)

$$\Phi_{xy}(s) = H(s)\Phi_{xx}(s)$$

which is indeed Eq. (14.32) if s is replaced by $j\omega$ to signify concern with frequency alone.

Equation (14.33) is an important relationship for the analysis of dynamic systems. One important use is in identifying the dynamic nature of a system through determining its impulse response $h(t)$ when the input ACF and the input-output CCF can be measured from input and output records (Sec. 15.4). A particularly useful situation arises, furthermore, when the input is effectively white noise, since $\phi_{xx}(\tau)$ is then an impulse function at $\tau = 0$, and Eq. (14.33) degenerates to

$$\phi_{xy}(\tau) = h(\tau) \tag{14.34}$$

Thatis, the CCF equals identically the impulse response of the system. It should be noted that "effective" whiteness of the input spectrum means only that it must be flat out to a break frequency which is large compared with the system's highest break frequency (say 10 times greater; see also Prob. 14.10).

It is instructive now to refer back to the visual-tracking results reported in Sec. 7.1 and illustrated in Fig. 7.6. In this case the unpredictable signal is synthesized from a set of nonharmonic sine waves, and the frequency response plots are obtained by "unscrambling" the response components at the various frequencies. As far as the eye is concerned, however, the unpredictable signal could just as well be a random signal with a continuous PSD and a corresponding ACF. The analysis could then proceed via Eq. (14.32) or Eq. (14.33) and would result in Bode plots or in an impulse response for the system, respectively. If the assumption of linearity is valid, the sine-wave analysis and the spectral and correlation analyses should yield the same results.

In practice the statistical method has not been widely used so far, since it involves much computational work. Nevertheless, a number of autocorrelation and cross-correlation applications have been reported, for example, by Barlow and colleagues on the EEG using a special analog correlation computer [14.17, 14.18]. Several electrode pairs are used and both the ACF for each pair and the CCF between pairs are computed. Figure 14.21 shows an example of these correlation functions for a subject with a tumor of the right cerebral hemisphere, the electrode placements being in the left and right parietal-occipital areas. In these applications, of course, there is no

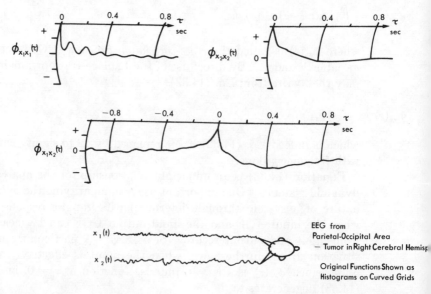

FIG. 14.21 Autocorrelation and cross-correlation of EEG. (*From Barlow et al.*)

established causal pathway in either direction, so that the application of Eq. (14.33) cannot produce a meaningful impulse response. Similar application of power spectral analysis using Eq. (14.32) has also begun to be reported. Thus, for example, Barnett and Michener [14.19] have estimated aortic impedance in anesthetized dogs by simultaneously measuring blood pressure and flow, where impedance is defined as the *s*-domain ratio of two variables and is therefore identically the transfer function.

14.4   Averaging Computations

(A) INTRODUCTION

All biological data exhibit statistical fluctuations either because the signal we seek is buried in some corrupting noise or because it has its own inherent statistical variability. In this context, noise can then be usefully defined as that part of the signal in which we are not currently interested. This section is concerned with averaging techniques which reduce the noise level relative to the signal (that is, increase the signal-noise ratio), with the underlying presumption that the resulting data can aid the discovery of cause-effect relations. This averaging is necessarily the same general process as filtering, although the detailed differences of emphasis are now pointed out. It should be remembered that averaging computers were introduced in Sec. 10.7 but that quantitative attention has necessarily been deferred until now.

The averaging problem arises equally with discrete and with continuous statistical signals, and some applications of each are discussed below. In each case the same general principle of averaging is applied, that of successively adding repetitive sequences of record— for example, electrocardiograms which are triggered at an appropriate event during successive cycles, and stimulus-response tests in which the stimulus provides the obvious trigger.

Averaging computers have been developed by individual experimenters during the last decade or so [14.20–14.23] but are now commercially available. Similarly, the more sophisticated form of correlation computer has also been developed for (example, [14.17, 14.24, 14.25]) and is commercially available (for example, [14.26]).

(B) DISCRETE SIGNALS

An example which neatly illustrates the nature of the averaging problem in obtaining a probability description of dynamic response to a repeated stimulus is available from recent work on the CNS. Burns and Smith [14.27] have recorded extracellularly from continually active cortical neurons in suprasylvian and other gyri in the unanesthetized but neurologically isolated cat's forebrain.

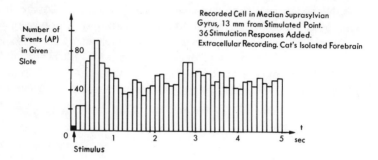

FIG. 14.22 *Averaging of discrete signal; stimulus-response tests.* (*From Burns and Smith.*)

The tests are of the stimulus-response type in which the stimulus is either a 100-msec flash of light onto the stationary retina or direct cortical stimulation (of 10 bipolar 0.5-msec-duration shocks in 100 msec, given to the brain surface in a region 13 mm away from the responding cell). In general, most cortical cells are continually firing in a pseudorandom fashion similar to that discussed in Sec. 14.2. Furthermore, although the mean frequency varies from cell to cell, that of any one cell remains apparently constant for several hours of recording in the undisturbed brain; thus, the mean frequency of firing is presumably not the main medium of information transfer between these cortical cells, in contrast to the situation with receptors. On the other hand, the pattern of firing after the response occurs is shown to be altered considerably on the average.

Consider first the unstimulated case. If an arbitrary interval of time (say 1 sec) is divided into a number of slots (say 10), and the number of firings which occur in each slot over a number of intervals (say 100 sec) is successively added, one would expect that the count in each slot should be approximately the same. Furthermore if the number of intervals (averaging time) is extended toward infinity, the rectangular histogram of Fig. 14.7*b* should be approached. For a finite number of intervals, however, Fig. 14.8 would be more representative of the experimental data, and the χ^2 test can be used to estimate the degree of confidence justified for the assumed distribution function. Of course, if by some unlikely chance the cell happens to be firing with a periodicity which synchronizes with the arbitrarily chosen period, the fit to a rectangular PDF would be very bad.

Consider now the stimulated case. The experimental data of Fig. 14.22 show that there is a visually obvious change from the rectangular pattern of firing probabilities that obtains before stimulation. These results are for successive addition, or averaging, of 36 stimulations. In this particular test, the stimulus was the electrical one and the periodicity was made long enough (5 sec) so that the complete response pattern could be established; it is seen to last at least 3 sec. If one postulates a model for the response pattern, the data's goodness of fit can again be tested by χ^2, but in this case the authors were only concerned to show that the χ^2 value was significantly larger than that of the unstimulated control. The relevant values are

Unstimulated control $\chi^2 = 1.05\bar{n}$
Stimulated test $\chi^2 = 2.49\bar{n}$

where $\bar{n}$ is the mean number of firings per slot. Let us, in addition, test the hypothesis of a rectangular distribution in the unstimulated control condition. Proceeding as in Sec. 14.2, we note that $\bar{n} \approx 46$ and $\nu = 45$, and we can thus compute the confidence level of the two values. For the control, the χ^2 value is 48 and there is about 35 percent probability of the data from a rectangular distribution exceeding this χ^2 value. For the test case, the χ^2 value is about 114 and there is an infinitesimally small probability ($\approx 10^{-5}$) that these data could be produced from a rectangular distribution.

This example shows that by averaging discrete periodic statistical data, response patterns can be inferred which would not be evident from single records. However, we have not specified quantitatively how many repetitions of the averaging process are needed for a given improvement in the regularity. Indeed, no fully satisfactory statistical technique seems to be presently available for this. Furthermore, it should be pointed out that simple statistical tests such as χ^2 do not test for shape of distribution, in the sense of providing hints about what the correct distribution might be.

(C) CONTINUOUS SIGNALS

There are many irregularly periodic continuous biological signals for which analysis may be aided by averaging of the waveform, for example, the electrocardiogram already mentioned. However, a more powerful application of the technique is provided by stimulus-

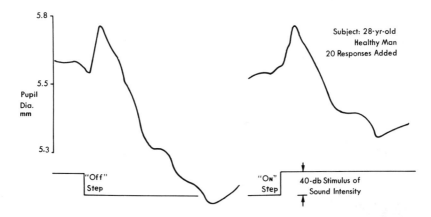

FIG. 14.23 *Averaging of continuous signal; stimulus-response tests.* (*From Clynes.*)

response tests in which the response is so buried in ongoing noise as not to be discernible at all from a single test.   Thus, after sufficient averaging it is possible to improve the signal-noise ratio of the response by 100 times or more.   With continuous signals the application to stimulus-response tests in EEG provides an obvious analogy to the discrete signals of single cells described in Sec. 14.4*B*.   As another example, therefore, Fig. 14.23 presents the small but clear pupil reflex to auditory signals after averaging of 20 responses [9.4].   The on and off steps of sound intensity are 40 db (change by a factor of 100) above threshold.

15 Biological Performance Criteria and Adaptive Control

15.1 Introduction

Living systems are termed *goal-seeking* or *purposive* because within limits they clearly choose or adapt their courses of action out of many possible alternatives, so as to sustain their existences under varying environmental disturbances. This statement applies equally to all levels of a hierarchical set of living systems from the cell to the ecology. A question therefore arises about what are the relevant criteria of performance for such systems in determining their courses of action. One obvious overall criterion is evolution itself, namely, that systems which operate in the most fitting way tend to survive, and an ecological example of this has already been briefly presented (Sec. 12.1). However, the evolutionary principle does not provide the immediate criterion for ongoing adjustment in such lower hierarchical levels as organ systems, and here some energy-related criteria seem to have evolved, as discussed in Sec. 15.2.

The block diagrams presented so far for biological and engineering control systems have implied a simple goal-seeking performance criterion, namely, reducing to zero as fast as possible the error of the system between its desired and achieved performances. In principle, the speed and accuracy of this error reduction can be designed to be very high, but at the same time a compensatory disadvantage appears in that large values of manipulated variables are required in such "high-performance" systems (Chap. 11). Therefore, one general principle of systems optimization is to attribute "costs" or penalties to various aspects of the system's operation, and then to seek that mode or course of action which minimizes the overall cost. Such a trade-off for the control-system configuration typically involves costs attributable to:

1. Existence of error (for example, hand not on target during a manual-tracking task)
2. Use of manipulated variable (metabolic cost of muscle forces)
3. Capital investment in process (the energy expenditure in resting metabolism, neural control, and cellular maintenance of the muscle mass)

In practice, much of the analyst's difficulty usually lies in discovering the appropriate weighting to apply to these variables in combining them as component costs of the performance criterion, for example, how to weight off-target error of the hand against the correcting muscle forces necessary.

When such a composite performance criterion is implemented, rather than a simple-minded fast reduction of error, the overall control system is termed *adaptive*, and the addition of at least one further outside loop is usually involved. Adaptive control is being actively studied and designed in engineering systems, although often with rather limited data on which to base performance criteria. On the other hand, the structure adopted is relatively simple and clear. In biological systems, adaptive control seems to be the universal rule; for example, the athlete enlarges the mass of muscles needed in his particular sport during his training, which increases the maximum actuator force available and also the relevant "capital investment." The performance criterion inferred from experimental data often seems to be a rather simple measure; on the other hand, the functioning of the biological loops by which it is put into effect is seldom at all clear. In Sec. 15.3 the functional nature of adaptive control systems is explored in some detail.

The task of designing adaptive or optimizing systems is simplified if the dynamic characteristics of the process being controlled can be identified, and this subject is discussed in Sec. 15.4. It is relevant that in the domain of man-machine systems the human operator has been shown to interrogate subconsciously the machine which he is controlling, presumably in order to identify his controlled process and hence to better his performance. On the other hand, the system designer must also be able to identify man's dynamic characteristics as a controller so that optimal, or at least satisfactory, overall operation may be obtained from complex man-machine systems.

15.2 Biological Performance Criteria

In this section some speculations and experimental data are presented concerning the performance criteria of biological systems. First of all, the performance requirements for large ecological systems are considered, the performance criterion presumably being related to population stability. The factors of power output and efficiency of individual organisms are then considered in relation to evolution, leading to an efficiency criterion for normal operations when maximum power output has no survival advantage. The data implying an energy-minimization criterion in walking and breathing are then presented, and some further ideas are discussed regarding performance criteria.

(A) ECOLOGICAL SYSTEMS

The biomass of the earth, comprising many ecological systems, is in constant flux and yet obviously possesses remarkable overall stability. The theory of evolution provides a satisfactory hypothesis to explain the nature of the dynamic interactions within and between species, but cannot by itself predict the resulting biomass of different species within ecological systems. Consequently, one can only guess at the performance criterion which is relevant to the system's operation.

The dynamics of ecological systems arise basically from different rates of population growth and from the phenomenon of predation in the broadest sense. Because of the basically explosive nature of the positive-feedback population loop, and assuming that all hierarchies of the food pyramid cooperate, it might be expected that the biomass would increase up to the largest amount which could be carried by the photosynthesizing base of the pyramid. In this case, the performance criterion of the ecological system would be that of maximizing its biomass. However, owing to unavoidable fluctuations in the various levels of the food pyramid resulting from natural catastrophes, etc., such an ecological system might well run the risk of extinction, or at least violent fluctuations. It is interesting, therefore, to find experimental evidence that ecological systems tend to exhibit greater stability than their nonliving environments [15.1]. Furthermore, Dunbar suggests that the operation of natural selection causes such systems to evolve in this direction of increasing stability of biomass, which in turn implies increasing probability for continued survival of the system [15.2]. By comparing the characteristics of the more stable, more adapted tropical marine ecologies with those of temperate and polar regions, Dunbar notes that such adaptation seems to involve a lowering of birthrates in some of the ecosystem's species.

Ecological systems are obviously interesting in their own right, but it is also worth noting that an organism such as the human is analogous in many respects to such a system. In particular, rela-

tive growth rates of different tissues and organs must be controlled to
achieve a harmoniously functioning system.   In this case, however,
a performance criterion for a single organism can be incorporated
centrally, presumably in the genetic structure, whereas in the ecosystem it must be distributed.   Thus, control of the ecosystem is
passive, in the terms of Chap. 2, whereas control in the human
organism is *active*.   We now limit our attention concerning performance criteria and adaptive control to individual organisms
and their subsystems.

(B) POWER OUTPUT AND EFFICIENCY

Each organism may be viewed as an engine which converts the chemical energy of its nutritional "fuel" into electric, thermal, mechanical, or other chemical energy.   In the last case the different materials
are either "information-rich" or "energy-rich" (DNA, RNA, ATP,
enzymes, etc.).   Energy conversions, whether in the physical or in
the biological domain, are inevitably less than perfectly efficient in
the sense that some proportion of energy always appears as heat
(Chap. 4).   Parenthetically, however, it should be noted that nature
has made a remarkable achievement in packaging efficient engines
which can produce the desired outputs of electricity, force, heat,
etc., by "burning" the chemical fuel at about 37°C, instead of at the
high temperatures of the engineer's combustion engines.

All transduction engines can be characterized to some extent by
two rather basic parameters: useful power output and efficiency,
where efficiency equals the ratio of useful power output P_0 to power
input P_1.

$$\eta = \frac{P_0}{P_1} \tag{15.1}$$

A general property is that the maximum-efficiency condition occurs
at a lower output than the maximum-power-output condition.   For
example, consider the shortening of muscle where the force F and
speed of shortening V (corresponding to θ in Fig. 12.18) are the
variables coupling with the load being driven.   The F-V characteristic for the human arm flexors is shown in Fig. 15.1.   The mechanical power output equals their product:

$$P = FV \tag{15.2}$$

and, as shown on the figure, passes through a maximum at about one-
third of maximum speed and one-fourth of maximum force.   On the
other hand, it has been shown [15.4] that the maximum efficiency
obtainable (in the frog muscle) is between 20 and 25 percent and
occurs at a lower speed of about one-quarter the maximum value,
when force is about one-half of maximum.   (Note that maximum
force occurs at zero speed and maximum speed at zero force.)   This
type of power-efficiency characteristic is also typical of energy-conversion devices in engineering (for example, Fig. 2.4a).

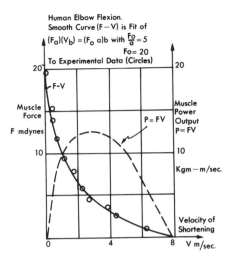

FIG. 15.1 *Force and power output of muscle as functions of velocity.* (*After Wilkie.*)

A high-maximum-power output of muscles enhances either the predatory or the escape ability of an organism, and therefore has presumably constituted an important evolutionary performance criterion [15.5, 15.6]. However, such demand for maximum-power output usually endures only briefly, and in addition, muscle correspondingly has a maximum-power characteristic which decreases as the duration of its demand is increased [15.7]. Furthermore, maximum-power output is only called for infrequently, and during the long intervals of low power output a maximum-efficiency operation would provide an evolutionary advantage. Thus it seems probable that both the maximum-power and maximum-efficiency conditions are included in nature's evolutionary performance criterion, although their relative weightings cannot easily be inferred from natural data. An engine designer similarly tries to provide an optimum design compromise between conditions of maximum power and "cruise" at maximum efficiency. Unfortunately, since such examples of overall performance criteria are hard to document in living systems, the examples of Sec. 15.2C are restricted to the suboptimizing performance criterion of operation at maximum efficiency. Even here the actual neurophysiological loops which achieve such adaptive control are not well understood, although such understanding obviously provides one key for tackling the integrative processes of the CNS.

(C) POWER–MINIMIZING PERFORMANCE CRITERIA

Walking. A very simplified model for the walking process has been considered in Sec. 8.2 in order to produce a mathematical model for the free-swinging limb. Since it assumed that no work is neces-

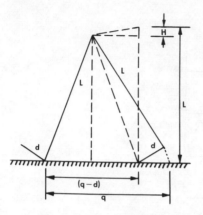

FIG. 15.2   Stiff-leg walking diagram.   *(From Cotes and Meade.)*

sary to swing each limb past the other, the power input necessary for walking should be limited to that needed by muscles which maintain balance and posture.   However, in practice the body's center of gravity must be raised and lowered during each step by raising the hips in order to allow the leg to clear the ground during its swing. In the present development we follow Cotes and Meade [15.8], who show that the vertical lift of the center of gravity H during each step is proportional to the squared difference between the pace length q and foot length d (Fig. 15.2):

$$H = \frac{(q - d)^2}{8L} \qquad\qquad (15.3)$$

where L is the length of the leg.   This lift work is initially stored in the body as PE.   However, if it is then dissipated as the hip falls during completion of the step, the necessary power becomes proportional to the product of H and the step frequency n:

$$P = \frac{MV^2}{8L}\left[n\left(\frac{1}{n} - \frac{d}{V}\right)^2 \right] \qquad\qquad (15.4)$$

where M is the weight of the body and V equals nq.   Experimental data regarding lift work do indeed show that P is proportional to V^2, and in addition that $n(1/n - d/V)^2$ has a constant value for natural walking in the range 2 to 4 mph.   This confirms Eq. (15.4) but also allows it to be simplified to

$$P = K\frac{MV^2}{8L} \qquad\qquad (15.5)$$

where K is a constant.   Cotes and Meade also find that the power consumption of walking as measured by oxygen consumption has the following form:

$$P_{02} = b + aV^2 \qquad\qquad (15.6)$$

Since the aV^2 term equals about 3.5 times the corresponding value of Eq. (15.5), and since muscle efficiency is in the region of $1/3.5$, it may be concluded that essentially all the vertical lift power is indeed dissipated as heat. Furthermore, since the constant b represents about twice the oxygen consumption necessary for resting metabolism, an additional power charge for walking may be assumed which is independent of velocity.

Our present interest is to determine whether a maximum-efficiency velocity exists, for which the relevant performance criterion is presumably that of least energy expenditure per unit distance covered W. (Of course, under different circumstances the organism is often prepared to pay an energy penalty for increased speed.) In this analysis, generally different results are obtained according to whether or not the resting metabolism r is first subtracted to yield that amount W chargeable to walking; thus, by using Eq. (15.6),

$$W = \frac{P_{02} - r}{V} = \frac{aV^2 + (b - r)}{V} \qquad (15.7)$$

The optimum walking speed is then that value of V which makes the slope dW/dV equal to zero, namely,

$$V_{opt} = \left(\frac{b - r}{a}\right)^{\frac{1}{2}}$$

The predicted optimal velocity from empirical values of the constants is 2.25 mph when resting metabolism is subtracted, but 3.5 mph when it is not. The value of 2.6 mph from simple pendulum analysis in Sec. 8.2 therefore acquires more confidence.

A general property of simple optimizations emerges if Eq. (15.7) is rewritten to show the presence of two component energy costs:

$$W \triangleq W_1 + W_2 = aV + \frac{b - r}{V} \qquad (15.8)$$

The W_1 component representing vertical lift work increases with V, whereas the W_2 component representing a constant metabolism decreases—in this particular case, as straight line and hyperbola, respectively (Fig. 15.3). The optimum condition occurs when their slopes have equal magnitude but opposite signs, so that

$$W_{opt} = 2\left(\frac{b - r}{a}\right)^{\frac{1}{2}} = 2aV_{opt}$$

Of course the performance criterion may equally well be one which should be maximized rather than minimized, so that the general term for optimality is that the performance criterion should be at an *extremum*.

The relevant energy expenditure per unit distance may be calculated from the data of Cotes and Meade for one subject walking

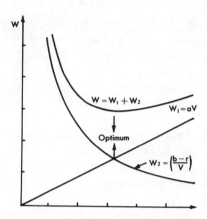

FIG. 15.3 *Optimality condition for "cost function" with two components.*

naturally on a flat surface at different speeds, and Fig. 15.4 shows that there does tend to be an optimum condition around 80 m/min, or about 3 mph. Furthermore, it is interesting to consider their data as the step frequency is changed with speed held constant. The vertical lift per minute, and hence the vertical lift power, decreases as step frequency is increased [Eq. (15.4) and experimental data of Fig. 15.5]. On the other hand, since the total power increases again after exhibiting a minimum at a step frequency of about 92 per minute, presumably the muscle power to drive the legs at higher than their natural frequency must account for the increasing component. Again a trade-off is exhibited between two cost functions, as in Eq. (15.8) and Fig. 15.3.

These empirical data disclose something about the actual behavior of the system and the performance criterion that is optimized, but not about how it is achieved. We know of course that walking in

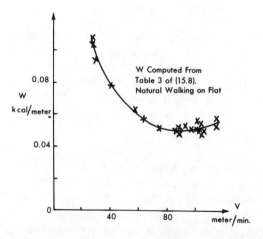

FIG. 15.4 *Energy expenditure per unit distance walked.*

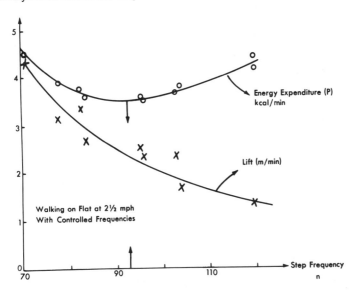

FIG. 15.5 *Constant speed walking at various frequencies.* (*From Cotes and Meade.*)

what is now seen to be an optimum range of about 2.25 to 3.5 mph is "effortless," and that walking faster than this does require more effort, which usually involves monitoring or training. However, this only begs the question of how the organism searches to find the optimum operating condition given a set of variable parameters such as step length and frequency.

Respiration. Respiration provides a second example of a system which operates so as to minimize an energy-based criterion under varying environmental conditions. The complete neuroanatomical block diagram for the inspiration-expiration cycle of respiration is complex, since several sets of muscle proprioceptors and chemo-receptors are involved. If the exploration is limited to the adaptive nature of respiratory control, however, a fairly simple closed-loop system suffices. In particular, factors that we shall presently ignore are the use of expiratory muscles to provide active lung contraction during high ventilation rates, the presence of tonic activity in both sets of muscles, and the other uses of the system, such as coughing and talking.

In normal breathing the respiratory cycle can largely be viewed as a relaxation oscillation. The viscoelastic lungs are expanded as a result of appropriate muscle action, and this expansion causes increased inhibitory feedback from proprioceptors. However, reciprocal inhibition within certain cell groups in the respiratory center tends to maintain the motoneuronal discharge for a short period. After this period the discharge ceases, the muscles contract passively like a stretched spring, and the proprioceptive discharge

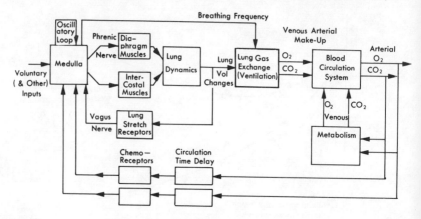

FIG. 15.6 *Simplified block diagram of respiratory system.* (*Adapted from Horgan and Lange.*)

decreases.   The cycle then recommences.   The inspiratory effect is largely produced by the combined actions of the diaphragm innervated by the phrenic nerve and of the intercostal muscles (between ribs) innervated by the intercostal nerves (Fig. 15.6).   The major feedback pathway mentioned is believed to be from stretch receptors in the lung tissues and carried along the vagus nerve.   The control or respiratory center is largely in the medulla, where the picture becomes complicated.   In particular the respiratory center comprises inspiratory and excitatory sets of motoneurons which reciprocally inhibit each other to produce a bistable control element ("on" or "off") [15.9].   Thus, respiration continues when the vagus nerve is cut, although it is slower and deeper.   (Similarly, the heart's pacemaker can maintain its own rhythm, although it is normally inhibited by the vagus nerve.)

   The purpose of the respiratory system is presumably to maintain the desired concentration of oxygen and carbon dioxide (and pH) in the blood, and there are chemoreceptors in the carotid and aortic bodies and medulla to provide the relevant information.   Figure 15.6 shows a simplified overall block diagram for the system, which is taken in part from Horgan and Lange [15.10].   It should be mentioned that these authors have particularly studied Cheyne-Stokes respiration, which exhibits a slow oscillation (of the order of 1 cycle/ min) in the ventilation rate, such that breathing may cyclically cease.   They show that the time delay in the loop due to blood circulation is a critical parameter in the control of respiratory rate. Although this disorder of a homeostatic system is only peripheral to the present interest of adaptive control, it helps to emphasize the several hierarchical levels involved.   Thus, first, the medullary oscillator produces an expiratory-inspiratory cycle whose frequency is modulated by feedback of lung volume.   Second, the actual ventilation rate results from the frequency and depth of respiration and, through chemoreceptor feedback, is used as the effector for homeostatic control of the blood's oxygen and carbon dioxide. The adaptive control to be discussed involves optimal mutual adjustment of the two variables, frequency and depth of respiration, under a given demand of ventilation rate. It may therefore be considered functionally intermediate in hierarchical level between the two just discussed, although neurophysiologically they may well be inextricable.

The viscoelastic nature of the lungs is important to an understanding of the results to be presented concerning optimization of the breathing rate. When the lungs expand as a result of muscular action they act in part as a spring and therefore store PE (Sec. 4.6); in part, however, they also act as viscous-type dissipators of energy since there is resistance to tissue-deformation velocity and to inspiratory airflow through the passages. The total work done during inspiration is therefore the sum of these components. During normal breathing when expiration is assumed to be passive, the PE of the spring effect is returned to the system, where it is dissipated in turn by the tissue and airflow resistances of expiration. Hence the input work of each cycle may usefully be considered as composed of elastic and nonelastic components, although by the end of the cycle both have in fact been dissipated. Since the lungs' parameters can be calculated after conducting certain tests, these two work components can be estimated for given conditions [15.11]. In this case, for example, it is pertinent to consider the curves representing the power of the elastic, nonelastic, and combined components as the frequency of respiration varies at a constant ventilation rate (Fig. 15.7). This situation is analogous to that of Fig. 15.5 for the walk-

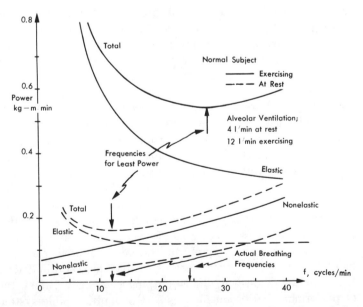

FIG. 15.7 Breathing-power expenditure as function of frequency. (*From Christie.*)

ing example, in which the lift power is analogous to the elastic power of breathing, and in which the (unplotted) dissipative power is analogous to the nonelastic power. Otis et al. [15.11] demonstrate that the total breathing power should pass through a minimum at some frequency for any given ventilation rate, and Christie [15.12] provides experimental evidence that humans do indeed breathe at a frequency close to the calculated optimum, under differing lung and environmental conditions. Thus, Fig. 15.7 shows that a particular normal subject's breathing rate changed from 12 breaths/min at rest to 24 breaths/min during exercise, while the calculated values were 12 and 27, respectively. Since Christie also shows that similarly optimal breathing rates are achieved by patients with various cardiac and pulmonary diseases for whom the system's parameters are widely altered, presumably this adaptive behavior is not preprogrammed into the system in open-loop fashion by hereditary factors. As another hint, it has also been shown that changes in tone of the airway muscles are utilized to minimize breathing work, by altering both airflow resistance and anatomical dead space [15.13]. Finally, it should be commented that although no detailed explanation is yet possible regarding the implementation of this adaptive behavior, it presumably must involve the information channels already shown on Fig. 15.6.

On Performance Criteria. It has also been suggested that the relevant performance criterion for respiratory control is minimizing average inspiratory muscle force rather than power [15.14]. On the other hand, this does not seem to change the general shape of curves such as that in Fig. 15.7 [15.13], presumably in part because some power input to muscle is necessary in any case to maintain a force output, whereas no mechanical power output is obtained until there is an overall velocity of contraction. Posture provides a significant example where the animal automatically tends to adjust limbs into a "comfortable" configuration in order to perform a given task, and it has been postulated that this is achieved by a performance criterion of minimizing the total muscular effort necessary [15.15].

Probably other performance criteria exist for biological control systems than those based on energy and force. For example, the focusing task involves photoelectric changes in the retinal cells due to image contrast changes, and presumably the criterion involves maximum "sharpness" of the image in some way, so that it must also be of the extremum type, in this case falling off on either side of the maximum-sharpness condition. In the more complicated task of visually tracking a moving object, the control functions of eye movement, focusing, and vergence are all involved in a coordinated fashion. Unfortunately, however, almost nothing is yet known about the form of performance criterion for this sort of multivariable system. In turn, the visual-tracking task merely represents one component in a visual-manual motor-coordination task, and similarly there is little knowledge of which criterion the CNS uses in programming its predictive operations and its constant-effort control segments.

Model of Plastic Neuron. To round out the treatment of biological performance criteria, we now mention some computer results by Griffith [15.16], arising from use of a plastic neuron model. In this example, the synaptic weightings are adaptively adjusted on the basis of a built-in performance criterion which essentially represents neuronal power.

Many artificial neuron networks have been assembled recently with the aim of developing machines which can recognize patterns, etc. Conventionally, such machines must learn by a stimulus-response procedure in which a teacher feeds back information to the machine regarding its performance, before it can perform any useful tasks itself. Such a procedure corresponds to some phases of human learning but presumably not to the plasticity exhibited by neuromuscular systems in limb reflexes, etc. Griffith has set up a small neuron network to simulate the basic muscle pair operating a limb, including proprioceptive feedback and tendon organ but excluding the γ system. The neuron model itself is analog in nature and has a built-in performance criterion to control its behavior. Each neuron autonomously adjusts the weighting of its afferent synaptic connections so as to minimize the function

$$P = f_0{}^2 + \sum_{i=1}^{n} f_i{}^2$$

where f_i is the input firing frequency at its ith synapse and f_0 is the output firing frequency.

If it is assumed that each action potential requires one unit of metabolic energy, the performance criterion P is related to power (although actually to its square), and it is thus generally compatible with the criteria inferred experimentally for walking and respiration. A fairly complicated hill-climbing procedure is implemented in the model by which it "learns" the optimal connections, a subject which is discussed in Sec. 15.4. The experimental results so far show that the system rapidly develops a proper physiological stretch reflex in response to alternating enforced movements of the limb, and also a jackknife reflex for large forces.

Summary. In summary, there is some evidence that biological control systems adaptively adjust their operating parameters in such a way as to minimize the power consumption of a given muscular task; this generally involves trading off one power function which increases as a function of the variable parameter, against another which falls. This adaptive behavior has evolutionary advantage for normal operations, and does not conflict with the idea that organisms also gain evolutionary advantage by increasing their maximum available power output during short periods.

15.3   Nature of Adaptive Control

Adaptive control occurs ubiquitously in biological systems, as common experience demonstrates.   The subject is also of great current interest in many disciplines (for example, cybernetics, control engineering, computer mathematics, and management) because systems that possess adaptive abilities can show rudimentary intelligence.   Consequently, considerable terminological confusion has arisen with such terms as *adaptive, self-adaptive, self-organizing, cybernetic, optimizing, goal-seeking, homeostatic, learning,* and *artificially intelligent,* which have considerable overlap and even contradiction.   No attempt is here made to distinguish among them, and only a few basic terms are used in the attempt of this section to outline some basic ideas of adaptive control.

(A) SEARCHING THE PERFORMANCE CRITERION

Adaptive control has been exemplified in the previous section by walking and respiration, for which the inferred performance criterion is that of power expenditure or some related function.   This power function typically passes through an extremum, in this case a minimum value, as some relevant manipulatable parameter of the system is altered, for example, breathing or walking frequency.   In general the performance criterion for adaptive control is therefore an even function of one or more appropriate manipulatable parameters, in that it either increases from a minimum value on either side of an optimum value of each parameter, or decreases from a maximum value (Fig. 15.8*b*).   In this light, then, can the conventional error-activated nonadaptive control system be viewed as having a performance criterion?   Certainly yes, because the present value of the error (or some function of it) is the criterion used to determine action aimed at reducing the error to zero, and this performance criterion is an odd function of its appropriate system parameter (error) (Fig.

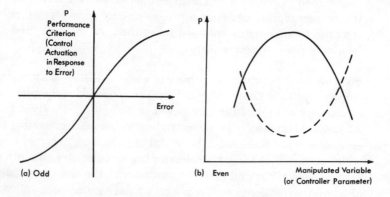

FIG. 15.8   Odd and even performance criteria.

15.8a). The pupillary-light reflex discussed in this book exemplifies nonadaptive control with an odd performance criterion, whereas the eye's accommodation in focusing exemplifies adaptive control with an even performance criterion [15.17]. However, the fact that both systems have performance criteria does not mean they are functionally similar. Thus, a basic difference is that the nonadaptive system attempts to minimize system error on the basis of predetermined and fixed values of controller parameters, while the adaptive system searches for those values of controller parameters which allow a usually more complex performance criterion to be optimized.

A little consideration reveals that the adaptive system must compare present performance with that at an earlier time in order to determine the present trend and hence the adjustment to be made to the system—which means that the system must learn from its past performance in an elementary way. In general, therefore, a process of hill climbing (or valley descending) is needed, and this is often implemented by using the steep-ascent procedure in which one climbs the hill in least distance by going up the steepest slope, as discussed in Sec. 15.4. When the optimum condition is reached by this searching action, hunting usually ensues since the presence of an optimum cannot be discovered without first going past it. If the optimum condition drifts more or less continuously in time, a continuous hunting action is needed, as exemplified by the control of shower temperature when there are frequent disturbances due to hot and cold water being run off elsewhere in the system.

(B) ADAPTIVE CONTROL STRUCTURE

The structure of an adaptive control scheme depends to some extent upon the nature of the relevant performance criterion, and we first consider the type of performance criterion which is the composite total of some compatible penalty terms. This situation arises, for example, when the penalty of permitting system error must be traded off against the penalty of actuating energy to eliminate it. The situation is shown for a general case involving the main loop variables in Fig. 15.9. In this regard, reconsider the manual-tracking task (Secs. 11.8 and 12.3) in which the input command and achieved performance are both instrumented visually for the operator. It can be plausibly presumed that the CNS controller attaches penalty both to the energy supply necessary for the actuating muscles and to the presence of error which indicates bad control, although of course their relative weighting can vary with the importance of precision, duration, etc., of the particular task. Such a composite performance criterion can seldom be predicted analytically as a function of the system's performance, but in principle it can be measured easily and continuously without special input signals being necessary since it exists automatically from ongoing activity. This information must then be incorporated into some sort of hill-climbing pro-

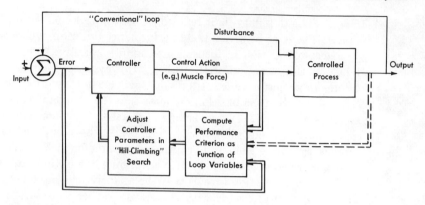

FIG. 15.9   *Adaptive control with composite performance criterion.*

cedure in order that the optimum but unknown value of the performance criterion may be reached (Sec. 15.4).

In other control systems, simple single-component performance criteria may be relevant; for example, such a consideration as stability or speed of response may be far more important than the energy cost of control actuation incurred. Therefore, in order to analyze the structure of adaptive control further, consider a simply structured system such as an aircraft with either a man or an autopilot in control. In order to determine the best values of autopilot parameters to produce desired handling characteristics for a given aircraft, the system designer must first predict or evaluate the aircraft response characteristics over a wide range of these parameters. In essence he maps out a pertinent performance criterion (for example 5 percent overshoot in the aircraft's response to step input of elevators), and from these picks the optimum controller settings permanently. Unfortunately for such nonadaptive design-stage optimizations, modern aircraft are required to fly over a great range of speed and altitude, with the result that the parameters of the aircraft's transfer function vary so widely that fixed controller parameters do not maintain constant handling characteristics. Consequently, adaptive autopilots have become necessary whose parameters are changed compensatorily with speed and altitude changes so that the handling characteristics do remain essentially invariant. Such adaptive autopilots are instrumented by an extra hierarchical loop that identifies handling performance (Fig. 15.10) and acts continuously, instead of by the "one-shot" design study of conventional autopilots. The problem of identifying the process dynamics themselves is taken up separately in Sec. 15.4.

Consider the functioning of the aircraft pilot as controller of this same system. It is clear that he operates adaptively in his role; in particular, he has been observed to perform the identification task subconsciously by providing frequent small pulse inputs to the control actuator in order to check aircraft-handling characteristics.

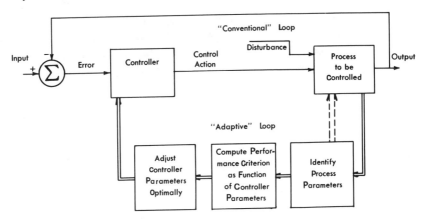

FIG. 15.10   Basic adaptive control configuration.

Therefore, it can be plausibly assumed that he changes loop gain and other control parameters as necessary to maintain the performance criterion he has set himself.   This strong analogy demonstrates that control of machines by man and by automatic means may indeed be studied together fruitfully.

Since the previously constant parameters of the controller are varied by the adaptive loop, it becomes uncertain whether they should preferably be called *system parameters* or *system variables*.   In any case, nonlinear operations such as multiplication are inevitably introduced, so that all adaptive control systems are nonlinear in common control-engineering terms.

The adaptive control configurations of Figs. 15.9 and 15.10 both show a conventional inner loop with outer adaptive loops.   This hierarchical configuration is not necessary as far as "hardware" paths are concerned, although it is conceptually convenient.   Thus, in applying computer control to large industrial processes, the trend is toward elimination of conventional inner loops in favor of direct control by the computer (Fig. 15.11).   However, it should be emphasized that this does not eliminate any of the information-processing requirements just discussed, but rather that they must all be per-

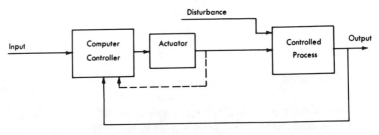

FIG. 15.11   Canonical configuration for computer control.

formed by the computer. These aspects suggest the question: Do living organisms achieve adaptive control preferentially through hierarchical loops, or through the canonical "direct" form?

15.4 Identification of Dynamic Systems

(A) GENERAL

The identification of the nature and parameter values of dynamic processes has provided one of the themes for this book. It is often called the *identification* or *transfer-function-discovery problem*, and in Sec. 10.2 the difference between it and the other two dynamic problems of forward analysis and input discovery was pointed out. Identification may be performed in both the time and frequency domains; for example, the technique of extracting exponential decay curves for multicompartment models (Sec. 10.4) is in the time domain, whereas the experimentally determined frequency responses of visual tracking (Fig. 7.6) identify this system as being in the frequency domain. In these examples the necessary calculations are normally performed off-line, that is, after the experiment is completed. On the other hand, a basic feature of adaptive control is that on-line control adjustments must be made, frequently or continuously, based upon recently updated knowledge of the dynamic performance. Since the relevant measure of this dynamic performance is the performance criterion (Figs. 15.9, 15.10), one major requirement of adaptive control is to implement its on-line identification. Of course, the identification of a composite performance criterion such as that illustrated in Fig. 15.9 involves only a direct calculation, but on the other hand, the identification of process dynamics as in Fig. 15.10 is not so trivial and usually involves hill-climbing procedures. Therefore in Sec. 15.4*B* we consider methods of identifying process dynamics, especially in the restricted form of identifying the parameter values of a differential-equation model whose form is assumed a priori. For illustrative purposes, this development centers around the identification of human-operator dynamics in a manual-tracking task using a quasi-linear model (Sec. 8.11), but in principle the method can be extended to more complex sampled-data models (Secs. 11.8, 12.3). Incidentally, even though the adaptive control configuration of Fig. 15.9 has a directly determined performance criterion, hill climbing is nonetheless required in adjusting the controller parameters, so that our considerations are relevant here also.

Identification methods are also available for which no form of model need be assumed, and one basic example involving determination of the impulse response from input-output records using statistical signals is briefly discussed in Sec. 15.4*C*. For details of other techniques, the reader should consult [15.18] to [15.20], for example.

(B) IDENTIFICATION OF SYSTEM PARAMETERS

Consider the problem of identifying a dynamic system or black box in which there is ongoing input-output activity and for which the model's structural form may be assumed to be either a differential equation (linear or nonlinear) or a transfer function (linear only). The problem is then one of adjusting the parameter values of this model until the model's output matches the actual system's output when subject to the ongoing input (Fig. 15.12), a process which is sometimes aided by adding suitable specific signals to the input. Note that the black box may equally well be a component of a control system or the complete system itself including all feedback loops. In the particular development to follow, the black box is taken to be the human operator, and his model is assumed to have the transfer function [essentially Eq. (8.73)]

$$H_m(s) = \frac{y}{x}(s) = \frac{(1 + a_1 s)e^{-Ts}}{b_0 + b_1 s + b_2 s^2} \tag{15.9}$$

which is equivalent to the differential equation

$$b_0 y(t) + b_1 \dot{y}(t) + b_2 \ddot{y}(t) = x(t - T) + a_1 \dot{x}(t - T) \tag{15.10}$$

where T is a time delay arising in neural pathways and the CNS.

Instantaneous Identification of One Parameter. In some cases of interest there may only be one parameter which normally varies significantly. Thus, for example, one might wish to study the variation of a_1 in the above model, since it represents the weight or emphasis with which the operator anticipates or differentiates the input signal [Eq. (15.10)]. As such it characterizes to some extent the operator's attempt to stabilize the system with phase lead when the process is sluggish, and one would expect its value to vary with the dynamics of each controlled process. A single parameter can be

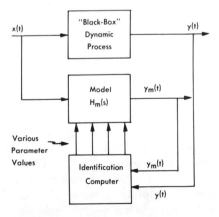

FIG. 15.12 *Conceptual configuration for identification.*

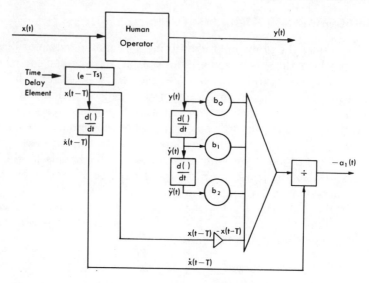

FIG. 15.13   Identification of one parameter.

determined continuously and instantaneously by performing certain operations on the input and output variables (see method of Corbin [15.21]). Thus, Eq. (15.10) can be rewritten to solve for a_1 as a function of time:

$$a_1(t) = \frac{b_0 y(t) + b_1 \dot{y}(t) + b_2 \ddot{y}(t) - x(t - T)}{\dot{x}(t - T)} \qquad (15.11)$$

which is shown implemented for an analog computer in Fig. 15.13. However, differentiation is a notoriously noisy process, whereas integration tends to average out any noise. Therefore, greater stability of solution can be obtained if two integrations are first performed on each side of Eq. (15.10).

$$b_0 \left[\int_0^t dt \int_0^t y(t)\, dt \right] + b_1 \left[\int_0^t y(t)\, dt - t y(0) \right] + b_2[y(t) - y(0) - t\dot{y}(0)]$$
$$= \int_0^t dt \int_0^t x(t - T)\, dt + a_1 \left[\int_0^t x(t)\, dt - t x(0) \right] \quad (15.12)$$

which can be written more compactly by defining some appropriate new variables:

$$b_0 y_{-2} + b_1 y_{-1} + b_2 y = x_{-2} + a_1 x_{-1}$$

This equation is clearly analogous to Eq. (15.10), and may also be solved for a_1; however, integration operations now replace the differentiation operations of Fig. 15.13 (see Prob. 15.1). This introduction of integration operations means that the identification system cannot follow changes of $a_1(t)$ as rapidly as before [15.20]. This is normally no disadvantage, however, and indeed the scheme has been shown to work well.

Identification of Complete Transfer Function.   A single equation can be solved explicitly for one parameter if all other variables and parameters are continuously known, but if it is necessary to solve for more than one parameter at a time, auxiliary equations must be introduced.   These equations are of the hill-climbing form, and operate so as to force the error between actual and model outputs to zero in the steady state.   The procedure is now not instantaneous, and indeed the overall configuration is identical with that of closed-loop control, as will be demonstrated.

When the model parameters of Eq. (15.10) do not equal those of the black box, the left- and right-hand sides of this equation cannot generally be equal; their difference can obviously be defined as the error of the identification procedure:

$$\epsilon(t) = x(t - T) + a_1(t)\dot{x}(t - T) - b_0(t)y(t) - b_1(t)\dot{y}(t) - b_2(t)\ddot{y}(t) \quad \textit{(15.13)}$$

A class of procedures defined as *methods of steep descent* (or steep ascent) is then available to specify how this error information should be used to adjust the several parameters simultaneously, so that a stable and fast descent to the condition $\epsilon = 0$ is ensured (see, for example, chap. 18 of [15.22]).   The reduction of the error to zero thus constitutes optimization of the performance criterion for this situation.

The performance criterion itself should first be defined as a general function of error ϵ:

$$p = p(\epsilon)$$

Some suitable specific relations are

Error magnitude	$p = \lvert \epsilon \rvert$	*(15.14)*
Quadratic error	$p = \epsilon^2$	*(15.15)*

since both of these reduce to zero when $\epsilon = 0$, and also since both penalize positive and negative errors impartially (Fig. 15.14).

The method of steep descent states that in order to adjust parameters of the performance criterion

$$p(t) = p[\epsilon(t)] = p[a_1(t), b_0(t), b_1(t), b_2(t)]$$

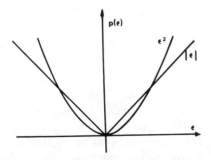

FIG. 15.14   *Possible performance criteria.*

so as to drive p to its minimum, the time rates of parameter adjustment should be made proportional to the corresponding partial spatial derivatives. This means that the trajectory moves down the *gradient* of the surface p, which is a multidimensional function of the a and b parameters; that is,

$$\frac{da_1(t)}{dt} = -k\frac{\partial p}{\partial a_1}$$
$$\frac{db_j(t)}{dt} = -k\frac{\partial p}{\partial b_j} \quad j = 0, 1, 2 \qquad (15.16)$$

where k is a rate constant. For manipulative purposes the partial differentiation expressions are rewritten usefully in this case as

$$\frac{\partial p}{\partial b_j} = \frac{\partial p}{\partial \epsilon}\frac{\partial \epsilon}{\partial b_j} \quad \text{etc.} \qquad (15.17)$$

For example, consider the quadratic error criterion [Eq. (15.15)]; in this case

$$\frac{\partial p}{\partial \epsilon} = 2\epsilon$$

and from Eq. (15.13)

$$\frac{\partial \epsilon}{\partial b_j} = -\frac{d^j y}{dt^j}$$

so that Eq. (15.17) becomes

$$\frac{\partial p}{\partial p_j} = -2\epsilon\frac{d^j y}{dt^j} \qquad (15.18)$$

Therefore, the actual rates of steep descent which must be instrumented are as defined by Eqs. (15.16) and (15.18):

$$\frac{da_1}{dt} = -2k\epsilon(t)\dot{x}(t - T) \qquad \frac{db_1}{dt} = +2k\epsilon(t)\dot{y}(t)$$
$$\frac{db_0}{dt} = +2k\epsilon(t)y(t) \qquad \frac{db_2}{dt} = +2k\epsilon(t)\ddot{y}(t) \qquad (15.19)$$

These four extra equations, together with Eq. (15.13) defining ϵ, allow a closed loop to be set up whose steady-state condition solves for the correct parameter values, on the assumption that T is known. The complete analog-computer implementation is shown in Fig. 15.15, and it clearly involves a nonlinear multiloop control scheme. In our conventional terms, integral controllers are used, since the parameters a and b are obtained by integrating certain functions of the system error [Eq. (15.19)]. In consequence, the system will track any variations of parameter values with a certain sluggishness. The scheme is not limited, in general, to any particular order of transfer function or to any particular input function.

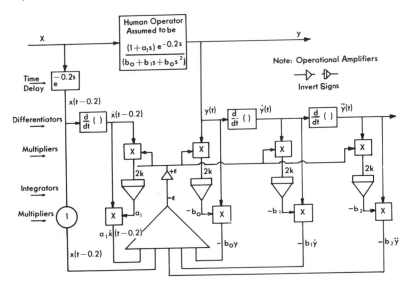

FIG. 15.15 Identification of transfer-function parameters. *(Con-figuration in part after Ornstein.)*

An obvious criticism of the scheme is that again successive derivatives of the input and output signal must be taken. This difficulty can be alleviated if the input waveform is under the experimenter's control because sinusoidal waveforms yield relatively noiseless derivatives. Unfortunately, however, it seems that at least one independent sine wave is needed for each pair of parameters being determined (in order to provide sufficient "information" in the signal) so that the alleviation is limited when the number of unknown parameters is large. A more basic alternative is to use the integrating technique previously described for single-parameter identification, although the extra integrations result in the overall system being more sluggish.

Human-operator Tracking Characteristics. We now briefly present the results of Ornstein [8.6], who has developed this technique with particular reference to the human operator's transfer function in manual tracking. In detail, he used an error-magnitude criterion [Eq. (15.14)], in consequence of which the circuitry involves relays also (Prob. 15.2). The operator uses a control stick in a compensatory task; that is, only the actual error between his desired and actual outputs is displayed to him (Fig. 15.16). The dynamics of the mechanism M between the operator's manipulated variable (his control stick) and the system output θ_0 can be varied by the experimenter, which usually results in the operator compensatorily modifying at least one of his parameters. The steep-descent circuit typically stabilizes on a set of parameter values within about 30 sec, although an iterative technique with at least two runs is used to improve overall accuracy.

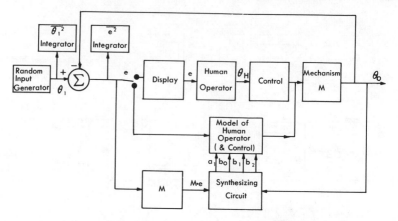

FIG. 15.16   *Configuration for human-operator identification.* (*From Ornstein.*)

When the parameters are repetitively obtained from the beginning of a new task, they do exhibit a learning curve.   After stable values are reached, it is possible to switch the operator out and his model in for brief periods without the operator realizing he is not in fact controlling the process; this presumably means that the modeling is satisfactory.   Ornstein also reports on tests to elucidate the dictum that man should be required to act no more complexly than as a proportional amplifier in designing man-machine control systems [11.1].   Of course, we have observed that in the tracking task he usually inserts predictive or derivative action, presumably because this helps optimize his personal performance criterion for the task. If, then, the display itself is quickened by feeding back the first, and possibly the second, derivatives of the output (Fig. 15.17), the operator would presumably need to provide less anticipation and so would "turn down the gain" on his a_1 coefficient.   This effect is indeed strongly evidenced by the tests.   Furthermore, when the process is made more sluggish, the value of a_1 is significantly increased. Finally, it is this same constant which particularly increases during the learning process.

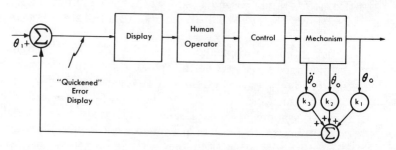

FIG. 15.17   *"Quickening" the error display.* (*From Ornstein.*)

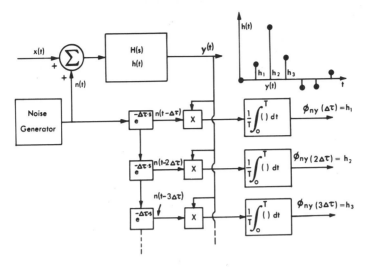

FIG. 15.18 *Impulse response indentification by cross correlation.* (*From Anderson.*)

(C) IDENTIFICATION OF THE IMPULSE RESPONSE BY CORRELATION TECHNIQUE

Chapter 14 has shown that some use has been made of correlation techniques in the biological field. As a further example, we now briefly outline a method of determining the impulse response of a system by using the convolution relation between autocorrelation and cross-correlation functions [Eq. (14.33)], which has been successfully utilized in the adaptive autopilot problem discussed in Sec. 15.3 [15.23]. In this example, wide-bandwidth almost-white noise $n(t)$ is generated and inserted with normal ongoing activity $x(t)$ into the aircraft-control actuators (Fig. 15.18). The output of the aircraft $y(t)$ is then cross-correlated with the noise input in a special-purpose computer, so that the effect of ongoing activity averages out. The period of the averaging is T, and the noise is of a particular discrete type which repeats itself over every T. Since the input is effectively white noise, the sampled cross-correlation function $\phi_{ny}(k\,\Delta\tau)$ which results is directly equivalent to the sampled impulse response h_k, where k represents the number of unit time shifts $\Delta\tau$ of the calculation. Therefore the system has been identified, although not in analytic form.

Problems

2.1 For the car-velocity example of Sec. 2.3 and Fig. 2.4a:

(a) Replot $F_r(V)$ and $F_e(V = 0) - F_e(V)$ against V on log-log coordinates, and show that the following analytical relationships apply:

$$F_r(V) = aV^\alpha$$
$$F_e(V) = F_e(V = 0) - bV^\beta$$

Note that $Fe(V)$ is the broken curve labeled $Fe(x_1, V)$ in Fig. 2.4a.

(b) Using the relevant differential equation, complete the symbolic block diagram corresponding to Fig. 2.4c. Note that the nonlinear nature of the functions F_e and F_r makes the problem of obtaining an analytic solution much easier than for the population-growth example of Sec. 2.5.

(c) Estimate graphically the initial acceleration at t_1 from Fig. 2.4b, and then estimate the car's mass by using Newton's second law and the appropriate F value from Fig. 2.4a. If it is assumed that $V(t)$ exponentially decays to V_2 from V_1, estimate the value of the time constant $1/k$ in Sec. 2.5.

2.2 Cells and other organisms control their concentrations of various materials at generally different levels from the bathing interstitial fluid.

(a) Considering only the diffusion phenomenon, draw an information-flow diagram for the dynamic relations between flow rates and concentration of, for example, sodium ions in the cell; ignore the more complex feature of flow dependence upon transmembrane potential. Show that this system has passive control.

(b) Incorporate the "sodium-pump" effect by which sodium ions are pumped out of a cell against a concentration gradient. Assume that a desired sodium concentration exists for the cell, and that the pumping action is controlled by the difference between the actual and desired levels. Redraw the information-flow diagram and show that active control exists.

(c) Model Probs. 2.2a and b mathematically by some simple linear differential equations. Then convert the information-flow diagram of Prob. 2.2b to a symbolic block diagram.

2.3 Prepare a crude information-flow diagram for the main transmission blocks used in the television process. The inputs to the system are to be taken as the visual and acoustic stimuli from an actor in the studio, and the outputs as the visual and acoustic perception of this actor by the viewer of a television set. Are there any obvious feedback paths? Modify the diagram to include tape recording of the program, and subsequent playback.

2.4 For the population-growth model of Sec. 2.5:

(a) Obtain estimates of world population data in recent history, and by plotting it semilogarithmically evaluate the growth-rate constant k. Estimate the world population in 2000 A.D. on this basis. How accurate does this linear model seem to be?

(b) For $k = 0.01$ and 0.03 how long does it take the population to double? To increase tenfold?

2.5 Lanchester [P.1] noted that when "modern" warfare techniques evolved, the old-style individual-combat pattern was replaced by collective action in which the casualty rate on one side at any time was proportional to the current firepower, and hence the current size, of the other side.

(a) Write the differential equations for this system where $x(t)$ and $y(t)$ denote the current size of each army.

(b) Draw corresponding information-flow and block diagrams.

(c) By eliminating the variable $y(t)$, show that the overall differential equation for $x(t)$ is

$$\frac{d^2x}{dt^2} - kx = 0$$

(d) The block diagram does constitute a single causal loop. Is the feedback effectively negative or positive?

2.6 The dynamics of an insect population in a closed environment are to be modeled for a short period as follows: The population P increases at a rate proportional to population (proportionality constant k_1) and decreases at a rate k_2 proportional to the mass W of a certain toxic substance. The mass of this toxic substance increases at a rate k_3 proportional to the population and decreases through decay at a rate k_4 proportional to its own mass.

(a) Write down the describing differential equations.

(b) Draw corresponding information-flow and block diagrams.

(c) Is this system actively or passively controlled?

2.7 For the closed loop with pure time delay of Fig. 2.26, plot the output response when the input is:

(a) A pulse of width $T/2$.

(b) A step input, with the time delay having static gains of $\frac{1}{2}$, 1, and 2.

(c) A square-wave input, of amplitude ± 1 and frequency $1/2T$, with the time delay having static gains of $\frac{1}{2}$, 1, and 2.

3.1 A simple linear model is to be prepared for the passive temperature control of poikilothermous animals. Assume that:

(a) Metabolic rate is proportional to the amount by which body temperature exceeds some constant value θ_0.

(b) Heat loss from the surface is governed by Newton's law of cooling [Eq. (3.15)].

(c) All the body can be lumped into a single homogeneous unit.

Derive the describing differential equation and draw the elementary block diagram. Discuss the stability of this simple model, and any improvements to the model suggested by this analysis.

3.2 The cardiovascular system has a complex of regulators, but in particular controls its average blood pressure. Identify the anatomy of this closed-loop control. From the mathematical modeling of the components, you will appreciate that the analysis is not simple. However, something is known about the transducer characteristics and about the fluid-flow phenomenon. What can you say about "lumping" versus wave propagation models in this case?

3.3 Obtain the Taylor's series expansion of $f(x + \Delta x)$ for the following functions of $f(x)$:

$$\sin x \qquad \cos x \qquad e^x$$

Hence obtain the following Maclaurin's series, by placing $x = 0$, and then by replacing Δx with x (since the series converge for all finite values of Δx):

$$\sin x = x - \frac{x^3}{3!} + \frac{x^5}{5!} - \frac{x^7}{7!} \cdots \qquad \cos x = 1 - \frac{x^2}{2!} + \frac{x^4}{4!} - \frac{x^6}{6!} \cdots$$

$$e^x = 1 + x + \frac{x^2}{2!} + \frac{x^3}{3!} \cdots$$

3.4 The expression $(1 + \delta)^4$ arises in connection with the radiation example. Multiply this out in the normal way, and show that the resulting terms equal identically those given by the binomial expansion [Eq. (3.19)]. Note that the number of terms is finite.

3.5 In turbulent flow of fluid past an orifice (see sketch), the defining mathematical relation for the volume rate of flow q is

$$q = cA \sqrt{\frac{2gp}{\rho}}$$

where A = orifice area
  c = coefficient of orifice, <1
  g = gravity constant ($= 981$ cm/sec², for example)
  ρ = fluid density
  p = drop in pressure across the orifice

(*a*) Calculate a family of steady-state curves for q against several values of orifice area (A, parameter) and pressure drop (p, abscissa) significant in the vascular system, using suitable dimensional units and realistic values for the other constants. (Take $c = 0.6$.)

(*b*) Actually, blood flow is normally *laminar* in nature, that is, the flow rate is described by a linear pressure relation, but in disease this may not remain true. In any case overplot some relevant linear relation on the turbulent figure. For what vascular conditions is it invalid to treat the blood as a continuous fluid?

(*c*) Treating q as a function of the two variables p and A, perform linearization both graphically and analytically. In particular, show that the analytical form can be expressed as

$$\frac{\Delta q}{q_0} = \frac{\Delta A}{A_0} + \frac{1}{2}\frac{\Delta p}{p_0}$$

where A_0, p_0, and q_0 are arbitrary equilibrium conditions.

4.1 Consider the dynamics and energetics of a short-distance runner. Assume that he accelerates at constant rate from the start until he reaches his full speed after 2 sec, thereafter maintaining this speed. Assume further that he runs a 100-m course in 10 sec. His weight W is 70 kg and his height is 1.9 m.

(*a*) What is his top speed? Hence show that 10 m/sec is an acceptable approximation.

(*b*) How much KE does he store at full speed? Use kg-m energy units and note that the relevant mass m for the formula is given by W/g, where g is the gravity constant equal to 9.81 m/sec^2.

(*c*) How much is his *momentum* p at full speed?

(*d*) How large is the effective accelerating force during the 2 sec of constant acceleration? Alternatively if the acceleration decreases linearly in time during the 2 sec at the end, what initial value of acceleration is then needed for the same top speed?

(*e*) At what angle must he lean forward during each acceleration program? Estimate the largest muscular forces necessary for these accelerations. Guess or find the necessary anatomical data.

(*f*) In the acceleration programs of Prob. 4.1*d*, at what average rate is he doing external work during the last $\frac{2}{10}$ sec of the accelerating period? (That is, what is his power output?) It will probably help to draw out time histories of force, acceleration, velocity, KE, etc., during this period. Do these curves suggest that any of the postulates are unlikely? In particular, can constant accelerating force be maintained, or is some such relation as the smooth decrease of accelerating force physically required?

(*g*) Estimate the rate of excess heat flow in the body due to muscular inefficiency. At what rate is ATP being used up initially? Take 1 cal = 4.18 joules = 4.27 kg-m. Also take the ideal available energy output of ATP as 7.73 kcal/mole, where mole equals molecular weight in grams, which equals 504.

(*h*) From the results of Prob. 4.1*g* and by obtaining or guessing appropriate muscle data, estimate the rates at which the relevant muscles heat up during acceleration. Take the specific heat $c = 1$ cal/g. Do the rates of change need estimating at different times during the 2 sec or would a total temperature jump from averaged data over the accelerating period be sufficiently accurate?

4.2 A pole-vaulting high jumper provides an example of almost conservative conversion of energy from the kinetic to the gravity potential kind [Eq. (4.16)].

(*a*) What characterizes maximum jumping efficiency for a given velocity? Draw diagrams to clarify the process.

(*b*) At what minimum speed must he run to clear 5 m? Neglect any inertial effects of the pole in this model.

(*c*) What effect is likely to limit how high man can jump this way? What tricks are possible, and practiced, to increase the height jumped? (For experimental data see [P.2].)

(*d*) Does he recover his KE after the jump? If so, what happens to it when he lands?

4.3 Electric eels can store electrical energy in special electrical organs, are able to produce potential differences as high as 250 volts, and can discharge up to about 6 kw for a few milliseconds (see [1.4], for example). Assuming a period of discharge of 5 msec at constant maximum power, what total energy must be stored? By guessing at a bioelectric conversion efficiency, how much ATP is needed to recharge the organs?

4.4 The following partial data is taken from Ackerman [1.4] for heart activity and blood flow.

Quantity	At Rest	Active
$\bar{p}$ = time-averaged pressure difference across heart	100 mm Hg	100 mm Hg
q = total blood flow	3.5 liters/min	35 liters/min
Hydrostatic power	1 watt	?
Kinetic power	0.13 watt	?
Total heart power	1.13 watts	?

(*a*) Fill in the missing data using the following relations: (1) Hydrostatic power represents the rate of supplying hydrostatic energy into the blood at the aorta, and is proportional to the product $\bar{p}q$, in compatible units. This energy input is dissipated as heat in the blood vessels due to their resistance to fluid flow, but in the aorta the energy is stored in fluid pressure of an equivalent gravity column [Eqs. (4.16), et seq.]. (2) Kinetic power represents the direct rate of energy supply as KE in the blood at the aorta, and hence has the characteristics of Eq. (4.23), when suitably modified for rate of flow; that is, replace m by ρq.

(b) Compare the heart power with thermoregulatory power, and with power for the running man in Prob. 4.2.

5.1 In the thermal lumped element of Sec. 5.2, remove the restriction that q_{cm} is zero, by considering both θ_c and θ_m as variables, and hence modify Eqs. (5.3) and (5.4). What is the new value of muscle time constant? In the development of Sec. 5.2, $\theta_m = \theta_c$ is the condition implicitly assumed when q_{cm} is made zero; under this condition, should not the two masses of muscle and core actually be lumped together? Does this produce a different time constant?

5.2 In the two-compartment model of Sec. 5.3:

(a) Show that the first-order differential equation for the concentration C_2, assuming the active flow of Eqs. (5.12) and (5.13) and consumption rate M, is given by

$$T \frac{dC_2}{dt} + C_2 = C_1 - \frac{k_{21}E + M}{k_{12} + k_{21}}$$

(b) Hence show that when $M = 0$ the steady-state condition is

$$C_2 = C_1 - E$$

provided

$$k_{21} \gg k_{12}$$

(c) Is the time constant of Prob. 5.2a smaller or larger than in Eq. (5.8)? Comment.

(d) Are E and M likely to exist together for the same material? Is the model helpful? What about the coupling between membrane potential and concentration differences?

5.3 The two-compartment model of Sec. 5.3 is to be generalized by making the supply compartment a variable system also, of volume V_1 and concentration C_1 with supply rate q_{01} from the environment (the independent input).

(a) Write out the two resulting first-order differential equations using the flow law of Eq. (5.7) between V_1 and V_2.

(b) Draw the block diagram.

(c) Under the special circumstances of a rapid hypodermic injection of q_{01}, this input may be looked upon as an equivalent concentration input obeying

$$C_{01} = \frac{1}{V_1} \int_0^{\Delta t} q_{01} \, dt$$

where Δt is the short injection period. In this case the block diagram can be redrawn with only one integration, showing that the two energy storages are not independent and that the complete system is in fact first order. Take $M = 0$ and integrate each of your equations once with respect to time to see how this is done. Show this analytically also by solving the equations for C_1 (or C_2). Show that the time constant is $(1/k)/(1/V_1 + 1/V_2)$.

5.4 For the nanoplankton respiration model of Sec. 5.4:

(a) Show that $C/(1/R + 1/R_b)$ in Eq. (5.15) has the dimensions of a time constant and that this is always smaller than RC.

(b) Write the relevant equations for nighttime respiration, evaluate the time constant, and compare the rise of f_R in sunlight with its decay at night.

(c) Can you estimate the probable ratio $R:R_b$? Under what condition would the nighttime and daytime constants be essentially the same?

(d) Which area on the daylight curves represents cellularly stored labile material? Since this material is all respired at night, which area in the night curve equals this?

(e) By comparing Eqs. (5.15) and (5.16), show that $f_c \to 0$ in the steady state during daylight.

5.5 Note that the nanoplankton system has two loops. Find the transfer function f/e_b by successive application of Eq. (5.28), as follows:

(a) By suitable use of Eq. (5.28) reduce Fig. 5.3b to the form

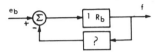

(b) Reduce the above figure to a single open-loop transfer function, and then check with Eq. (5.15).

▼

6.1 By direct integration of the forward Laplace transform show that an exponential input function $e^{\alpha t}$ has the following transform pair:

$$e^{\alpha t} \rightleftharpoons \frac{1}{s - \alpha}$$

Compare this result with the population model of Sec. 2.5 for k positive.

6.2 Reconsider the relation which defines the unit impulse function (Fig. 6.1a). Take the value of a to be finitely large so that the pulse is the superposition of a positive step of strength $1/a$ followed by a negative step of strength $1/a$. Calculate the time response when this is the input to a first-order system defined by $T/(Ts + 1)$. By superplotting responses for several small a/T ratios show quantitatively that as $a \to 0$ the response to the rectangular pulse approaches the response to the unit impulse input. Hint: Either (a) find the step response, and then subtract the corresponding negative one with the necessary time shift (superposition), or (b) calculate the positive step response until the pulse input ceases at $t = a$, and then treat the further response as an initial-condition problem without input.

6.3 Assume that when some particular biochemical material is released into the bloodstream, it is carried with time delay T to a chemoreceptor which responds to this biochemical with a pulse frequency $f_0(t)$ proportional to the concentration. By direct use of the Laplace transform show that when the concentration undergoes a step change at the release point, the chemoreceptor output function $f_0(s) = e^{-Ts}/s$. Hence show that the transfer function of the bloodstream in this situation is $H(s) = e^{-Ts}$, which is often defined as the *pure time delay, latency,* or *transmission delay.*

6.4 For the step response of Eq. (6.17) calculate the proportion by which $y(t)$ still falls short of y_∞ at $t = nT$ for significant nonintegral values of n. Check your results on Fig. 6.3c.

▼

6.5 Consider the lead-lag system of Fig. 6.6:

(a) Derive the unit step response of the lead-lag system [Eq. (6.40)] by using the partial-fraction technique.

(b) Derive the stated transfer function of the circuits of Fig. 6.6b, viz., $e_0/e_1(s)$, assuming the output load current is negligible (infinite output impedance). Hence confirm the α and T relations given in the figure. Note that the generalized impedance is essentially a transfer function and is the effort-flow ratio

$$Z(s) = \frac{e(s)}{f(s)}$$

so that for a capacitor $Z = 1/Cs$, for example (see Chap. 4).

6.6 Rework the thermoregulatory lump problem of Sec. 6.7 using Δ variables, and hence show that the overall transfer function is given by Eq. (6.46) rather than by Eq. (6.48). Redraw the responses of Fig. 6.7 with these new variables.

▼

6.7 In the first-order cascade system of Sec. 6.7:

(*a*) Check the impulse response $z(t)$ of the second first-order system, given [Eq. (6.52)], by both the partial-fraction technique and the transform table.

(*b*) Set $k_1 = 1 = k_2$ and $T_1 = T_2/2$, and then plot the overall impulse response $h(t)$ [$\equiv z(t)$] by drawing the indicated exponentials and subtracting them graphically.

(*c*) Repeat Prob. 6.7*b* for the case of repeated roots, when $T_1 = T_2 = T$, so that the impulse response is $h(t) = (t/T^2)e^{-t/T}$. Establish the maximum value of $h(t)$ and its time of occurrence by differentiating $h(t)$ and setting this derivative equal to zero.

▼

6.8 Perform approximate graphical convolution for a basic first-order system having the impulse response $h(t) = (1/T)e^{-t/T}$, with an input function of the same form: $x(t) = (1/T)e^{-t/T}$. Use squared graph paper of coarse grid (four or five lines per inch) and draw $h(t)$ on a transparent template. It can then be manually folded back and slid to successive locations to provide $h(t - \sigma)$. Perform the integration by "counting squares." Compare the resultant plot with the analytically obtained plot of Prob. 6.7*c*.

7.1 Basic trigonometry and series evaluations:

(*a*) Find the t in the expression ωt which produces a phase of 25°, 30°, $\pi/4$ rad, 60°, $\pi/2$ rad, 135°, 225°, $\tfrac{3}{2}\pi$ rad, and 315°, when ω equals 2 rad/sec.

(*b*) Evaluate $\sin \omega t$ and $\cos \omega t$ at the above values using conventional trigonometric tables.

(*c*) Evaluate $\sin \omega t$ and $\cos \omega t$ approximately by using the Maclaurin's series expansions of Prob. 3.3. Sketch the curves for about 90° by successively adding in terms of the series until adequate accuracy is achieved.

7.2 In phase-plane presentations the output $y(t)$ and its first derivative $dy(t)/dt$ are plotted as abscissa and ordinate, respectively, and the resulting curve is often called a *trajectory*.

(*a*) Show that if $y(t) = a \sin \omega t$, the phase-plane trajectory is a unique ellipse whatever the initial condition and frequency. For what value of ω is the special case of a circle achieved?

(*b*) Plot the phase-plane trajectory for $y(t) = y_0 e^{-t/T}$ from initial condition y_0.

(*c*) Lissajous figures are obtained if the input-output data of frequency response are shown as abscissa and ordinate, respectively, on an oscilloscope. What shapes are the figures if the AR = 1 and $\phi = 0$, $-\pi/2$, and $-\pi/4$, respectively?

▼

7.3 (*a*) Prove the validity of Euler's theorem [Eq. (7.15)]

$$e^{j\alpha} = \cos \alpha + j \sin \alpha$$

by use of the Maclaurin's series expansion for these three functions.

(*b*) From Eq. (7.14) prove that $|z| = r$ by using Eq. (7.10) and the trigonometric identity

$$\sin^2 \alpha + \cos^2 \alpha = 1$$

7.4 DeMoivre's theorem states that

$$(\cos \alpha + j \sin \alpha)^n = \cos n\alpha + j \sin n\alpha$$

where n is a positive integer.

(*a*) Use the self-evident identity $(e^{j\alpha})^n \equiv e^{j\alpha n}$ to prove this.

(*b*) Show that DeMoivre's theorem specifies the phase of a vector z^n defined in terms of the basic vector $z = re^{j\alpha}$. What is the corresponding magnitude relation? Thus show the vectors z and z^3 on the complex plane in polar coordinates, when $z^3 = -2 + j$ in rectangular coordinates.

7.5 By use of the partial-fraction technique show that the term specified by Eq. (7.31) is the only one which does not decay when the linear system is subjected to a sinusoidal input of angular frequency ω_0. Argue for the particular case of the basic first-order system $H(s) = 1/(Ts + 1)$.

7.6 Show that the polar plot of $H(j\omega) = 1/(j\omega T + 1)$ is a unit semicircle.

7.7 (*a*) Sketch the Bode plots for a three-unit and a four-unit cascade of first-order elements where each element is the same, namely, $H_i(s) = 1/(Ts + 1)$. (Note that the error of the asymptotes at $\omega T = 1$ can become quite large.) Hence show that static gains of 8 and 4, respectively, are needed to produce the closed-loop instability condition discussed in Sec. 7.7. Redraw the Bode open-loop plots with these static-gain values, that is, for the systems

$$H(s) = \frac{8}{(Ts + 1)^3} \quad \text{and} \quad \frac{4}{(Ts + 1)^4}$$

Note that the absolute value of T makes no difference in this case to the static-gain values, although it does affect the oscillatory frequencies. What are they?

(*b*) Sketch the polar plots for Prob. 7.7a to illustrate the aspects discussed. Do not worry about high accuracy.

7.8 (*a*) Sketch the Bode plots for a cascade of first-order units as follows:

$$H_1(s) = \frac{2}{s + 1} \qquad H_2(s) = \frac{3}{\dfrac{s}{2} + 1} \qquad H_3(s) = \frac{1}{\dfrac{s}{10} + 1}$$

(*b*) Sketch the Bode, polar, and transient response plots for a lead-lag network defined by

$$H_4(s) = \frac{\dfrac{s}{5} + 1}{\dfrac{s}{50} + 1}$$

(*c*) Sketch the combined Bode plots when the systems for Probs. 7.8a and b are placed in series. Show that the $-180°$ point is now exceeded at a higher frequency than in Prob. 7.8a. Does this represent better dynamic performance? For example, would the step response now be completed earlier?

8.1 Consider Li's torsional spring with ends clamped onto each thigh so that gravity-restoring torque is now augmented by an amount

$$T_s = -k_s\theta$$

where k_s is spring torque per radian.

(*a*) Hence show that Eq. (8.12) still applies but that the undamped natural frequency ω_{n1} is now increased in the ratio

$$\left(\frac{\omega_{n1}}{\omega_n}\right)^2 = 1 + \frac{2k_s}{mgL}$$

For the standard man, what k_s value is necessary to increase the natural frequency by 50 percent? Work in kg-m/rad.

(*b*) Regarding the energy aspects of this new system, does the 50 percent speed improvement involve extra muscle power—both ignoring and including friction?

(c) Consider the different modification of stilts. Assume unit density (1 g/cc), thin (10-cm-diameter) wooden poles 3 m long swinging from the hips. What is the new value of ω_n? Assuming the same maximum angular deflection during the stride, what is the stilting speed?

(d) How fast would a man walk naturally on the moon, presuming that his moon suit does not modify his thin-cylindered legs?

▼

(e) Since various animals have differing leg lengths, Prob. 8.1c suggests that a generalized prediction of natural walking speeds should be possible. Assuming that all legs are thin columns, the only comparative anthropological data needed is that of leg lengths. Comment on the implications and limitations of such an analysis.

8.2 (a) A spinning skater accelerates (or decelerates) his angular velocity by hugging (or throwing out) his arms. Estimate the relevant J in the withdrawn-arm case by assuming the body is a simple vertical cylinder; the axis of rotation is then of course coincident with the cylinder's long axis. Estimate the new J when the arms are flung out radially, assuming they are simple cylinders weighing 6 kg and are 1 m long.

(b) The change of angular velocity between the two conditions of (a) can readily be found from the conservation-of-energy principle where, in the rotational case, the stored energy is $E = \frac{1}{2}J\omega^2$. Find the percentage change in ω, and also show the validity of this energy-storage relation.

▼

8.3 Plot the characteristic function of the leg model [Eq. (8.19)],

$$f(s) = \left(\frac{s}{\omega_n}\right)^2 + 2\zeta \frac{s}{\omega_n} + 1$$

for $\zeta = 0, \frac{1}{2}, 1$, and 2, and for $\omega_n = 4$ rad/sec. Plot for a sufficient range of s to establish the behavior of the curves. How do the results correlate with the definition of roots in the text? Locate as many of the poles graphically as possible.

8.4 Plot the impulse response of the semicircular-canal transfer function [Eq. (8.22)], for the given constants, by the graphical method of Fig. 8.10. Use only sufficient accuracy to show the significant results.

8.5 The saccadic motion of the eye has been best-fitted as the step response of a basic linear second-order transfer function, with $\omega_n = 120$ rad/sec and $\zeta = 0.7$.

(a) Sketch the impulse and step responses in this case. (Do not use dimensionless time). What are the values of settling time, response time, and peak overshoot time? How large is P?

(b) Assuming a steady-state gain of unity, what is the bandwidth? How well do the correlations of Sec. 8.10 fit these data?

8.6 Refer back to Prob. 2.5 concerning warfare.

(a) Write down the transfer function, and appreciate that an unforced initial-condition problem exists as stated. Would the flow of reinforcements constitute a valid input? Evaluate the poles of this system and plot them upon the complex plane, thus showing that they are both real, but that one is in the right half-plane.

(b) Sketch the system response from initial conditions

$x(0) = 10^6$
$y(0) = 0.5 \times 10^6$

There is a physical constraint in this problem which ensures that the usual unstable response of a system with a right-half-plane pole cannot develop. Comment on this.

(c) Consider Fig. 7.1 showing the establishment of the steady-state conditions necessary for frequency response measurements. Hence appreciate that the right-half-plane pole prevents the frequency response from being found physically, although it can be analytically plotted easily enough from definition. Do this.

8.7 Refer back to the insect population (Prob. 2.6), which provides a second-order model but again without input as specified. Consider now an experiment to obtain frequency response data on this system, whereby we start with zero population P and zero waste W. The input rate is q in actual insects per second placed into the system. This is what the experimenter can control in sine waves, or any other function shape.

(a) Show that the transfer function is given by

$$\frac{P}{q}(s) = \frac{1}{s^2 + (k_4 - k_1)s + (k_2 k_3 - k_1 k_4)}$$

Using Eq. (8.42) and any other necessary relations of Chap. 8 satisfy yourself that all roots lie in the left half-plane if $k_4 > k_1$ and $k_2 k_3 > k_1 k_4$.

(b) In particular, assume

$$k_1 = 0.5 \times 10^{-6} \text{ sec}^{-1} \qquad k_2 = 1 \times 10^{-6}$$
$$k_3 = 2 \times 10^{-6} \qquad k_4 = 2 \times 10^{-6}$$

and hence sketch the Bode and polar locus plots.

(c) What are M_p, ω_n, ω_d, ω_R, and ω_{bw}?

9.1 Consider the two transfer functions between F_e and M_m as inputs and θ_c as output for the thermoregulatory example, using Eqs. (3.30) to (3.35), but to simplify tedious manipulations, set $q_{cs} = 0$ [Eq. (3.35)], also F_r, M_b, M_x, F_c, and F_{rad}.

(a) Show that the transfer functions are of the form

$$\theta_c = \frac{k_1 F_e}{b_3 s^3 + b_2 s^2 + b_1 s + b_0} + \frac{k_2(T_s s + 1)M_m}{b_3 s^3 + b_2 s^2 + b_1 s + b_0}$$

(b) Show that all the b coefficients of Prob. 9.1a are positive. Are there always three real poles for this passive system?

9.2 (a) Evaluate the lag-delay parameters (Fig. 9.4) for the nth-order equal-time-constant cascade

$$H(s) = \frac{1}{(1 + Ts)^n} \qquad \text{for } n = 2, 4, 8$$

Proceed by first plotting the relevant impulse responses (Fig. 9.6) and then by graphically integrating them (counting squares) to obtain the step responses. Overplot the relevant lag-delay models' responses and comment upon the fit. Sketch and compare exact and approximate Bode plots.

▼

(b) Evaluate the T_m and T_s values from Eqs. (9.21) and (9.22) for the n values in Prob. 9.2a. Sketch the indicated Gaussian impulse responses and compare with the exact responses by overplotting. Also sketch and compare Bode plots.

9.3 The second-order Padé polynomial to approximate the pure time delay is given by

$$H(s) = \frac{1 - \dfrac{T}{2}s + \dfrac{(Ts)^2}{12}}{1 + \dfrac{T}{2}s + \dfrac{(Ts)^2}{12}}$$

(a) Compare its Bode plots with the idealized requirement.

(b) What are ζ and ω_n for the characteristic function? Sketch the pole-zero plot.

▼

(c) Find and sketch the step response.

9.4 Sketch the transient and frequency response curves in similar manner to Figs. 9.10 to 9.14, for the second-order-denominator URS element

$$H(s) = \Omega \frac{ks}{(Ts + 1)^2}$$

9.5 Consider the heat-exchanger model of Eq. (9.42) as applying to blood circulation through the skin. Assume that steady-state temperature conditions exist, $\frac{\partial \theta}{\partial t} = 0$, so that blood temperature is defined by $\theta(x)$ as a function only of path length through the skin, and the skin temperature itself is assumed constant at ϕ.

(*a*) Show that blood temperature decays exponentially with distance, and that the "spatial transfer function" corresponds to a first-order lag as previously met in the time domain.

▼

(*b*) Now permit the skin to heat up along the path by virtue of heat convection being necessary to remove all heat from the skin's surface. Repeat the analysis allowing for skin capacity, but again in the steady state.

10.1 A frequency response test provided the following experimental data:

Frequency ω, rad/sec	0.01	0.05	0.10	0.20	0.30	0.40	0.50	0.60	0.80	1.2	2.0	4.0
AR	2.1	2.15	2.2	2.8	3.7	5.3	2.7	1.5	0.70	0.27	0.08	0.02
Phase ϕ, deg	0	-2	-6	-14	-35	-89	-138	-158	-167	-175	-182	-193

Prepare the Bode plots and hence "discover" the transfer function; in particular evaluate steady-state gain, time constants, or ζ and ω_n, and any other parameters which may be necessary. Assuming that the phase data are good, is there any evidence for a non-minimum-phase element?

10.2 Reconsider the simple population-growth model of Sec. 2.5 and Prob. 2.4. Assume that as the population increases the birth rate decreases as $k_b = k_{b0}e^{-P(t)/P_1}$, whereas the death rate stays constant at $k_d = k_{d0}$.

(*a*) Show that a steady-state world population is now reached:

$$P(t = \infty) = P_1 \log_e \frac{k_{b0}}{k_{d0}}$$

and evaluate it for $k_{b0} = 0.03$ per year, $k_{d0} = 0.02$ per year, and $P_1 = 3 \times 10^{10}$.

(*b*) Write the overall differential equation for the system in the form $dP/dt = f(P)$. Plot the $f(P)$ against P for the constants of Prob. 10.2a, and then carry out slope-line integration to solve for the steady state resulting from an initial condition of $P_1 = 3 \times 10^9$ (as in 1964). Select your Δt for the accuracy you think necessary. How long before the steady state is essentially reached?

(*c*) Solve similarly by the Euler finite-difference recurrence formula.

10.3 Reconsider the multicompartment model of Fig. 10.6 and note that the three exponential decays derived in the text can be explained as the independently first-order dynamics of communication among each of the other three compartments. Clearly, therefore, the impulse response of compartment 1 must indeed have a finite value of $0+$, as shown in Fig. 10.5. For exemplary purposes, assume that the constants k_{ij} of Fig. 10.6 specify linear flow rates

$$q_{ij} = k_{ij}C_i$$

and that the moles stored in a compartment Q_i depend upon the compartment's concentration C_i and volume V_i:

$$Q_i = C_iV_i$$

(a) Prove that the transfer function relating external flow rate q_{01} in moles per second as input to concentration C_1 as output must have a second-order numerator in the transfer function as well as a third-order denominator, in order for this initial response to be achieved. Failing all other techniques, the initial-value theorem shows this directly (Sec. 6.8).

(b) Sketch the response of compartment 4 to the same impulse input into 1; in particular, compare the initial response with that of 1. Since the denominator of the transfer function is inevitably the same third-order denominator as in Prob. 10.3a, what order must the numerator now be?

(c) Write down the describing differential equations and draw up an analog-computer circuit with $q_{01}(t)$ as an arbitrary flow input into compartment 1. What minimal numbers of integrators and summers are needed?

(d) Modify the model so that compartment 4 exchanges only with compartment 3 (k_{34} and k_{43}); then repeat Probs. 10.3b and c.

10.4 The following plasma-activity-curve data are reported in [10.4]:

t	0	1	2	3	5	24	30	47	70	94	118	143
$x(t)$	1	0.9	0.86	0.78	0.76	0.51	0.41	0.35	0.28	0.24	0.23	0.21

Graphically perform the exponential-decay extraction procedure of Sec. 10.4, obtaining three separated roots.

10.5 It has been noted [P.3] that the impulse response to a one-shot input of advertising (sales per unit time after insertion) is similar to the typical plot of the spread of an epidemic (for example, measles or poliomyelitis in new cases per unit time). Indeed, there seems a plausible analogy between these examples regarding such features as incubation period, spread of infection by carrier, and "invasion of a susceptible body by an infecting organism." In a particular situation the following numbers arose:

t	0	1	2	3	4	5	6	7	8	9	10	12	14	16	18
$h(t)$	0	0	40	175	260	225	175	90	65	50	40	25	15	9	5

Plot both $h(t)$ and log $h(t)$ versus t. Estimate the incubation period. Apply the root-extraction technique as far as possible. However, after extracting the first exponential a new situation appears because some of the new plotted points are negative. Can you find any new ways to extract the remaining dynamics? Refer particularly to Fig. 8.10 regarding impulse responses which are initially zero, and show that the root extraction still applies in this case, with one minor modification.

10.6 Prepare an unscaled analog-computer wiring diagram for the semi-circular canal:

(a) As in Fig. 10.10c except that the cupula acceleration $\ddot{\phi}$ is to be made explicitly available as a computer variable.

(b) From the equivalent block diagram of Fig. 8.6b and c.

10.7 A mathematical model is to be prepared for rotational eye movements, based upon the idea that the vitreous and aqueous humors (Fig. 2.11) are only coupled to the cornea via linear viscous and elastic forces. Assume also that corneal movement relative to the skull generates viscous and elastic restoring forces.

(a) Write out the relevant differential equations and find the transfer function from applied muscle torque to corneal rotation. Estimate the J constants.

(b) Prepare an unscaled analog-computer diagram for solution of Prob. 10.7a.

10.8 In the input-discovery problem of Sec. 10.3, show that an alternative way to generate $H_2^{-1}(s)$ of Fig. 10.3 is to place the original function $H_2(s)$ in the feedback path around a high-gain operational amplifier.

10.9 Return to the finite-difference solution for the step response of the second-order standard form of the semicircular-canal model [Eqs. (10.30) and (10.31)], using the parameter values of the text.

(*a*) Carry out the solution for about ten increments using $\Delta t = 10^{-3}$ sec, and a 6-digit word with roundoff. Estimate the exact answer by any shortcuts (for example, Taylor's series expansions) and compare. What pattern of response is developing after the initial phase?

(*b*) Repeat (*a*) with $\Delta t = 1$ sec and hence show that the resulting solution is indeed unstable.

10.10 Reconsider one-dimensional transient heat conduction, for which the partial differential equation is given by Eq. (3.14).

(*a*) For digital-computer solution the time derivative must also be finite-differenced. Do this using both of the following forms:

$$\frac{d\theta_n}{dt} \approx \frac{\theta_n - \theta_{n-1}}{\Delta t} \approx \frac{\theta_{n+1} - \theta_{n-1}}{2\Delta t}$$

and hence obtain and compare the overall recurrence formulas for θ_{n+1} which result. Is one of the above derivative approximations more accurate than the other? Draw chords on a curve like Fig. 2.7 to illustrate the processes.

(*b*) What is the significance of two data points θ_n, θ_{n-1} being needed in the finite-difference formula to make one step forward to θ_{n+1}? How does one get the problem started?

10.11 The differential equation for the swinging-leg model [Eq. (8.15)] is to be prepared for solution by finite-difference calculations. Obtain the standard form of the equation. Estimate Δt and carry out as much of the numerical solution as interests you for an initial-deflection unforced case, with $\zeta = 0.2$.

11.1 Elimination of steady-state error: Consider the following closed-loop configuration:

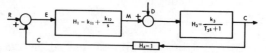

(*a*) Show that the proportional-plus-integral control scheme of H_1 is equivalent to

$$H_1 = k_{11}\frac{T_1 s + 1}{s}$$

Sketch its step, impulse, and asymptotic Bode frequency responses.

(*b*) Using Eqs. (11.2) and (11.3) obtain the closed-loop transfer functions for C and E, respectively.

(*c*) Using Prob. 11.1*b*, show that the numerator polynomials of the transfer functions for E as output have no s term of lower order than s^1, and therefore that the steady-state error to both reference and disturbance steps is zero. Show that a compatible conclusion can be drawn from the transfer functions for C.

(*d*) Set $k_{12} = 0$, but at the same time modify the process to

$$H_3 = \frac{k_3}{s(T_3 s + 1)}$$

H_3 specifies the one-integrator plant, or controlled process. Repeat Probs. 11.1*b* and *c* for this plant and hence deduce the general rule that one integrator is needed in the forward path between the input and output of interest, in order to obtain zero steady-state error to step inputs.

(*e*) Repeat the manipulation of Prob. 11.1*b* with $H_4 = 1/(T_4 s + 1)$.

11.2 Reconsider the car-dynamics problem (Figs. 2.4, 10.4).

(*a*) Initially assume that F_e does not decrease from 100 lb as a function of V, and hence estimate the linearized time constant for the car at 20 ft/sec, 57 ft/

sec, and an average over this interval, by drawing appropriate tangents and chords on Fig. 10.4a.

(b) Include the decreasing $F_e(V)$ function also, and again estimate the time constants for the velocity conditions of Prob. 11.2a.

(c) Consider the relevance of these results to the dynamic analysis of car-speed control (Sec. 11.3), particularly to the reduction of T as the loop gain is increased.

11.3 Repeat the block-diagram manipulations of Fig. 11.7 to obtain the transfer functions of the same system but have F_e and E as outputs; hence confirm Eqs. (11.17) and (11.20).

11.4 Second-order closed-loop:

(a) For the elementary block diagram of Fig. 11.11b obtain the closed-loop transfer function [Eq. (11.28)] by block-diagram manipulation as in Prob. 11.3.

(b) Consider more deeply the alternative second-order system of Eq. (11.32). Obtain the transfer expressions for manipulated variable and for error as outputs. Sketch some relevant step and frequency responses in order to examine how the dynamics have been changed. In particular, show that the manipulated-variable step response demand is less severe for the reference step than in the second-order model of the text, but that it is similar for the disturbance step.

▼

11.5 Stability:

(a) Show that the characteristic function for proportional control of a first-order cascade process with $n = 4$ [Eq. (11.43)] is $(Ts)^4 + 4(Ts)^3 + 6(Ts)^2 + 4(Ts) + (1 + K)$.

(b) Apply Routh's test to Prob. 11.5a and hence confirm the instability condition of Fig. 11.22d; furthermore, confirm that the stable range for K is $-1 < K < 4$.

(c) Repeat the above steps for proportional control of the basic overdamped second-order process used as an example in the text, for which $T_1 = 2T_2$ (Fig. 11.13).

11.6 Compensation:

(a) In Fig. 11.11b, incorporate velocity feedback k_v as well as position feedback. Manipulate both the equations and the block diagram to show that the k_v and b effects are additive. Satisfy yourself of the stabilizing effect which results, by sketching comparatively the Nyquist plots when k_v is increased.

(b) Prepare the closed-loop Bode plots of the lead-lag compensated system in Fig. 11.19 for the $k_l = 4$ condition. Repeat for $k_l = 14$.

▼

(c) Apply a lag-lead network with $\alpha = \frac{1}{2}$ to the position-control system of Fig. 11.19 but without the lead-lag network. Place the "nose" frequency of the phase plot $1/(\sqrt{\alpha}\, T)$ at a sufficiently low frequency on Fig. 11.19 that the phase is essentially not affected in the resonance area, say

$$\frac{1}{\sqrt{\alpha}\, T} = \frac{\omega_n}{10}$$

Show that the open-loop plot is depressed in the region of resonance, and consequently that the loop gain for the same value of M_p can be increased. Why is this desirable?

Investigate the effect of this lagging network by the root-locus method (for example, compare with Fig. 11.23). Can you predict the effect on the transient response?

▼

11.7 (a) Write down the forms of open-loop transfer function corresponding to the log AR Bode plot of Fig. 11.17, for all three options at low frequency.

(b) Plot the open-loop poles and zeros from Prob. 11.7a on the complex plane. Hence, using the root-locus technique, sketch in some of the interesting details of the closed-loop root loci.

11.8 The curves for linear control of a double-integrator plant are to be compared with those for maximum-effort control (Figs. 11.25 and 11.26). Assume unity-gain position and variable-gain velocity feedback from the plant, where the latter's transfer function is $x/F = 1/ms^2$ [see Eq. (11.48)]. Also assume that the controller output F is proportional to error, where the constant of proportionality is defined so that the maximum value of F equals the F_M of Fig. 11.25. Plot comparative results for closed-loop damping ratios of 0.5, 0.707, and 1, and comment on the comparison between linear and maximum effort control.

12.1 One revealing experiment which illustrates how the semicircular canal can fail in its task of providing a drive for compensatory eye movement is as follows: The subject undergoes a constant angular acceleration $\dot{\Omega}$ (say 4 rad/sec²) until some predetermined velocity (say 10 rad/sec) is reached. Then Ω remains constant for a "long" period (say 40 sec), after which it is decreased rapidly to zero with constant deceleration (say 10 rad/sec²). The subject is thereafter "dizzy" for a considerable period. What happens to the eyes during the dizzy period? Sketch the cupula's quantitative time response to this forcing input function Ω, and also the eye's approximate response pattern. Why are the canals said to have failed?

Ballet dancers avoid this problem, partly by adaptation and partly by using a particular technique of head movement. What is the technique? Perform a detailed analysis of the corresponding cupula deflections to establish why the problem is thus overcome.

12.2 Sampled-data model of eye movement, with head stationary:

(*a*) Sketch the expected eye-movement response to successively longer pulses of target position than that shown in Fig. 12.9*a*, and note any significant pulse lengths.

(*b*) Sketch the expected eye-movement response to a square-wave target input of $\frac{1}{10}$, 1, 2, and 10 cps, making guesses as necessary about the predictability of the input.

(*c*) Sketch the expected eye-movement responses to an exponentially increasing target position.

13.1 The step response of the cockroach mechanoreceptor has been fitted [13.8] by the function

$$f(t) = 24 + 165e^{-12t} + 120e^{-1.1t} + 42e^{-0.125t}$$

(*a*) Show that the corresponding transfer function of the receptor is given by

$$H(s) = \frac{351s^3 + 2,548s^2 + 1,211s + 55.5}{s^3 + 13.3s^2 + 15.5s + 2.3}$$

(*b*) Compare the Bode plots of this transfer function with those of Fig. 13.6. Use ingenuity to avoid unnecessary calculation—for example, the frequency response behavior as $\omega \to 0$ and $\omega \to \infty$ can be quickly evaluated.

13.2 Refer to MacKay's model of receptors and intensity perception (Sec. 13.4).

(*a*) Show that Stevens' law is still obtained if the frequency relations of receptor and organizer are themselves power laws:

$$f_R = k_R(I - I_0)^{\alpha_R}$$
$$f_O = k_O(\psi - \psi_0)^{\alpha_O}$$

(*b*) Prepare a modified flow diagram for cross-modal matching in which the subject perceives one receptor input and matches with it, by manipulation of a control, another environmental variable which also provides receptor input. For example, as noise intensity is varied, the subject adjusts the room's illumination to the same perceived intensity. Examine whether MacKay's model provides for a power-law relation between modes in this case, as Stevens does find experimentally.

14.1 Using Eq. (14.13) for the exponential PDF, find the zeroth, first, and second moments; hence show that Eq. (14.3) is satisfied, that the first moment is μ, and the second moment about $x = 0$ is $2\mu^2$.

14.2 Extreme-value statistics: By referring to tables, or otherwise, find the number of standard deviations beyond the mean which correspond to 10, 5, and 1 percent probability levels for the exponential, Gaussian, and rectangular PDFs. In the extreme-value statistics problem, one is concerned with the very low probability, say less than $\frac{1}{10}$ percent, of catastrophic events (very severe winters, earthquakes, droughts, epidemics). Since accurate records are often not available for more than perhaps 100 years, what can you say of the confidence in using a priori models (such as the Gaussian) whose parameters have been determined from these experimental data?

14.3 Although probability distributions are basically defined for statistical signals, they can be computed for deterministic signals. Compare the bar-chart distributions for symmetric regular and symmetric randomly switched square waves (± 1 amplitudes); modify the latter case for when the probability of state $+1$ is p (not just $\frac{1}{2}$). Find the "PDF" for a symmetric sawtooth wave and a sinusoid. (Note that in any small amplitude interval Δx the proportional time spent there is inversely related to the slope.)

14.4 Prepare the bar chart of probability for simultaneous tossing of 3 dice (by counting outcomes, or otherwise) and note that it is beginning to look like the Gaussian curve. Overplot the graphs for $n = 1, 2$, and 3 dice using a dimensionless variable $\tau = (x - \mu_n)/\sigma_n$, where μ_n and σ_n are separately calculated for each n. Overplot the normalized Gaussian PDF also.

14.5 The following experimental data on event frequencies $n(x)$ are to be fitted with a Gaussian PDF, and the χ^2 test is to be applied regarding goodness of fit:

x	$n(x)$	x	$n(x)$
0.0–0.2	1	0.8–1.0	72
0.2–0.4	5	1.0–1.2	24
0.4–0.6	10	1.2–1.4	6
0.6–0.8	48	1.4–∞	3

Note that the data for the slots at each end should be combined with their respective neighboring slots for the χ^2 test see [Eqs. (14.19)].

14.6 Find the best least-squares line to the following data [14.6], regarding the estimation of the mean number of responding units m, such as in Fig. 14.12. Also compute the correlation coefficient.

m_1	m_2
3.36	3.22
2.89	3.19
2.64	2.72
2.33	2.40
1.86	1.95
1.50	1.48
1.15	1.08
1.04	1.08
0.58	0.59
0.38	0.31

14.7 The following data from Fatt and Katz [14.2] specify the frequencies with which various time periods between MEPP occurred, in a total of 800 intervals. The first number represents the number of events in the interval from 0 to 0.04 sec, and successive numbers refer to successive 40-msec slots. Fit an exponential PDF to these data and perform the χ^2 test for significance:

125, 123, 96, 90, 60, 53, 50, 33, 23, 26, 19, 14, 24, 6, 5, 8, 8, 11, 5, 3, 8, 1, 1, 3, 0, 0, 0, 2, 0, 0, 1, 0, 0, 1, 1

14.8 The following data [14.9] refer to the probability distribution histogram for the thresholds of primary auditory neurons at their characteristic frequencies, for two different frequency ranges. Fit the higher-frequency data with a Gaussian PDF; the lower-frequency data, however, are bimodal and need two separate Gaussian curves, which should also be fitted.

Range of Threshold, Negative db	*Number of Neurons Responding* (9–20 kcps)	*Number of Neurons Responding* (0.6–1 kcps)
10–15	1	2
15–20	1	4
20–25	2	5
25–30	3	2
30–35	9	1
35–40	12	3
40–45	14	3
45–50	14	10
50–55	13	14
55–60	14	21
60–65	10	15
65–70	8	18
70–75	5	13
75–80	3	10
80–85	2	5
85–90	4	5
90–95	1	0
95–100	1	0

14.9 The highest frequency present in the ventilation record of Fig. 14.14*e* is $\frac{1}{20}$ cps [14.12]. (The record is actually smoothed from data sampled every 10 sec.) Assuming that one wishes a resolution of $W = \frac{1}{1,000}$ cps in the PSD calculation, how long should the maximum time shift τ_m be, and how many sample data points does this represent? If the probable nonstationarity of the process limits one to 5 hr of recording, how many degrees of freedom are available? What are the 90 percent confidence limits on the spectral density estimate so obtained? Approximately how many different products must the computer form in performing the indicated ACF calculation?

▼

14.10 Linear filtering of statistical signals:

(*a*) Spectral densities: White noise, of (constant) spectral density Φ_0, is to be filtered by a basic first-order lag of transfer function $H(j\omega) = 1/(Tj\omega + 1)$. Proceeding as in Fig. 14.20 with logarithmic coordinates, show that $|H(j\omega)|^2$ has a high-frequency asymptote of -2, and hence that the output spectral density is defined by

$$\Phi_{yy}(j\omega) = \frac{\Phi_0}{(\omega T)^2 + 1}$$

This procedure illustrates the fact that if a given PSD can be defined analytically, it can always be considered as generated by some specific linear filter operating on white noise.

Let $\Phi_{yy}(j\omega)$ be further filtered through a second first-order system, $H(j\omega) = 1/(\alpha T j\omega + 1)$, where α is a constant. Draw the output spectrum $\Phi_{zz}(j\omega)$ on the previous figure and show that it is defined analytically by

$$\Phi_{zz}(j\omega) = \frac{\Phi_0}{[(\omega T)^2 + 1][(\alpha\omega T)^2 + 1]}$$

(b) *Mean square value:* Equation (14.31) specifies the mean square value of a statistical signal after linear filtering. Furthermore, Prob. 14.10a has shown that any $\Phi_{xx}(j\omega)$ may be defined in the form

$$\Phi_{xx}(j\omega) = \Phi_0 |H_0(j\omega)|^2$$

where $H_0(j\omega)$ is a conceptual filter to produce $\Phi_{xx}(j\omega)$ from white noise Φ_0. Hence Eq. (14.31) may always be written as

$$\overline{y^2} = \frac{\Phi_0}{2\pi} \int_{-\infty}^{+\infty} \left| \frac{c_m(j\omega)^m + \cdots + c_0}{d_n(j\omega)^n + \cdots + d_0} \right|^2 d\omega$$

where the c and d polynomials derive from the constants of $H(j\omega)$, $H_0(j\omega)$, etc., and where $n \geq m + 1$. The solutions for this integral are tabulated for values of n up to tenth order as functions of the c and d coefficients, for example [14.16]. Here we write out the first two only since they become cumbersome with increasing n:

$$n = 1 \qquad \overline{y^2} = \Phi_0 \left(\frac{c_0{}^2}{2 d_0 d_1} \right)$$

$$n = 2 \qquad \overline{y^2} = \Phi_0 \left(\frac{c_1{}^2 d_0 + c_0{}^2 d_2}{2 d_0 d_1 d_2} \right)$$

(1) Using the solution for $n = 1$, show that the mean square value for y is $\overline{y^2} = \Phi_0/2T$.

(2) Using the solution for $n = 2$, show that the mean square value for z is $\overline{z^2} = \Phi_0/[2T(\alpha + 1)]$.

(3) Now consider the effect of the filter of time constant αT on $\Phi_{yy}(j\omega)$. Clearly, as the value of αT becomes small with respect to T ($\alpha \ll 1$), this second filter has very little effect on the mean square value $\overline{z^2}$, which $\approx \overline{y^2}$. This confirms quantitatively the comment in the text that as long as the break frequency of a near-white input signal is large compared with the natural or break frequency of the filtering dynamic system, this input spectrum may be considered white.

15.1 Draw up the analog-computer diagram to solve Eq. (15.12) for parameter a_1, in similar manner to that of Fig. 15.13. Note that a time base must be generated, by integrating a step function in order to implement the initial-condition expressions. What minimum numbers of integrators, summers, multipliers, and dividers are necessary?

15.2 Show that when the absolute-value-of-error performance criterion [Eq. (15.14)] is used instead of the quadratic value [Eq. (15.15)], Eqs. (15.16) yield, for the steep-descent relations,

$$\frac{db_j(t)}{dt} = -k \frac{\epsilon}{|\epsilon|} \frac{d^i y}{dt^i}$$

Hence prepare an analog-computer implementation, as in Fig. 15.15, where relay switches are used to implement the function $\epsilon/|\epsilon|$ which only takes on values ± 1. This is the circuit reported by Ornstein [8.6].

References and Bibliography

References

1.1 Wiener, N.: "Cybernetics," The M.I.T. Press, Cambridge, Mass., and John Wiley & Sons, Inc., New York, 1961.

1.2 Beer, S.: "Cybernetics and Management," John Wiley & Sons, Inc., New York, 1959.

1.3 Casey, E. J.: "Biophysics; Concepts and Mechanisms," Reinhold Publishing Corporation, New York, 1962.

1.4 Ackerman, E.: "Biophysical Science," Prentice-Hall, Inc., Englewood Cliffs, N.J., 1962.

1.5 Ruch, T. C., and J. F. Fulton: "Medical Physiology and Biophysics," W. B. Saunders Company, Philadelphia, 1961.

1.6 Stacy, R. W., et al.: "Essentials of Biological and Medical Physics," McGraw-Hill Book Company, New York, 1955.

1.7 Oncley, J. L., et al.: "Biophysical Science—A Study Program," John Wiley & Sons, Inc., New York, 1959.

1.8 Maxwell, J. Clerk: On Governors, *Proc. Roy. Soc.*, **16**:270–283 (1867).

2.1 Ashby, W. R.: "Design for a Brain," John Wiley & Sons, Inc., New York, 1960.

2.2 Malthus, T. R.: Mathematics of Population and Food, "The World of Mathematics," vol. II, pp. 1189–1199, Simon and Schuster, Inc., New York, 1956.

2.3 Ruch, T. C., and J. F. Fulton: "Medical Physiology and Biophysics," W. B. Saunders Company, Philadelphia, 1961.

2.4 Lowenstein, O.: (i) Pupillography—Methods and Diagnostic System, *AMA Arch. Ophthalmol.*, **55**:565–571 (April, 1956).
(ii) Electronic Pupillography—Why, How and When, *Eye, Ear, Nose Throat Monthly*, **38**:549–558 (July, 1959).

2.5 Stark, L.: Stability, Oscillations and Noise in the Human Pupil Servo-mechanism, *Proc. IRE*, **47**(11) (November, 1959).

2.6 Clynes, M.: Computer Dynamic Analysis of the Pupil Light Reflex, *Proc. 3d Intern. Conf. Med. Electron.*, International Federation of Medical Electronics & Biomedical Engineering, and Institute of Electrical Engineers (London), pp. 356–358, 1960.

2.7 Jones, R. W.: Some Properties of Physiological Regulators, "Automation and Remote Control," vol. II, Butterworth Scientific Publications, London, 1961.

2.8 Wagman, I. H., and L. M. Nathanson: *Proc. Soc. Exp. Biol. Med.*, **49**:466–470 (1942).

2.9 Lettvin, J. Y.: Form-Function Relation in Neurons, Quart. Progr. Rept. 66, p. 333, Research Laboratory for Electronics, M.I.T., Cambridge, Mass., July, 1962.

2.10 Clynes, M.: Nonlinear Biological Dynamics of Unidirectional Rate Sensitivity, Symposium on Rhythmic Functions in the Living System, *Ann. N.Y. Acad. Sci.*, 1963.

2.11 Sandberg, A. A., and L. Stark: Analog Simulation of the Human Pupil System, Quart. Progr. Rept. 66, pp. 420–428, Research Laboratory for Electronics, M.I.T., Cambridge, Mass., July, 1962.

2.12 Alpern, M., and F. W. Campbell: The Behavior of the Pupil during Dark-adaptation, *J. Physiol.*, **165**:5–7 (1962).

3.1 Ruch, T. C., and J. F. Fulton: "Medical Physiology and Biophysics," W. B. Saunders Company, Philadelphia, 1961.

3.2 Lyman C. P.: Hibernation in Mammals and Birds, *Am. Scientist*, **51**(2): 127–138 (June, 1963).

3.3 Stolwicjk, J.: unpublished lecture, Pierce Foundation, Yale University, New Haven, Conn., 1962.

3.4 Crosbie, E. J., J. D. Hardy, and E. Fessenden: Electrical Analog Simulation of Temperature Regulation in Man, *IRE Trans. Bio-Med. Electron.*, **BME-8**(4) (October, 1961).

3.5 Benzinger, T. H., C. Kitzinger, and A. W. Pratt: The Physiological Regulation of Human Body Temperature, *Proc. 5th U.S. Navy Sci. Symp.*, Office of Naval Research, ONR-9, vol. II, April, 1961.

3.6 Burton, A. C.: "Temperature," Operating Characteristics of the Human Thermoregulatory Mechanism, pp. 522–528, Reinhold Publishing Corporation, New York, 1941.

3.7 Iberall, A.: Human Body as an Inconstant Heat Source and Its Relation to Clothes Insulation, papers 59-IRD-7, 8, American Society of Mechanical Engineers, New York, January, 1960.

3.8 Granit, R.: "Receptors and Sensory Perception," Yale University Press, New Haven, Conn., 1955.

3.9 Carlson, L. D.: Maintaining the Thermal Balance in Man, *IRE Intern. Conv. Record*, pt. 9: 94 (1962).

3.10 Jones, R. W.: Physiological Control Systems, *Proc. 16th Ann. Conf. Eng. Bio. Med.*, ISA-IEEE, 1963.

3.11 Hardy, J. D.: Physiology of Temperature Regulation, *Physiol. Rev.*, **41**(1): 521–606 (January, 1961).

3.12 Van Beaumont, W., and R. W. Bullard: Sweating; Its Rapid Response to Muscular Work, *Science*, **141**(3581): 643–646 (1963).

3.13 Burington, R. S.: "Handbook of Mathematical Tables and Formulas," McGraw-Hill Book Company, New York, 1948.

3.14 Benzinger, T. H.: The Diminution of Thermoregulatory Sweating during Cold Reception at the Skin, *Proc. Natl. Acad. Sci.*, **47**:1683–1688 (1961).

3.15 Hensel, H., and Y. Zotterman: *Acta Physiol. Scand.*, **23**:291 (1951).

3.16 Zotterman, Y.: Thermal Sensations, "Handbook of Physiology," sec. 1, vol. 1, chap. 18, American Physiological Society, Washington, D.C., 1959.
3.17 Zotterman, Y.: *Ann. Rev. Physiol.*, **15**:357 (1953).

4.1 Keenan, J. H.: "Thermodynamics," John Wiley & Sons, Inc., New York, 1941.
4.2 Paynter, H. M.: "Analysis and Design of Engineering Systems," The M.I.T. Press, Cambridge, Mass., 1960.
4.3 Crandall, S. H.: "Engineering Analysis," McGraw-Hill Book Company, New York, 1956.

5.1 Giese, A. C.: "Cell Physiology," W. B. Saunders Company, Philadelphia, 1962.
5.2 Edelman, I. S.: Transport through Biological Membranes, *Ann. Rev. Physiol.*, **23** (1961).
5.3 Pace, W. H., Jr.: An Analog Computer Model for the Study of Water and Electrolyte Flows in the Extracellular and Intracellular Fluids, *IRE Trans. Bio-Med. Electron.*, **BME-8**(1) (January, 1961).
5.4 Goresky, C. A.: A Linear Method for Determining Liver Sinusoidal and Extravascular Volumes, *Am. J. Physiol.*, **204**(4) (April, 1963).
5.5 Horgan, J. D., and D. L. Lange: Analog Computer Studies of Periodic Breathing, *IRE Trans. Bio-Med. Electron.*, **BME-9**(4) (October, 1962).
5.6 Cole, D. F., and J. F. Meredith: Target Organ Interaction; Considerations of the Effects of Sodium and Water Retaining Hormones on Renal Function, *Bull. Math. Biophys.*, no. 19:23–39 (1957).
5.7 Odum, H. T., R. J. Beyers, and N. E. Armstrong: Consequences of Small Storage Capacity in Nanoplankton Pertinent to Primary Production in Tropical Waters, *J. Marine Res.* 1963.

8.1 Cotes, J. E., and F. Meade: The Energy Expenditure and Mechanical Energy Demand in Walking, *Ergonomics*, **3**(2) (April, 1960).
8.2 Mayne, R.: The Dynamic Characteristics of the Semicircular Canals, *J. Comp. Physiol. Psychol.*, **43**(4) (1950).
8.3 van Egmond, A. A. J., J. J. Groen, and L. B. W. Jongkees: The Mechanics of the Semi-circular Canal, *J. Physiol.*, **110**:1–17 (1949).
8.4 Jones, G. Melvill, and J. H. Milsum: Spatial and Dynamic Aspects of Visual Fixation, *IEEE Trans. Bio-Med. Electron.*, **BME-12**(2) (April, 1965).
8.5 Hixon, W. C., and J. Niven: Application of the System Transfer Function Concept to a Mathematical Description of the Labyrinth, Naval School of Aviation Med., Rept. 57/NASA order R1, Pensacola, Fla. (1961).
8.6 Ornstein, G. N.: The Automatic Analog Determination of Human Transfer Function Coefficients, (i) Rept. NA61 H-1, North American Aviation, Inc., Columbus, Ohio, January, 1961.
(ii) *Med. Electron. Bio. Eng.*, **1**(3) (August, 1963).

9.1 Truxal, J. G.: "Automatic Feedback Control System Synthesis," McGraw-Hill Book Company, New York, 1955.
9.2 Paynter, H. M.: On an Analogy between Stochastic Processes and Monotone Dynamic Systems, *Proc. Conf. Control Tech.* (*Heidelberg*), Butterworth Scientific Publications, London, September, 1956.
9.3 Grodins, F. S.: "Control Theory and Biological Systems," Columbia University Press, New York, 1963.
9.4 Clynes, M.: The Non-linear Biological Dynamics of Unidirectional Rate Sensitivity Illustrated by Analog Computer Analysis, Pupillary

Reflex to Light and Sound, and Heart Rate Behavior, *Ann. N.Y. Acad. Sci.*, **98**:art. 4 (Oct. 30, 1962).

9.5 Clynes, M.: Unidirectional Rate Sensitivity: A Biocybernetic Law of Reflex and Humoral Systems as Physiologic Channels of Control and Communication, *Ann. N.Y. Acad. Sci.*, **92**:art. 3 (July 28, 1961).

9.6 Goresky, C.: A Linear Method for Determining Liver Sinusoidal and Extravascular Volumes, *Am. J. Physiol.*, **204**(4) (April, 1963).

10.1 Silverman, M., and A. S. V. Burgen: Application of Analog Computer to Measurement of Intestinal Absorption Rates with Tracers, *J. Appl. Physiol.*, **16**(5) (September, 1961).

10.2 Paynter, H. M.: Analyzing Control Systems Graphically, *Control Eng.*, February, 1955.

10.3 Macmillan, R. H.: "Non-linear Control Systems Analysis," The Macmillan Company, New York, 1962.

10.4 Matthews, C. M. E.: The Theory of Tracer Experiments with ^{131}I-labelled Plasma Proteins, *Phys. Med. Biol.*, **2**(1):36 (July, 1957).

10.5 Rescigno, A.: *Acta Biochim. Biophys.*, **21**:111 (1956).

10.6 Reeve, E. B., and H. R. Bailey: Mathematical Models Describing the Distribution of I^{131} Albumin in Man, *J. Lab. Clin. Med.*, **60**(6):923–943 (December, 1962).

10.7 Walker, W. G., R. S. Ross, and J. D. S. Hammond: Study of the Relationship between Plasma Volume and Transcapillary Protein Exchange Using I^{131} Labelled Albumin & I^{125} Labelled Globulin, *Circulation Res.*, **8**(5) (September, 1960).

10.8 Shepherd, C. W.: "Basic Principles of the Tracer Method," John Wiley & Sons, Inc., New York, 1962.

10.9 Van Liew, H. D.: Semilogarithmic Plots of Data which Reflect a Continuum of Exponential Processes, *Science*, **138**(3541) (Nov. 9, 1962).

10.10 Gardner, D. G., J. C. Gardner, G. Laush, and W. W. Meinke: Method for the Analysis of Multicomponent Exponential Decay Curves, *J. Chem. Phys.*, **31**(4):978–986 (October, 1959).

10.11 Worsley, B.: Analysis of Decay-type Data, *Commun. ACM*, January, 1964.

10.12 Crandall, S. H.: "Engineering Analysis," McGraw-Hill Book Company, New York, 1956.

10.13 Wooldridge, D. E.: "The Machinery of the Brain," McGraw-Hill Book Company, New York, 1963.

10.14 Cooper, T., T. Pinakatt, and A. W. Richardson: The Use of the Thermal Dilution Principle for Measurement of Cardiac Output in the Rat, *Med. Electron. Biol. Eng.*, **1**(1) (January, 1963).

11.1 Birmingham, H. P., and F. V. Taylor: A Human Engineering Approach to the Design of Man-operated Continuous Control Systems, Naval Research Laboratory, Rep. 4333, Washington, D.C., April, 1954.

11.2 Goldman, S.: Further Considerations of Cybernetic Aspects of Homeostasis, "Self Organizing Systems," Pergamon Press, New York, 1961.

11.3 Smith, O. J. M.: Nonlinear Computations in the Human Controller, *IRE Trans. Bio-Med. Electron.*, **BME-9**(2) (April, 1962).

11.4 Wilde, R. W., and J. H. Westcott: The Characteristics of the Human Operator Engaged in a Tracking Task, *Automatica*, **1**(1) (January, 1963).

12.1 Errington, P. E.: The Phenomenon of Predation, *Am. Scientist*, **51**(2) (June, 1963).

12.2 Patten, B. C.: Competitive Exclusion, *Science*, **134**(3490) (1961).

12.3   Curtis, H. J.: Biological Mechanisms Underlying the Ageing Process, *Science*, **141**(3582) (Aug. 23, 1963).

12.4   Maruyama, M.: The Second Cybernetics: Deviation—Amplifying Mutual Causal Processes, *Am. Scientist*, **51**(2) (June, 1963).

12.5   Pritchard, R., W. Heron, and D. O. Hebb: Visual Perception Approached by the Method of Stabilized Images, *Can. J. Psychol.*, **14**:67–77 (1960).

12.6   Jones, G. Melvill, and J. H. Milsum: Spatial and Dynamic Aspects of Visual Fixation, *IEEE Trans. Bio-Med. Electron.*, **BME-12**(2) (April, 1965).

12.7   Young, L. R.: A Sampled-data Model for Eye-tracking Movements, doctoral dissertation, Instrumentation Laboratory, M.I.T., Cambridge, Mass., 1962.

12.8   Westheimer, G.: (i) Mechanism of Saccadic Eye Movements, *AMA Arch. Ophthalmol.*, **52**:710 (1954).
(ii) A Note on the Response Characteristics of the Extraocular Muscle System, *Bull. Math. Biophys.*, **20**:149 (1958).

12.9   Vossius, G.: Der Sogenannte "innere" Regelkreis der Wilkur Bewegung, *Kybernetik*, **1**:28 (1961).

12.10   Robinson, D. A.: The Mechanics of Human Saccadic Eye Movement, *J. Physiol.*, **174**:245–264 (1964).

12.11   Jones, G. Melvill, and S. Mishkin: Evidence for Non-compensatory Vestibular-ocular Reflex, 5th Canadian Federation of Biological Societies, June, 1963.

12.12   Morgan, C. T., et al.: "Human Engineering Guide to Equipment Design," McGraw-Hill Book Company, New York, 1963.

12.13   Eccles, J. C., R. M. Eccles, I. Iggo, and M. Ito: Distribution of Recurrent Inhibition among Motoneurons, *J. Physiol.*, **159**:479–499 (1961).

12.14   Eldred, E., R. Granit, and P. A. Merton: Supra-spinal Control of the Muscle Spindles and Its Significance, *J. Physiol.*, **122**:498–523 (1953).

12.15   Hammond, P. H., P. A. Merton, and G. G. Sutton: Nervous Gradation of Muscular Contraction, *Brit. Med. Bull.*, **12**(3):214 (1956).

12.16   Partridge, L. D., and G. H. Glaser: Adaptation in Regulation of Movement and Posture: A Study of Stretch Response in Spastic Animals, *J. Neurophysiol.*, **23** (May, 1960).

12.17   Hammond, P. H.: An Experimental Study of Servo Action in Human Muscular Control, *Proc. 3d Intern. Conf. Med. Electron.*, IEE (London), 1960.

12.18   Stark, L., et al.: The Dynamic Characteristics of a Muscle Model Used in Digital-computer Simulation of an Agonist-Antagonist Muscle System in Man, Quart. Progr. Rept. 64, Research Laboratory for Electronics, M.I.T., Cambridge, Mass., January, 1962.

12.19   Granit, R.: Neuromuscular Interaction in Postural Tone of the Cat's Isometric Soleus Muscle, *J. Physiol.*, **143**:387 (1958).

12.20   Wilkie, D. R.: The Mechanical Properties of Muscle, *Brit. Med. Bull.*, **12**:177 (1956).

12.21   Jewell, B. R., and D. R. Wilkie: An Analysis of the Mechanical Components in Frog's Striated Muscle, *J. Physiol.*, **143**:515 (1958).

12.22   Wilkie, D. R.: The Circuit Analogue of Muscle, *Electron. Eng.*, October, 1950, p. 435.

12.23   Coulter, N. A., and J. C. West: Non-linear Passive Mechanical Properties of Skeletal Muscle, Rept. WADD 60-636, U.S.A.F.

12.24   Lippold, O. C. J., J. W. T. Redfearn, and J. Vuco: The Effect of Sinusoidal Stretching upon the Activity of Stretch Receptors in Voluntary Muscle and Their Reflex Responses, *J. Physiol.*, **144**:373 (1958).

12.25   Davey, M. R., and T. D. M. Roberts: A Transfer Function Relating

Deformations to Impulse Frequency for a Muscle Spindle in the Frog's Toe, *J. Physiol.*, **144**: 16, (1958).

12.26 Houk, J. C., Jr., V. Sanchez, and P. Wells: Frequency Response of a Spindle Receptor, Quart. Progr. Rept. 67, p. 223, Research Laboratory for Electronics, M.I.T., Cambridge, Mass., October, 1962.

12.27 Granit, R., B. Holmgren, and P. A. Merton: The Two Routes for Excitation of Muscle and Their Subservience to the Cerebellum, *J. Physiol.*, **130**:213 (1955).

13.1 Katz, B.: Depolarization of Sensory Terminals and the Initiation of Impulses in the Muscle Spindle, *J. Physiol.*, **111**:261–282 (1950).

13.2 Matthews, B. H. C.: The Response of a Single End Organ, *J. Physiol.*, **71**:64–110 (1931).

13.3 Fex, G.: Auditory Activity in Centrifugal and Centripetal Cochlear Fibers in Cat—A Study of a Feedback System, *Acta Physiol. Scand.*, **55**(189) (1962).

13.4 Van Bergeijk, W. A., J. R. Pierce, and E. E. David: "Waves and the Ear," Science Study Series, Doubleday & Company, Garden City, N.Y., 1960.

13.5 Bronk, D. W.: The Mechanism of Sensory End Organs, *Res. Publ. Assoc. Res. Nervous Mental Disease*, **15**:60–82 (1934).

13.6 Chapman, K. M., and R. S. Smith: A Linear Transfer Function Underlying Impulse Frequency Modulation in a Cockroach Mechanoreceptor, *Nature*, **197**(4868):699–700 (Feb. 16, 1963).

13.7 Landgren, S.: *Acta Physiol. Scand.*, **26**(1) (1952).

13.8 Pringle, J. W. S., and V. J. Wilson: *J. Exp. Biol.*, **29**:220 (1952).

13.9 Stevens, S. S.: "Sensory Communication" (W. A. Rosenblith, ed.), The M.I.T. Press, Cambridge, Mass., 1961.

13.10 MacKay, D. M.: Psychophysics of Perceived Intensity; A Theoretical Basis for Fechner's and Steven's Laws, *Science*, **139**(3560):1213–1216 (1963).

13.11 Jones, R. W., C. C. Li, A. U. Meyer, R. B. Pinter: Pulse Modulation and in Physiological Systems, Phenomenological Aspects, *IRE Trans. Bio-Med. Electron.*, **BME-8**(1) (January, 1961).

14.1 Fatt, P., and B. Katz: An Analysis of the End-plate Potential Recorded with an Intracellular Electrode, *J. Physiol.*, **115**:320–370 (1951).

14.2 Fatt, P., and B. Katz: Spontaneous Subthreshold Activity at Motor Nerve Endings, *J. Physiol.*, **117**:109–128 (1952).

14.3 DelCastillo, J., and B. Katz: Quantal Components of the End-plate Potential, *J. Physiol.*, **124**:560–573 (1954).

14.4 DelCastillo, J., and B. Katz: Biophysical Aspects of Neuro-muscular Transmission, *Progr. Biophys.*, **6**:121–170 (1956).

14.5 Boyd, I. A., and A. R. Martin: Spontaneous Subthreshold Activity at Mammalian Neuro-muscular Junctions, *J. Physiol.*, **132**:61–73 (1956).

14.6 Boyd, I. A., and A. R. Martin: The End-plate Potential in Mammalian Muscle, *J. Physiol.*, **132**:74–91 (1956).

14.7 Feller, W.: "An Introduction to Probability Theory and Its Applications," John Wiley & Sons, Inc., New York, 1950.

14.8 Hoel, P. G.: "Introduction to Mathematical Statistics," John Wiley & Sons, Inc., New York, 1954.

14.9 Rosenblith, W. A. (ed.): "Processing Neuroelectric Data," The M.I.T. Press, Cambridge, Mass., 1961.

14.10 Campbell, F. W., J. G. Robson, and G. Westheimer: Fluctuations of

Accommodation under Steady Viewing Conditions, *J. Physiol.*, **145**:579–594 (1959).

14.11 Farrar, J. T.: Use of a Digital Computer in the Analysis of Intestinal Motility Records, *IRE Trans. Med. Electron.*, **ME-7**(4) (October, 1960).

14.12 Goodman, L.: On the Dynamics of Pulmonary Ventilation in Quiescent Man, Rept. SRC 18-L-62-1, Systems Research Center, Case Institute of Technology, Cleveland, Ohio, May, 1962.

14.13 Barlow, J. S., M. A. Brazier, and W. A. Rosenblith: The Application of Autocorrelation Analysis to Electro-encephalography, *Proc. 1st Natl. Biophys. Congr.*, Yale University Press, New Haven, Conn., 1959.

14.14 Blackman, R. B., and J. W. Tukey: "Measurement of Power Spectra," Dover Publications, Inc., New York, 1959.

14.15 Bendat, J. S.: Interpretation and Application of Statistical Analysis for Random Physical Phenomena, *IRE Trans. Bio-Med. Electron.*, **BME-9**(1) (January, 1962).

14.16 Newton, G. C., L. A. Gould, and J. F. Kaiser: "Analytical Design of Linear Feedback Controls," appendix E, John Wiley & Sons, Inc., New York, 1957.

14.17 Barlow, J. S., and R. M. Brown: An Analog Correlator System for Brain Potentials, Tech. Rept. 300, Research Laboratory for Electronics, M.I.T., Cambridge, Mass., July, 1955.

14.18 Barlow, J. S., and M. Z. Freeman: Comparison of EEG Activity by Means of Auto-correlation and Cross-correlation Analysis, *Electroencephalog. Clin. Neurophysiol.*, **11** (1959).

14.19 Barnett, G. O., and J. C. Michener: Power Spectrum Analysis of Aortic Impedance, *Proc. 16th Ann. Conf. Eng. Med. Bio.*, Baltimore, 1963.

14.20 Dawson, G. D.: A Summation Technique for the Detection of Small Evoked Potentials, *Electroencephalog. Clin. Neurophysiol.*, **6**:65–84 (1954).

14.21 Barlow, J. S.: A Small Electronic Analog Averager and Variance Computer for Evoked Potentials of the Brain, *Proc. 2d Intern. Conf. Med. Electron.*, Iliffe and Sons, London, June, 1959.

14.22 Smith, G. K., and B. D. Burns: A Biological Interval Analyzer, *Nature*, **187**:512–513 (1960).

14.23 Clynes, M., and M. Kohn: Portable 4-channel On-line Digital Average Response Computer, *Proc. 4th Intern. Conf. Med. Electron.*, New York, July, 1961.

14.24 Burns, B. D., G. Mandl, and G. K. Smith: An Auto- and Cross-correlator for Digital Information Processing, *Electron. Eng.*, April, 1963, pp. 220–228.

14.25 Casby, J. U., R. Siminoff, and T. R. Houseknecht: An Analog Cross-correlator for the Analysis of Nervous Activity, *Proc. 15th Ann. Conf. Eng. Bio. Med.*, Chicago, 1962.

14.26 Saper, L. M.: On-line Auto- and Cross-correlator Realized with Hybrid Computation Techniques, *IEEE Intern. Conv. Record*, pt. 9, March, 1963.

14.27 Burns, B. D., and G. K. Smith: Transmission of Information in the Unanaesthetized Cat's Isolated Forebrain, *J. Physiol.*, **164**:238–251 (1962).

14.28 Katsuki, Y., N. Suga, and M. Nomoto: Peripheral Neural Mechanisms of Hearing in the Monkey, *Proc. Bionics Symp.*, WPAFB, U.S.A.F., March, 1963.

15.1 Patten, B. C.: Preliminary Method for Estimating Stability in Plankton, *Science*, **134**(3484) (October, 1961).

15.2 Dunbar, M. J.: The Evolution of Stability in Marine Environments: Natural Selection at the Level of the Ecosystem, *Am. Naturalist,* **94**(875) (March, 1960).

15.3 Wilkie, D. R.: The Relation between Force and Velocity in Human Muscle, *J. Physiol.,* **110**:249–280 (1950).

15.4 Hill, A. V.: The Mechanical Efficiency of Frog's Muscle, *Proc. Roy. Soc.,* **13**(127): 434–451 (1939).

15.5 Lotka, A. J.: "Elements of Physical Biology," The Williams & Wilkins Company, Baltimore, 1925.

15.6 Odum, H. T., and R. C. Pinkerton: Time's Speed Regulator: The Optimum Efficiency for Maximum Power Output in Physical and Biological Systems, *Am. Scientist,* **43**(2) (April, 1955).

15.7 Wilkie, D. R.: Man as a Source of Mechanical Power, *Ergonomics,* **3**(1) (January, 1960).

15.8 Cotes, J. E., and F. Meade: Energy Expenditure and Energy Demand in Walking, *Ergonomics,* **3**(2) (April, 1960).

15.9 Burns, B. D.: Central Control of Respiratory Movements, *Brit. Med. Bull.,* **19**(1) (1963).

15.10 Horgan, J. D., and R. L. Lange: Digital Computer Simulation of the Human Respiratory System, *IEEE Intern. Conv. Record,* pt. 9, 1963.

15.11 Otis, A. B., W. O. Fenn, and H. Rahn: Mechanics of Breathing in Man, *J. Appl. Physiol.,* **2**:592 (1950).

15.12 Christie, R. V.: Dyspnoea in Relation to Visco-elastic Properties of the Lung, *Proc. Roy. Soc. Med.,* **46**(5) (May, 1953).

15.13 Widdicombe, J. G.: Regulation of Tracheo-bronchial Smooth Muscle, *Physiol. Rev.,* **43**(1):1-37 (January, 1963).

15.14 Mead, J.: Control of Respiratory Frequency, *J. Appl. Physiol.,* **15**:325–337 (1960).

15.15 Nubar, Y., and R. Contini: A Minimum Principle in Biomechanics, *Bull. Math. Biophys.,* **23** (1961).

15.16 Griffith, V. V.: A Model of the Plastic Neuron, *Proc. Bionics Symp.,* WPAFB, U.S.A.F., March, 1963.

15.17 Eykhoff, P.: Adaptive and Optimalizing Control Systems, *IRE Trans. Auto. Control,* **AC-5**(2) (June, 1960).

15.18 Sheridan, T. B.: Time-variable Dynamics of Human Operator Systems, (i) DACL Rept., M.I.T., Cambridge, Mass., March, 1960 (AFCRC-TN-60-169).
(ii) Doctoral dissertation in mechanical engineering, M.I.T., Cambridge, Mass., 1959.

15.19 Gabor, E., W. Wilby, and R. Woodcock: A Universal Non-linear Filter, Predictor and Simulator Which Optimizes Itself by a Learning Process, *IEE (London),* paper 3270 (July, 1960).

15.20 Milsum, J. H.: Hill Climbing and Computer Control, *Proc. Modern Control Theory Seminar,* NRC, Ottawa, Canada, 1963.

15.21 Mishkin, E., and L. Braun, Jr. (eds.): "Adaptive Control Systems," McGraw-Hill Book Company, New York, 1961 (see method of Corbin, p. 350).

15.22 Beckenbach, E. F. (ed.): "Modern Mathematics for the Engineer," McGraw-Hill Book Company, New York, 1956.

15.23 Anderson, G. W.: A Self-adjusting System for Optimum Dynamic Performance, *IRE Natl. Conv. Record,* 1958 (also chap. 9 of Mishkin and Braun, Bibliography).

P.1 Lanchester, F. W.: "The World of Mathematics," Mathematics in Warfare, vol. IV, pp. 2138–2157, Simon and Schuster, Inc., New York, 1956.

P.2 Fletcher, J. C., H. E. Lewis, and D. R. Wilkie: Human Power Output:
The Mechanics of Pole Vaulting, *Ergonomics,* **3**(1) (January, 1960).
P.3 Benjamin, B., W. P. Jolly, and J. Maitland: Operational Research and
Advertizing—Theories of Response, *Operations Res. Quart.,* **11**(4)
(1960).

Bibliography

BASIC INEXPENSIVE BACKGROUND TEXTS

Alger, P. A.: "Mathematics for Science and Engineering," McGraw-Hill Book
Company, New York, 1957; $2.95.   (Covers algebra, calculus, differential
equations, Laplace transforms, probability.)
Gruenberger, F. J., and D. D. McCracken: "Introduction to Electronic Com-
puters," John Wiley & Sons, Inc., New York, 1963; $2.95.
Welbourne, D.: "Analog Computer Methods," Commonwealth Library Series,
The MacMillan Company, New York, 1965.

ADAPTIVE AND OPTIMUM CONTROL THEORY

Britton, J. E., and H. J. Kushner: Reference Bibliography; Adaptive or Self-
optimizing Control Systems, Lincoln Laboratory, M.I.T., Cambridge,
Mass., February, 1961.
Chang, S. S. L.: "Synthesis of Optimum Control Systems," McGraw-Hill Book
Company, New York, 1961.
Gibson, J. E.: "Nonlinear Automatic Control," McGraw-Hill Book Company,
New York, 1963.
Leondes, C. T. (ed.): "Computer Control Systems Technology," McGraw-Hill
Book Company, New York, 1961.
Mishkin, E., and L. Braun: "Adaptive Control Systems," McGraw-Hill Book
Company, New York, 1961.
Peterson, E. L.: "Statistical Analysis and Optimization of Systems," John
Wiley & Sons, Inc., New York, 1961.
Schulz, W. C., and V. C. Rideout: Control System Performance Measures; Past,
Present and Future, *IRE Trans. Auto. Control,* February, 1961.

ANALOG AND DIGITAL COMPUTERS

Alt, F. L.: "Electronic Digital Computers," Academic Press Inc., New York,
1958.
Ashley, J. R.: "Introduction to Analog Computation," McGraw-Hill Book
Company, New York, 1963.
Borko, H.: "Computer Applications in the Behavioral Sciences," Prentice-Hall,
Inc., Englewood Cliffs, N.J., 1962.
Gottlieb, C. C., and J. N. P. Hume: "High-speed Data Processing," McGraw-
Hill Book Company, New York, 1958.
Jackson, A. S.: "Analog Computation," McGraw-Hill Book Company, New
York, 1960.
Karplus, W. J.: "Analog Simulation," McGraw-Hill Book Company, New York,
1958.
McCormick, E. M.: "Digital Computer Primer," McGraw-Hill Book Company,
New York, 1959.
Nathan, R., and E. Hanes: "Computer Programming Handbook," Prentice-
Hall, Inc., Englewood Cliffs, N.J., 1961.

Rogers, A. E., and T. W. Connolly: "Analog Computation in Engineering Design," McGraw-Hill Book Company, New York, 1960.

Say, M. G., et al.: "Analog and Digital Computers," G. Newnes.

Scott, N. R.: "Analog and Digital Computer Technology," McGraw-Hill Book Company, New York, 1960.

Siegel, P.: "Understanding Digital Computers," John Wiley & Sons, Inc., New York, 1961.

Soroka, W. W.: "Analog Methods in Computation and Simulation," McGraw-Hill Book Company, New York, 1954.

Truitt, T. O., and A. E. Rogers: "Basics of Analog Computers," John F. Rider, Publisher, Inc., New York, 1960.

Warfield, J. N.: "Introduction to Electronic Analog Computers," Prentice-Hall, Inc., Englewood Cliffs, N.J., 1959.

Welbourne, D. "Analog Computer Methods," The MacMillan Company, New York, 1965.

AUTOMATIC CONTROL THEORY

"Automatic Control," Scientific American, Inc., New York, and Simon and Schuster, Inc., New York, 1955.

Campbell, D. P.: "Process Dynamics," John Wiley & Sons, Inc., New York, 1958.

Doebelin, E. O.: "Dynamic Analysis and Feedback Control," McGraw-Hill Book Company, New York, 1962.

Eckman, D.: "Automatic Process Control," John Wiley & Sons, Inc., New York, 1958.

Gibson, J. E.: "Nonlinear Automatic Control," McGraw-Hill Book Company, New York, 1963.

"Handbook of Automation, Computation and Control," vol. II, John Wiley & Sons, Inc., New York, 1958.

Holzbock, W. G.: "Automatic Control," Reinhold Publishing Corporation, New York, 1958.

Kuo, B. C.: "Automatic Control Systems," Prentice-Hall Inc., Englewood Cliffs, N.J., 1962.

Macmillan, R. H.: "Non-linear Control Systems Analysis," Pergamon Press, New York, 1962.

Raven, F. H.: "Automatic Control Engineering," McGraw-Hill Book Company, New York, 1961.

Smith, O. J. M.: "Feedback Control Systems," McGraw-Hill Book Company, New York, 1958.

Truxal, J. G.: "Automatic Feedback Control System Synthesis," McGraw-Hill Book Company, New York, 1955.

CONTINUOUS STATISTICAL SIGNALS AND THEIR LINEAR FILTERING

Blackman, R. B., and J. W. Tukey: "Measurement of Power Spectra," Dover Publications, Inc., New York, 1959.

Cherry, C.: "On Human Communication," John Wiley & Sons, Inc., New York, 1957.

Harman, W. W.: "Principles of the Statistical Theory of Communication," McGraw-Hill Book Company, New York, 1963.

Lee, Y. W.: "Statistical Theory of Communication," John Wiley & Sons, Inc., New York, 1960.

Newton, G. C., Jr., L. A. Gould, and J. F. Kaiser: "Analytical Design of Linear Feedback Controls," appendix E, John Wiley & Sons, Inc., New York, 1957.

Truxal, J. G.: "Automatic Feedback Control System Synthesis," McGraw-Hill Book Company, New York, 1955.

CYBERNETICS

Ashby, W. R.: (i) "Design for a Brain," John Wiley & Sons, Inc., New York, 1960. (ii) "Introduction to Cybernetics," John Wiley & Sons, Inc., New York, 1958.

Bell, D. A.: "Intelligent Machines—An Introduction to Cybernetics," Blaisdell Publishing Co., New York, 1962.

Cherry, C.: "On Human Communication," John Wiley & Sons, Inc., New York, 1957.

Guilbaud, G. T.: "What Is Cybernetics," William Heinemann, Ltd., London, 1959.

"Mechanization of Thought Processes," H. M. Stationary Office, London, 1958.

Pask, G.: "An Approach to Cybernetics," Harper & Row, Publishers, Incorporated, New York, 1961.

Wiener, N.: "Human Use of Human Beings," Houghton Mifflin Company, Boston, 1950.

Yovits, M. C., G. T. Jacobi, and G. D. Goldstein: "Self-organizing Systems 1962," Spartan Books, Washington, D.C., 1962.

DIFFERENTIAL EQUATIONS

Agnew, R. P.: "Differential Equations," 2d ed., McGraw-Hill Book Company, New York, 1960.

Ford, L. R., Sr., and L. R. Ford, Jr.: "Calculus," McGraw-Hill Book Company, New York, 1963.

Hildebrand, F. B.: "Advanced Calculus for Engineers," Prentice-Hall, Inc., Englewood Cliffs, N.J., 1949.

Kells, L. M.: "Elementary Differential Equations," 5th ed., McGraw-Hill Book Company, New York, 1960.

DISCRETE STATISTICS

Feller, W.: "An Introduction to Probability Theory and Its Applications," John Wiley & Sons, Inc., New York, 1950.

Ferguson, G. A.: "Statistical Analysis in Psychology and Education," McGraw-Hill Book Company, New York, 1959.

Hoel, P. G.: "Introduction to Mathematical Statistics," John Wiley & Sons, Inc., New York, 1954.

Mainland, D.: "Elementary Medical Statistics," W. B. Saunders Company, Philadelphia, 1963.

Steele, R. G. D., and J. H. Torrie: "Principles and Procedures of Statistics: With Special Reference to the Biological Sciences," McGraw-Hill Book Company, New York, 1960.

LAPLACE TRANSFORMS AND LINEAR ANALYSIS

Brown, R. G., and J. W. Nilsson: "Introduction to Linear Systems Analysis," John Wiley & Sons, Inc., New York, 1962.

Gardner, M. F., and J. L. Barnes: "Transients in Linear Systems," John Wiley & Sons, Inc., New York, 1942.

Savant, C. J., Jr.: "Fundamentals of the Laplace Transformation," McGraw-Hill Book Company, New York, 1962.

MATHEMATICS AND CONTROL

Grabbe, E. M., S. Ramo, and D. E. Wooldridge: "Handbook of Automation, Computation and Control," vol. I, Control Fundamentals; vol. II, Computers and Data Processing; vol. III, Systems and Components, John Wiley & Sons, Inc., New York, 1958.

Korn, G. A., and Theresa M. Korn: "Mathematical Handbook for Scientists and Engineers," McGraw-Hill Book Company, New York, 1961.

Riggs, D. S.: "The Mathematical Approach to Physiological Problems," The Williams & Wilkins Company, Baltimore, 1963.

Stibitz, G. R., and J. A. Larrivee: "Mathematics and Computers," McGraw-Hill Book Company, New York, 1956.

NEUROPHYSIOLOGY

Eccles, J. C.: "The Physiology of Nerve Cells," The Johns Hopkins Press, Baltimore, 1957.

"Handbook of Neurophysiology," American Physiological Society, Washington, D.C., 1960. Matthews, P. B. C.: Muscle Spindles and Their Motor Control, *Physiol. Rev.*, **44**:219–288 (1964).

NUMERICAL ANALYSIS

Hamming, R. W.: "Numerical Methods for Scientists and Engineers," McGraw-Hill Book Company, New York, 1962.

Hildebrand, F. B.: "Introduction to Numerical Analysis," McGraw-Hill Book Company, New York, 1956.

Milne, W. E.: "Numerical Solution of Differential Equations," John Wiley & Sons, Inc., New York, 1953.

Ralston, A., and H. S. Wilf: "Mathematical Methods for Digital Computers," John Wiley & Sons, Inc., New York, 1960.

Scarborough, J. B.: "Numerical Mathematical Analysis," 4th ed., The Johns Hopkins Press, Baltimore, 1960.

Stanton, R. G.: "Numerical Methods for Science and Engineering," Prentice-Hall, Inc., Englewood Cliffs, N.J., 1960.

Stibitz, G. R., and J. A. Larrivee: "Mathematics and Computers," McGraw-Hill Book Company, New York, 1956.

SAMPLED–DATA SYSTEMS

Jury, E. I.: "Sampled-data Control Systems," John Wiley & Sons, Inc., New York, 1958.

Kuo, B. C.: "Analysis and Synthesis of Sampled-data Control Systems," Prentice-Hall, Inc., Englewood Cliffs, N.J., 1963.

Ragazzini, J. R., and G. F. Franklin: "Sampled-data Control Systems," McGraw-Hill Book Company, New York, 1958.

Thaler, G. J., and R. G. Brown: "Analysis and Design of Feedback Control Systems," 2d ed., McGraw-Hill Book Company, New York, 1960.

Index